AF794634

Corrosion Policy Decision Making

Corrosion Policy Decision Making

Science, Engineering, Management, and Economy

Edited by

Reza Javaherdashti
Eninco Engineering B.V.
The Netherlands

WILEY

This edition first published 2022
© 2022 John Wiley and Sons, Inc.

All rights reserved. No part of this publication may be reproduced, stored in a retrieval system, or transmitted, in any form or by any means, electronic, mechanical, photocopying, recording or otherwise, except as permitted by law. Advice on how to obtain permission to reuse material from this title is available at http://www.wiley.com/go/permissions.

The right of Reza Javaherdashti to be identified as the author of the editorial material in this work has been asserted in accordance with law.

Registered Office
John Wiley & Sons, Inc., 111 River Street, Hoboken, NJ 07030, USA

Editorial Office
111 River Street, Hoboken, NJ 07030, USA

For details of our global editorial offices, customer services, and more information about Wiley products visit us at www.wiley.com.

Wiley also publishes its books in a variety of electronic formats and by print-on-demand. Some content that appears in standard print versions of this book may not be available in other formats.

Limit of Liability/Disclaimer of Warranty
In view of ongoing research, equipment modifications, changes in governmental regulations, and the constant flow of information relating to the use of experimental reagents, equipment, and devices, the reader is urged to review and evaluate the information provided in the package insert or instructions for each chemical, piece of equipment, reagent, or device for, among other things, any changes in the instructions or indication of usage and for added warnings and precautions. While the publisher and authors have used their best efforts in preparing this work, they make no representations or warranties with respect to the accuracy or completeness of the contents of this work and specifically disclaim all warranties, including without limitation any implied warranties of merchantability or fitness for a particular purpose. No warranty may be created or extended by sales representatives, written sales materials or promotional statements for this work. The fact that an organization, website, or product is referred to in this work as a citation and/or potential source of further information does not mean that the publisher and authors endorse the information or services the organization, website, or product may provide or recommendations it may make. This work is sold with the understanding that the publisher is not engaged in rendering professional services. The advice and strategies contained herein may not be suitable for your situation. You should consult with a specialist where appropriate. Further, readers should be aware that websites listed in this work may have changed or disappeared between when this work was written and when it is read. Neither the publisher nor authors shall be liable for any loss of profit or any other commercial damages, including but not limited to special, incidental, consequential, or other damages.

Library of Congress Cataloging-in-Publication Data

Names: Javaherdashti, Reza, editor.
Title: Corrosion policy decision making : science, engineering, management, and economy / edited by Reza Javaherdashti.
Description: First edition. | Hoboken, NJ : Wiley, 2022. | Includes bibliographical references and index.
Identifiers: LCCN 2021031559 | ISBN 9781119764311 (hardback) | ISBN 9781119764328 (adobe pdf) | ISBN 9781119764335 (epub)
Subjects: LCSH: Corrosion and anti-corrosives.
Classification: LCC TA462 .C65675 2022 | DDC 620.1/1223–dc23
LC record available at https://lccn.loc.gov/2021031559

Cover Design: Wiley
Cover Image: © Phonix_a Pk.sarote/Shutterstock

Set in 9.5/12.5pt STIXTwoText by Straive, Pondicherry, India

MIX
Paper from responsible sources
FSC® C013604
www.fsc.org

10 9 8 7 6 5 4 3 2 1

To:
Nikola Tesla
The man who changed the world

Contents

Preface

Corrosion Management (CM) has become a very hot topic these days. Many professional groups related to corrosion from the USA to Australia invite their corrosion experts to give lectures/webinars on this topic. There are two common points among all these lectures and webinars; one is that the lecturers speak about how to address corrosion in their own industries and presumably manage it to an audience who are themselves corrosion professionals, and the other is that the invited speakers are always corrosion professionals themselves. Management of matter, that in our understanding is equal to CM, is no less important than management of energy, and therefore it must be explained in such a way that non-corrosionists also understand and appreciate it. Namely, politicians, non-corrosionist engineers and professionals, economists and environmentalists must also realize the significance of corrosion and its proper management.

The theme of CM has become so repetitive and predictable that one can envisage almost the entirety of the content: a few words about the economy of corrosion (normally borrowed from figures given in some well-known websites); a few words about the systems(s) in which corrosion is of interest (or rather, what the speaker has some experience about); how the corrosion has been recognized and addressed technically in these systems; and then the section entitled "Lessons to Learn."

However, none of these webinars or lectures point out matters which are out of "the box of corrosion": there are no words about the impact of the thermodynamic nature of corrosion or its management and the difficulties it creates, or how to explain and control ecological measures imposed by corrosion and CM techniques, nor is there anything related to how creativity can be an essential issue in finding solutions for corrosion problems.

This book is first of its kind that looks at corrosion not just as a technical matter but as a multidisciplinary issue that involves many skills and faculties—mainly non-technical in the sense that engineering literature and nomenclature are considered technical.

In this book we have tried to show that while due to the thermodynamic nature of corrosion, it cannot be stopped. It is necessary to recognize many misunderstandings associated with corrosion (such as the difference between corrosion prevention and corrosion control) and redefine many basic concepts (such as the concept of failure within the context of introducing the terms pseudo-FFS and Zugzwang effect state) to understand it correctly from a practical point of view that can be called Management.

We would also like to call our approach toward CM "Smart Management of Corrosion" in the sense that it has the capability to adapt itself with all industries, not just a few, and also because management of corrosion benefits from contributions from both CM and Corrosion Knowledge Management (CKM).

We hope our readers will enjoy reading this book as much as we did writing it and applying its principles in our daily professional lives!

Reza Javaherdashti
The Netherlands
April 2021

Authors and Contributors

R. Javaherdashti

- General Manager, Eninco Engineering B.V., The Netherlands
- PhD (Materials science–Corrosion)
- More than 25 years of experience; successfully delivered 400+ projects in corrosion management, problem-solving, expert-witnessing, and Root Cause Analysis in various industries globally
- Certified corrosion and MIC lecturer by ASME (USA) and Society of Petroleum Engineers (USA)
- Performed 5000+ hours of teaching corrosion management and MIC to various industries (from oil & gas to mining, aviation, and chemical industries worldwide)
- Authored/co-authored several internationally referenced books on corrosion (published by Elsevier, Springer, CRC Press, etc.) as well as research papers and Root Cause Analysis reports
- E-mail: javaherdashti@enincoating.com
- LinkedIn profile: https://www.linkedin.com/in/dr-reza-javaherdashti-9a2a2415/

F. Javaherdashti

- Senior advisor at IMI (Industrail Management Institute), I.R. Iran
- PhD (Management)
- Senior advisor in business management, branding, and CRM
- More than 28 years of experience in national and international management consulting
- More than 10,000 hours of teaching management principles and consultancy for several industries
- E-mail: Farnak.javaherdashti@gmail.com

A. Ghanbarzadeh

- Research at RIPI (Research Institute of Petroluem), I.R. Iran
- M.Sc. (Chemical Engineering)
- Academic member and Head of research projects on protective coatings at RIPI
- 31 years of work experience
- Authored four books and 12 papers
- E-mail: ali.ghanbarzadeh@gmail.com
- LinkedIn profile: https://www.linkedin.com/in/ali-ghanbarzadeh-a5abba55/

M. Mostashar Nezami

- CEO, AtavArt, privately held co, I.R. Iran
- MA in General Economics and BS in Accounting
- More than 17 years of experience in the accounting, financial, and economic studies
- Carried out an 8-year research study in the economics of renewable energies, and introduced an applicable way for using renewable energies as resident's daily energy
- More than 400 hours teaching of Accounting, Finance Management, and Economics
- Attended "Model Thinking" traing course offerd by University of Michigan
- Wrote more than ten articles in economic analysis in the National Press
- E-mail: mahsanezami60@yahoo.com
- LinkedIn profile: https://www.linkedin.com/in/mahsa-mostashar-nezami-a539b578

M.R. Hamedghafarian

- Senior Process Engineer of Gas Plants Utilities & Off-Sites, Ministry of Petroleum, I.R. Iran
- M.Sc. in Advanced Chemical Engineering
- B.Sc. in Chemical Engineering (natural gas processing)
- Committee member / Gas Plant Corrosion
- Chairman Committee / Baseline Pollution Determination in gas plants (Wastewater Treatment)
- More than twelve years of experience in operation and design of gas processing plants (Reception Facilities and Utilities)
- Seven papers in classical thermodynamics, corrosion, environment, process additives management, and CO_2 capture in Iranian congresses
- Translation of two books on water treatment chemicals, and technical report writing in chemical engineering

- More than 1000 hours teaching Troubleshooting in Utility Processes, Corrosion, Heat Exchanger Design, Industrial Wastewater Treatment, Process Engineering Documentation, and Water Treatment topics in Refineries and Gas plants
- Associate member of technical staff in Iranian Corrosion Association
- Member of Iranian Society of Engineering Education
- E-mail: mohamedreza.hamedghafarian@gmail.com
- LinkedIn profile: https://www.linkedin.com/in/mohamedreza-hamedghafarian-955463a0/

M. Basirzadeh

- Board member at Kavosh Meyar Tabasum Pasargad Company
- Industrial Management MSC, Tehran Amirkabir University
- Iranian TRIZ teacher and consultant
- Author of four books in Farsi language about TRIZ and creativity
- E-mail: irantriz@gmail.com
- LinkedIn profile: https://www.linkedin.com/in/mehdi-basirzadeh-4980481

1

Introduction

Reza Javaherdashti

General Manager, Eninco Engineering B.V., The Netherlands

This book is a first-hand answer to two questions:

1) What corrosion management (CM) really means;
2) What to expect from CM.

Let us be more specific in describing what we mean by the above. For those of you who have been too engaged with your integrity management tasks that cannot find time to watch movies and particularly horror movies, I do recommend to watch the movie "Final Destination; Part 5," and watch it with the eyes of a corrosion/integrity management specialist. In this movie (and its previous parts) what is shown is that something of little importance goes wrong, and in a series of unpredictable, sad events, one (or sometimes more) of the actors are killed in a very dramatic, graphic way. As said, the main lesson learned from this movie is that one problem can lead to another and in the end, a catastrophic result occurs. The same is also true with neglecting corrosion in an integrity management plan; something that one would think is not that important will lead to another, and if it is in its "pseudo-FFS[1]" (Fit-For-Service State) stage, the result at its best could be arriving at a "Zugzwang effect" [1] stage, or simply said, failure.

In classical academic literature related to Management and its principles, perhaps one of the best definitions have been given by Henry Fayol as "To manage is to forecast and to plan, to organize, to command, to co-ordinate and to control [1]." On the other hand, corrosion has also a clear definition that can be simplified as *the chemical reaction between an electron donor (anode) and electron receiver*

1 Please study Chapter 3 for a better understanding of these terms.

Corrosion Policy Decision Making: Science, Engineering, Management, and Economy, First Edition.
Edited by Reza Javaherdashti.
© 2022 John Wiley & Sons, Inc. Published 2022 by John Wiley & Sons, Inc.

(cathode) via a medium that allows exchange of ions and a metallic path that allows electron transfer.[2] However, what is CM? In fact, when it comes to corrosion science, we know that we are talking about laboratories, white-collar researchers, academic environments, and the search for understanding fundamentals and mechanisms of corrosion processes. In this context, innovation and how innovative research must be carried out is of fundamental significance. When we talk about corrosion engineering, it is the way the accumulated science and knowledge about corrosion and corrosion processes will become applicable in the field. For instance, the use of cathodic protection to let structures survive longer, or use of corrosion inhibitors and biocides to chemically control corrosion. Figure 1.1 can serve to show the essential elements of corrosion engineering and how corrosion engineering and other engineering disciplines are interconnected with each other. We see that when talking about corrosion engineering, we are actually talking about a multidimensional topic that in its wholeness is more complex than other engineering disciplines not in based on the subject of focus, but on the methodologies that apply to address corrosion and its various aspects such as monitoring and treatment:

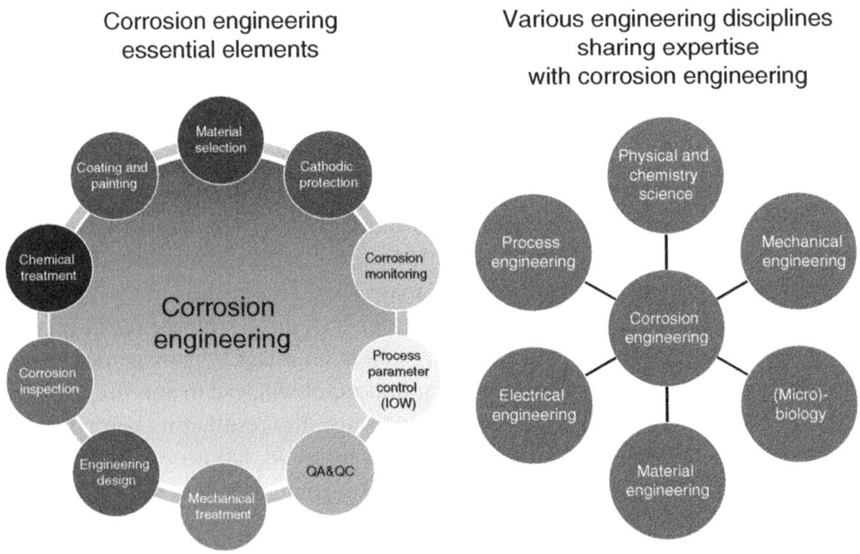

Figure 1.1 Corrosion engineering and its relation to other engineering disciplines (right), essential features and components of corrosion engineering (left) (Used by permission from Eng. Riky Bernardo – Qatar).

2 The metallic path when referred to corrosion under real life conditions has no meaning because corrosion already is taking place on metals.

All of these applications have their own codes and standards. When we talk about corrosion treatment, we know that we are talking about five strategies to deal with corrosion (physical measures, chemical measures, electrical measures, mechanical measures, and design/material selection measures). Corrosion monitoring is to address methods and technologies by which severity of corrosion within its course of action is studied by codes and measures pertaining to corrosion monitoring. All of the above bring to mind certain codes, working environments and specialties, and expertise. However, what is CM? It obviously has a part dealing with corrosion and a part dealing with management. The confusion arises from here; how can management which is seeming a non-technical issue, be matched with corrosion which is a highly technical issue?

All the materials written to date on the management of corrosion are just looking at the science/engineering, and to some extent economy of corrosion, without detailing with the actual requirements (for example, about the economic nomenclature that is needed to understand cost of corrosion). Some of the publications about CM do suffer from the defects mentioned above. Some examples of CM literature are:

A. Morshed, "An Introduction to Corrosion Management in Industry," NACE, USA, 2017.

A. Morshed, "An Introduction to Asset Corrosion Management in the Oil and Gas Industry, 2nd edition," NACE, USA, 2016.

A.S. Groysman, "Corrosion Problems and Solutions in Oil Refining and Petrochemical Industry," Springer, 2017.

"My Manual: Practical Corrosion Management," a manual published by IDC Technologies, 2009, Perth, Australia.

Even IMPACT report by ex-NACE (now AMPP) that was published in 2016 had some sections on management and economy without any focus on engineering and science aspect of corrosion.

No economist, environmental sciences specialist, or a non-technical manager is likely to read those books (and similar ones) or attend CM seminars and webinars. The reason is simple; corrosion has not been defined for these non-corrosionist professionals. Perhaps, the importance of corrosion has sometimes been flagged by some politicians,[3] as well as passing bills to allow the establishment of certain corrosion related entities, say, within the army, but apart from those individual sparks, no overall, systematic understanding about corrosion and its various aspects exists yet.

Our iconoclast approach toward CM is what we call "Smart Management of Corrosion" in the sense that (i) it is smart because this system can adapt itself with any

3 https://www.defenseindustrydaily.com/Sen-Tom-Coburn-Americas-Fiscal-Defense-Crisis-06412, last visited 15 February 2021.

industry and, contrary to the existing assumption that CM is to be discussed within the context of a certain industry (oil and gas), smart management of corrosion is not industry-specific; and (ii) it focuses of the management side rather+ than the corrosion side. Our understanding is that current CM systems mainly deali with the risk of corrosion, however it is essential to also deal with the cost of corrosion, whether economic cost or ecological cost. Management of corrosion is a term that acts as an umbrella to both CM (risk of corrosion) and corrosion knowledge management (CKM) (cost of corrosion) simultaneously and being as such, Management of Corrosion is a more general term than CM alone.

There are three very important features about corrosion:

1) Corrosion and Failure: Corrosion is inevitable and manageable; failure is inevitable and non-manageable. Being non-manageable does not mean that it cannot be prevented, it simply means that when failure happens, it happens! This distinguishment between these two terms if of vital importance; the prejudice of many who deal with corrosion issues is that they are entitled to find a way to prevent or control corrosion.[4] Corrosion occurs anyway due to the thermodynamic nature of it, it is our duty, though, to prevent failures. It is through failures and leaks that not only economic costs are imposed (from shut down costs to maintenance and replacement), but also environmental effects also occur. Study of corrosion processes ending in leak and failure is the task of a corrosion specialist, post-mortem forensic investigation of the failure is what a forensic investigator does to help build a root cause analysis report, but who is to measure the seriousness of the environmental effects thus produced? Who is going to estimate the economic loss (direct cost and indirect cost)?
2) Risk and Likelihood: If the likelihood of getting a certain type of corrosion-related failure is low, it does not necessarily mean that is risk is also low and vice versa. The confusion between risk and likelihood is with no doubt one of the most important sources of problems that can even reach to the level of disasters. It is the mindset that can be observed in industry quite frequently and it has the potential of being lethal.
3) Corrosion and Aging: Aging implies that a structure has been in service for quite a long time, whereas corrosion can actually occur in structures right after being put into service. Therefore, corrosion-related failures could happen in structures after a relatively short period of service (sometimes months).

In our professional judgment, management of corrosion needs to address all aspects of corrosion as we have tried to do in this book; corrosion and the algorithm of innovation (TRIZ; the Theory of Inventive Problem Solving) have been covered to the required extent in Chapters 2 and 6. In Chapter 2, some important

4 Corrosion prevention and corrosion control are not the same and cannot be used interchangeably. We will get back to this concept later in this book.

aspects of corrosion science is re-defined, that is, as much as needed. The engineers looking after corrosion issues may not themselves have attended pre-employment corrosion courses, and thus during their employment years they attend courses or conferences with specific titles, or even learn from each other and/or their colleagues verbally. There is a very small minority who, with a pre-employment background in corrosion, still keep their knowledge up to date. This author during his more than two decades of consulting, as well as education training courses, has seen these issues very clearly.

The executive force for any management of corrosion scheme is humans and not robots, although we often forget this. Along with it, we also happen to forget the golden wisdom "To err is human." In fact, a great number of industrial disasters—including those related to corrosion—stem from human error. One of the ways by which human error can be highly decreased is via training. In IMPACT report by AMPP, the cost for training is not mentioned as cost but as investment. Creating motivation and paying special attention to training is of paramount significance in a corrosion knowledge management (CKM) scheme, and is an integral part of smart management of corrosion. In addition to the science of corrosion, it is important to find the best way to invent solutions. With decades of splendid past, TRIZ can be an instrumental factor in creating innovation patterns. This author has shown the application power of TRIZ in some of his publications. Researchers need to be familiar with TRIZ to find better solutions which are innovative and feasible particularly in the field of management of corrosion.

The heart of this book which is Smart Management of corrosion is the topic of Chapter 3. Engineering aspects of corrosion is explained in Chapter 5. No matter what our interpretation about management of corrosion, understanding corrosion in action is a must.

Discussing enough about economics as required to define corrosion damage is addressed in Chapter 4, whereas environmental effects of corrosion has been addressed in Chapter 7.

This book is highly likely to experience a "Semmelweis Reflex" which essentially means "to stick to preexisting beliefs and to reject fresh ideas that contradict them [2]. No literature of CM known to us issues such details of economy, environment, TRIZ, and innovative interpretation of CM as we will cover in this text.

References

1 Aquinas, P.G. (2011). *Principles and Practices of Management*. Phagwara, India: Lovely Professional University.

2 Gupta, V. K., Saini, C., Oberoi, M. et al. (2020) "Semmelweis Reflex: An Age-Old Prejudice," *World Neurosurgery*. https://pubmed.ncbi.nlm.nih.gov/31837492, last visited 13 February 2021.

2

A Short Review of Some Important Aspects of the Science of Corrosion

Reza Javaherdashti[1*] *and Ali Ghanbarzadeh*[2†]

[1] *General Manager, Eninco Engineering B.V., The Netherlands*
[2] *Research at RIPI (Research Institute of Petroleum), I.R. Iran*

2.1 Introduction

Corrosion is a process that happens naturally. The way corrosion happens is a story that has been repeated in all books written on the subject, but it will still make sense if we repeat it here. The main reason for doing so is that the thermodynamics behind corrosion is not only important from a scientific/engineering point of view, but also from a corrosion management (CM) approach. (We have described the significance of the thermodynamic nature of corrosion and how it is important in engineering practices, and particularly CM and management of corrosion in Chapter 3.)

Corrosion is derived from a Greek word that may be translated as "bitten away." Some authors—including us—justify this word by implying that a corroded part seems to have been bitten. The invisible bite probably comes from the very nature of the metal because metals are not found in their pure form—that is without having oxides, sulfates, carbonates, and the like—in their natural form. Even what we call iron (that originally means "coming from heaven" probably due to it being mainly found in meteors) is not naturally found as a totally pure material free of non-metallic materials. Metals such as iron, copper, and the like are always found in their ores. Ore is a rock part that contains these metals as compounds and not elements. These rocks are extracted and crushed and the metallic compounds within them are then extracted. The second and the most important stage from a corrosion point of view, however, is the extraction of metal from the

*https://www.linkedin.com/in/dr-reza-javaherdashti-9a2a2415.
†https://www.linkedin.com/in/ali-ghanbarzadeh-a5abba55.

Corrosion Policy Decision Making: Science, Engineering, Management, and Economy, First Edition.
Edited by Reza Javaherdashti.
© 2022 John Wiley & Sons, Inc. Published 2022 by John Wiley & Sons, Inc.

"surrounding" non-metallic compounds around it. The core process in metal extraction is the reduction processes during which electrons are added to electrons already existing in the metal element itself. These reduction processes, collectively referred to as "extraction metallurgy processes" are basically those happening in metallurgy plants. It is by these processes that metals with high purity are being produced.

In this sense, the pure metal is not at rest when compared to its state when it was in its ore. In other words, the state of energy in two stages of "in-ore" and "as-extracted" will be far different from each other; Gibbs free energy of "in-ore" is much lower than that of "as-extracted." This means that any way by which this excess of energy can be released will favor corrosion.

Now that it is understood that the force by which corrosion is being driven comes from the very nature of materials, particularly metals, we need to understand some basic concepts upon which the whole electrochemical corrosion principles are based. Why electrochemical corrosion? It is because a number of corrosion processes ranging from corrosion under insulation to microbiologically influenced corrosion (MIC) to atmospheric corrosion are all gathered under the umbrella of electrochemical corrosion. Therefore, in this chapter, we will review and describe some fundamental aspects of electrochemical corrosion, what it is, and how it can be treated. As we will see in Chapter 3 about "smart management of corrosion," prevention of corrosion is different from corrosion control. Very briefly, prevention of corrosion can be applied to a Greenfield project related to corrosion, or equipment which is still too young in its service life so that options such as modification of design or materials selection/upgrade or even change and modification of process parameters have yet to be considered. Corrosion control, on the other hand, is applied to a Brownfield project or equipment which has been in service for a quite long time (perhaps more than a year), so that not only are any change or modification in any process parameters not permissible, but also options such as change of design or materials selection can be highly likely to be considered too costly to be applied. By "corrosion treatment" we address both corrosion control and corrosion prevention where applicable, per case.

We will therefore start with very definition of the elements of an electrochemical corrosion process and continue with available tools to predict corrosion, that is, to foresee if it will occur or not, and then we continue with five treatment approaches toward corrosion. The five approaches are (i) chemical treatment, (ii) electrical treatment, (ii) mechanical treatment, (iv) design/material selection, and (v) physical treatment. What we have to advise from the beginning is that the fourth treatment approach can still be an option even if the problem is categorized as a corrosion control issue. In this regard, it is the cost of application that will be the major playing factor if design/material selection is to be better applied for a corrosion control approach.

Our approach in this section may sound somewhat radical to those who have been used to reading materials in their usual context, because instead of being too academic, our understanding of corrosion theory and practice is directly focusing on its pragmatic, immediate benefits for the readers and whoever engaged in the so-called CM practice. We would not allow our readers to be, literally, buried under formulas, diagrams, and the like. We want to be straight-forward and write frankly!

2.1.1 Essentials of Electrochemical Corrosion

If corrosion is a natural process by which energy of the system is lowered, then any approach toward corrosion must be the same when facing any natural process. The main thing here, however, is that as long as corrosion has not been allowed to make a hole in an asset—the process we call through-wall pitting—it is safe. This approach may sound quite different from what we have been taught so far; any manual on CM starts with the unwritten assumption that the job for a corrosion engineer is to prevent corrosion. In fact, the job of a corrosion engineer—and for that matter an integrity management expert—is to prevent through-wall pitting occurrence as manifested by severe, localized corrosion.

Figure 2.1 shows an example of two assets that have been severely corroded through localized corrosion. In both of the figures there are three elements of corrosion that can be recognized. The anode is the spot at which anodic reaction occurs, and in cases similar to those shown in Figure 2.1, it is where sever pitting in the form of severe through-wall pitting happens. At the anode, electrons are being released and travel via the metallic material body toward the cathode, the cathode could be adjacent to the anode. However, the point is that in localized corrosion, contrary to uniform corrosion, anode and cathode spots do not move relative to each other, so that the anode is fixed and from the very spot the anode becomes fixed, pitting starts. It is also due to this feature that localized corrosion happens in one particular spot on the metallic structure. This may also be the main characteristic of localized corrosion that makes its prediction almost impossible; one cannot know where that attack will occur and with what severity, normally measured in terms of corrosion rate and expressed by pitting factor, it can occur. The moisture on the metallic surface that is basically the water vapor already available in the air will be the main electrolyte through which ion exchange can happen.

It is quite understandable if these very concepts that form electrochemical corrosion—or more generally referred to as aqueous corrosion—leave a lot of questions for the readers. For example, from a spatial point of view, while the position of anode is visible how can we decide where exactly the cathode adjacent to it is? Is it on the left side of the anode? Its right side? Where is it located? Another matter is how do we know if the anode and cathode "creep" relative to each other on a metallic surface? And, for that matter, how do we know if one particular spot and not another spot on the metal will become the anode?

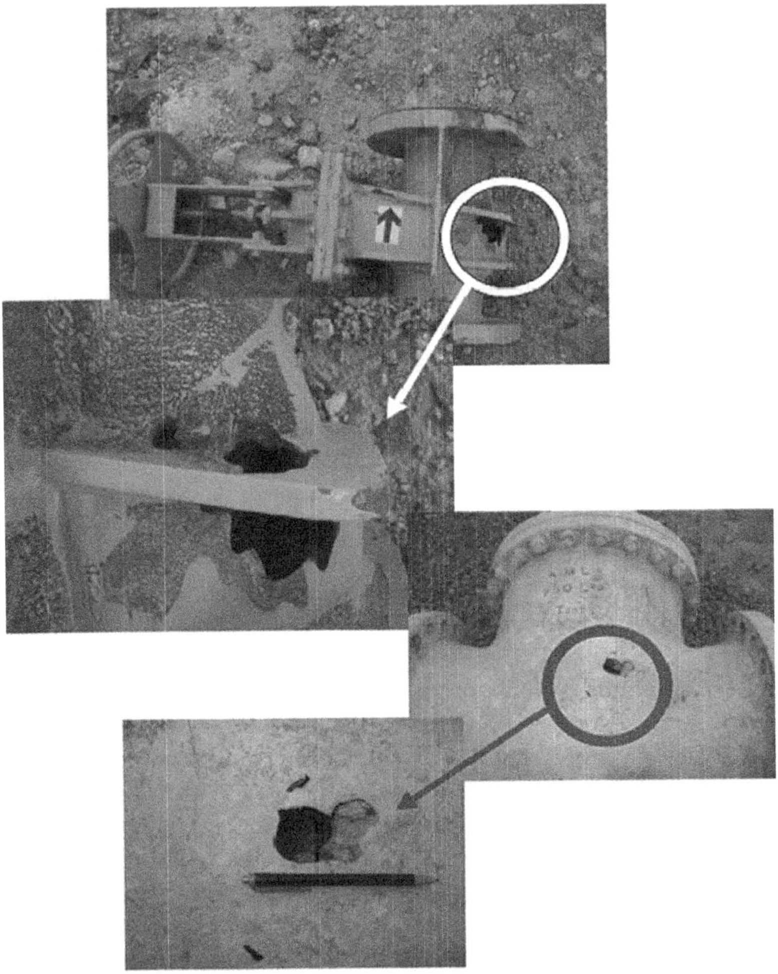

Figure 2.1 Two examples of severely corroded equipment leading into a through-wall localized failure. The anode is the point at which perforation has occurred (white arrow) and the cathode is immediately adjacent to it so that electron flow happens via the metallic surface (orange arrow), while ionic transfer happens through the moisture layer already formed on the metal. (*Source:* Taken from the collection of Dr. R. Javaherdashti.)

The above questions are actually some of fundamental importance should we be writing this book as a textbook. There are numerous books and papers that may become handy in answering these questions and even more puzzling ones. For us, however, the above questions carry no value at all. From a practical point of view, all we need to care about are the following five basic concepts:

1) At anode, anodic reactions happen. These reactions involve emitting the extra electron given to the metal through extractive metallurgy processes.
2) These electrons will be taken by the cathode. In other words, electrons emitted by the anode will be gained at the cathode, leading to the establishment of an electron flow (electricity) between anode and cathode.
3) The electron flow constitutes what we refer to as "current." The current thus produced is in fact the way by which voltage difference can be leveled between anode and cathode.
4) The moisture already established on the metal's surface acts as a carrier medium for the exchange of ions thus produced as a result of anodic and cathodic reactions. In other words, the electron emittance and electron gaining will serve to produce anion and cation ions which can be carried through the moisture layer (the electrolyte).
5) It follows, then, contrary to what textbooks on corrosion try to impose on our understanding saying that there need to be four elements (the anode, the cathode, the electrolyte, and the metallic path) for a electrochemical corrosion scenario to exist, there are actually only three elements necessary; these are the anode, the cathode, and the electrolyte. As corrosion is already happening on a metallic asset, we really do not require to make more confusion that that is already existing in the electrochemistry of corrosion. As some may already know, we may name a few of these confusions; one can name the concept of using the wrong term of oxidation instead of anodic reaction, the polarity signs of anode and cathode, the arguable term of under deposit corrosion, and even the term biofilm to explain the venue at which electrochemical cells are established.[1]

---

1 Examples in this regards can be that oxidation implies that oxygen is involved or must be involved in the corrosion process. While for many cases it is true, for an even larger number this may not necessarily be taken as a true statement (take, for example, MIC by anaerobic sulfate reducing bacteria). Therefore, it is always better to address electrons giving off reactions as "anodic reactions." In addition, "corrosion under deposit" does not say anything about the corrosion mechanism(s) involved, it just indicates where corrosion is taking place, which is, of course, under deposits. If the involved corrosion mechanism(s) are not specified, mere use of "under deposit corrosion" is a useless, pseudo-professional statement. Dissimilar metal corrosion is yet another technically wrong expression; this corrosion process that is correctly referred to as "galvanic corrosion" can happen on the same metals but with different conditions that will make their electrochemical properties become different. Welding a new pipe of smaller surface area (length) to the same size of the same pipe material that has been already in use for some time makes the former become anodic to the latter as the old pipe could actually act as a cathode, due to factors such as being covered with a layer of rust. Thus, never replace the right term "galvanic corrosion" with the wrong term "dissimilar metal corrosion." Biofilm implies a film that has a totally biological fabric. Both implications are incorrect; biofilm is neither a film nor is 100% biological. In our publications, we have proposed another term which is "temenos" to compensate for these misunderstandings and faults.

The above fundamental concepts explain both the science required to explain how corrosion occurs as well as the science needed to explain its prediction and treatment, as we will see in the next sections.

2.1.2 Prediction of Corrosion

We obviously need to know when corrosion happens, as this will enable us to prepare ourselves with the upcoming event and try to manage it as much as possible, in the best way achievable. As we have mentioned earlier, this is not a textbook about corrosion, and therefore many of the concepts that would have been covered in length in those books will be treated in the most practical approach possible to make these concepts as understandable for a corrosion specialist as possible.

There are mainly three tools that can be used to help us predict corrosion, in other words, to let us determine anodic and cathodic reactions and therefore find out about the possibility of corrosion.

These tools are:

a) standard hydrogen electrode scale (SHE)
b) galvanic series
c) Roubaix diagrams

By using these tools, it is possible to find out what chemical reactions can be considered as anodic and therefore giving off their extra electrons, and which reactions can be cathodic reactions to receive and gain those electrons. Obviously, the substrate on which these reactions will take place is a metallic surface (to allow electron transfer), and the environment will be water or water containing (the necessary electrolyte for ionic transfer).

2.1.2.1 Standard Hydrogen Electrode/Electrochemical Series

The main assumptions that need to be made are as follows.

1) It is assumed that the reduction voltage for hydrogen reaction is zero. In other words, it is assumed that the conversion of hydrogen ion (H^+) to hydrogen gas is zero. Needless to say that the actual voltage is not zero at all, but this assumption is accepted so that some chemical reactions can be placed above hydrogen evolution (due to their positive reduction voltage), and some reactions will be placed below that due to their negative potential. Those with positive potential will be noble, meaning that they are very highly likely to act as a cathode and those with negative potentials will be branded as active, meaning that they are very highly likely to act as an anode.
2) All of the substances taking part in the electrode reaction have unit activity.
3) Temperature is 25 °C.
4) Hydrogen pressure in the reference electrode is one atmosphere.

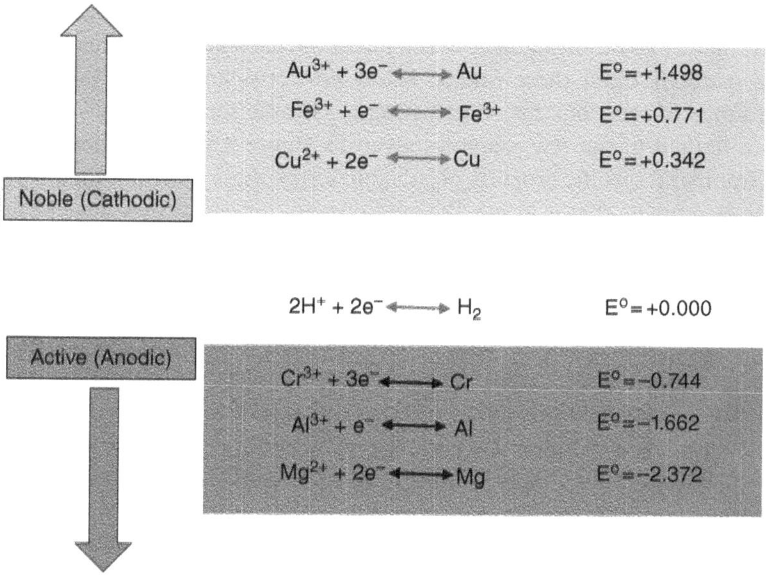

Figure 2.2 Schematic presentation of electrochemical series with some reactions as cathodic and anodic reactions. Standard potentials (E^0) are in volts vs. SHE (standard hydrogen electrode).

Figure 2.2 schematically shows an example of an electrochemical series:

However, it is evident that items ii–iv can only be achieved under strict laboratory conditions and under real life, industrial conditions it is not possible to main temperatures and pressures as required by the electrochemical series. It can be said that it is mainly due to these restrictions as dictated by laboratory-controlled conditions that industrial application of electrochemical series must be replaced with a more application-friendly option. Although, as we see later, standard hydrogen potential is a necessary element in constructing Pourbaix diagrams, which are very useful in applications.

2.1.2.2 Galvanic Series

Due to limitations of hydrogen electrode measurement for constructing electrochemical series, another option has been applied which is called "electrochemical series." The potentials used in a galvanic series are only valid in:

a) a given environment (electrolyte), and
b) at a given temperature (25 °C).

Temperatures can be corrected for any given environment, but the importance is to know what environment we are talking about. An example of a galvanic series for seawater at 25 °C can be seen in Figure 2.3.

As it can be seen from the figure, some reactions are still considered as cathodic and anodic; mainly those sitting near to the top of the galvanic series are considered noble (cathodic) and those ranked below are considered active (anodic). While the range of reactions for constructing a galvanic series for seawater at 25 °C is not limited to the few examples given in Figure 2.3, it is obvious that due to environment-specific property of galvanic series, these reactions hold true only for the electrolyte, that is to say, the environment for which they have been constructed.

It is also evident that we still have the restrictions imposed by the specific temperature that must be maintained in ordering anodic and cathodic reactions for the given environment. Furthermore, what is to be noticed with regards to the galvanic series given in Figure 2.3 the same as electrochemical series ranking, any reaction which is placed above another reaction is regarded cathodic to that reaction. For instance, while in Figure 2.2, aluminum reaction was more cathodic with respects to magnesium, in Figure 2.3, bronzes are regarded more anodic with regards to copper–nickel (70–30). It follows that while electrochemical and galvanic series may differ in many respects, it is still the ranking of a particular reaction with regards to the other one that determines if it is noble (cathode) or active (anode).

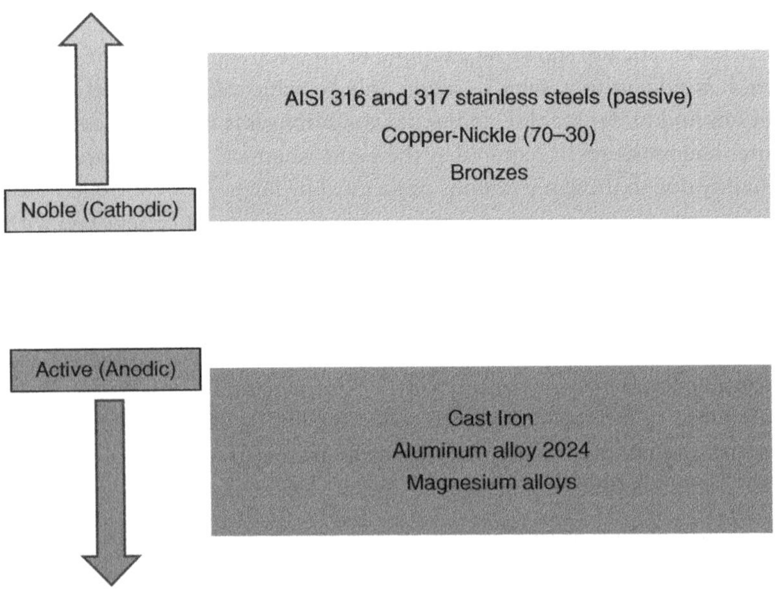

Figure 2.3 Some examples of active and passive metals in seawater at 25 °C for the specific environment seawater.

As we mentioned earlier, although galvanic series seem to be more practical than electrochemical series, there are still some limitations. Being environment- and temperature-specific is an inescapable limitation on all chemical reactions as they do happen under certain conditions, the most important of which are the electrolyte (that is, the environment), the temperature (as chemical reactions are kinetics highly dependent on temperature), and of course, pressure. However, there is yet another measure that can assist us in predicting corrosion behavior of metallic alloys and it was invented by a PhD student some decades ago, named Marcel Pourbaix.[2]

2.1.2.3 Pourbaix Diagrams

An example of a Pourbaix diagram is shown in Figure 2.4.

Pourbaix diagrams have two axes, one for corrosion potential as measured in hydrogen potential, and one for measuring acidity of the environment shown as pH. For a given set of potential-pH-environment, some "domains" will be created. These domains can be used to predict, under the given conditions for the three parameters mentioned above, if it is possible to expect corrosion or immunity to corrosion (in other words, passivity). Existence of some corrosive ions such as, but not limited to, chlorides, can somehow change the domains and thus shift the potentials in which corrosion or passivation can be expected.

One very important aspect of Pourbaix diagrams in addition to enabling us to predict safe and unsafe values of potential and pH with regards to corrosion for a given environment is to put emphasis on what is almost forgotten by a majority of non-corrosion expert professionals; it is normally assumed that as long as we know about the pH of the environment, it is safe to say if it is corrosive or not! The rule of thumb for these self-acclaimed corrosionists is that if pH is below 7 the environment is acidic and thus corrosive, if pH is 7 it is neutral, and if it is larger than 7 the environment is basic. This is wrong! In order to interpret correctly, one has to know both corrosion potential and pH. As an example, take the Pourbaix diagram in Figure 2.4 when chloride is present; at an acidic pH $= 6$ and potential $= -0.6$ V (red dashed line intersection in Figure 2.4), there is immunity against corrosion, whereas in the same system, but this time for a neutral pH $= 7$ and potential $= -0.4$ V, corrosion is highly likely to happen (black dashed line intersection in Figure 2.4). This alone can serve to show how powerful Pourbaix diagrams are in dealing with corrosion and predicting it. However, as noticed by our readers, the restrictions due to temperature do still remain.

---

2 In some sources, the reasons put forward for the rejection and creation of problems in of Purbaix's PhD thesis and graduation are described as "*The war and some confusion among the jury on the sign of electrode potential.*"

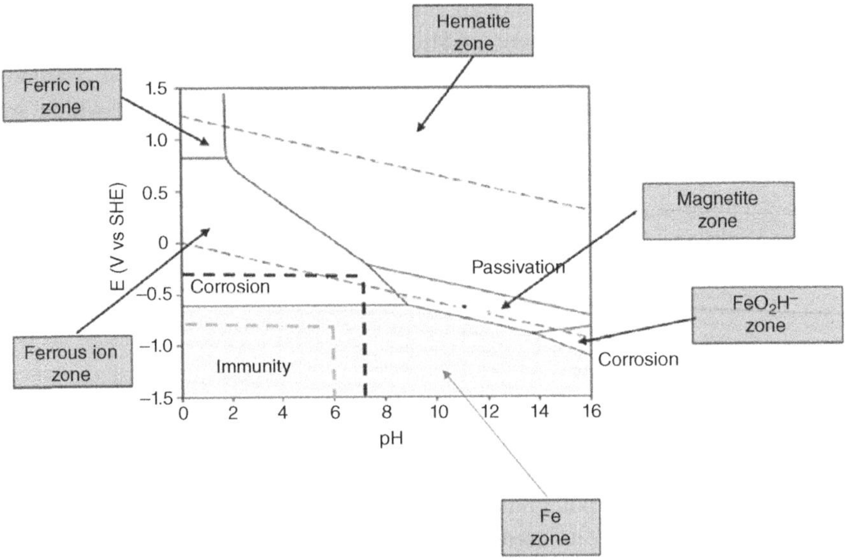

Figure 2.4 A typical Pourbaix diagram (simplified) for an Fe–water system at iron concentration of 10–6 mol/l and 25 °C. The region between the two dashed lines is the water stability zone. The two dashed lines, upper and lower, define the domains for oxygen and hydrogen stability; above the upper line water is oxidized to O_2, so oxygen is evolved above the upper line. Below the lower line, water decomposes to H_2 and thus, hydrogen will be liberated below the hydrogen line. The zones shown indicate the chemical species formed. Red and black dashed lines, show immunity under acidic and corrosion under neutral pH values, respectively, emphasizing the necessity for specifying and studying pH–voltage pair to decide about possible corrosivity scenarios. Hematite: Fe_2O_3; Magnetite: Fe_3O_4; Ferric ion: Fe^{+3}; Ferrous ion: Fe^{+2}.

2.2 Important Technical Treatment Strategies for Corrosion Treatment

Corrosion treatment strategies can be categorized into five categories:

1) Design modification-change/materials selection (change and/or modification of design including change/upgrading of materials).
2) Chemical treatment; use of various kinds of corrosion inhibitors (anodic inhibitors, cathodic inhibitors and mixed effect inhibitors, biocides either as oxidizing biocides or non-oxidizing biocides).
3) Electrical treatment (the most noticeable example in this context is cathodic protection).
4) Mechanical treatment (pigging).
5) Physical treatment (paints/coatings).

As we can see, the corrosion treatment strategies and techniques are indeed very limited in number. However, depending on the corrosion process in focus, there can exist some alterations. For example, if the corrosion case happens to be MIC, then yet another option can be added; the biological treatment of MIC. These alterations do not mean that the essential five strategies cannot be applied. For instance, in the case of MIC, physical treatment as well as electrical and mechanical treatment may also be applicable. The alterations are only applicable where, due to the very nature of the corrosion process, more treatment options may be made available.

Below we will briefly review these five treatment strategies. Once again, we should remind our readers that neither this chapter nor the book itself is about teaching corrosion science and the electrochemistry behind it in the way that many famous handbooks and textbooks do. Our main focus here is on knowing as much necessary for an engineer who happens to become responsible for developing a CM strategy and who may need to know some basic, important elements of the science of corrosion when trying to differentiate between CM various models. As we have described in full in Chapter 3 under the title of "Smart Corrosion Management Elements," having an efficient, realistic, smart model for CM is an integral part of any strategy that seriously considers management of corrosion.

2.2.1 Design Modification-change/Materials Selection

Contrary to what many engineers and operators may think, the as-is design of assets is not the last word when it comes to CM of the asset. In fact, many factors during fabrication, welding, coating, testing, and operation may become the points at which corrosion can be invited. For instance, it is usually recommended to use continuous welding instead of spot welding to avoid problems such as moisture entrance, or having as few branches as possible in a pipeline to avoid stagnation points. In addition, it is highly likely that MIC problems can occur in post-hydrostatic testing of pipelines, which is mainly applied to test the welding strength (contrary to pneumatic test which is essentially a leak test, hydrotest is both a strength and leak test).

Yet another issue that may arise is to have formed galvanic cells by joining two metals that each can have different electrochemical properties (such as welding together the same metals with and without coating, or coupling dissimilar metals to each other so that in addition to creating a potential difference, anode/cathode area ratios less than one will be created). Operation conditions may lead into initiation and development of MIC in the asset, an example can be seen for example, in underground firewater rings or within reverse osmosis membranes systems. Examples are numerous.

All that calls for being constantly alert to consider the option for a design change-modification. In this context, design modification means that the existing design is altered with minimum costs, whereas change may mean complete makeover, which would of course bring many costs.

The policy is that before opting for design change-modification/materials selection, do a thorough feasibility study to compare this option with the economy and possible outcome with the other four options (when applicable), and then decide if you want to continue with it or leave it, either temporarily or permanently.

Yet another strategy in this respect could be increasing the thickness of the metal. This technique is technically referred to as creating "Corrosion Allowance." As the very name implies, the method is essentially based on increasing the thickness of the metallic asset. This way we actually "allow" corrosion to work its way through a certain thickness of the asset before too much corrosion of the thickness occurs. It is important to note that corrosion is essentially effective on the thickness, as it is this dimension that on a larger scale carries the stress imposed on the material. Corrosion allowance is applied not to treat corrosion but to delay it. In addition, this method can be taken as a practically useful measure should the form of corrosion be uniform. However, as it may occur with many corrosion process types, localized corrosion in the formation of pits leads to deepening of these pits and results in through-wall leak, and failure of coalescence of the pits leads to cracks and thus lowers the threshold of tolerating applied stress that can occur. Corrosion allowance is not widely used except in occasions such as submarine (deep and ultra-deep) pipelines and pressure vessels.[3] On such occasions, the other four corrosion treatment strategies may not applicable all the time (such as installing cathodic protection [CP]). However limited the number of industries that use corrosion allowance, it is still an especially important option for a better corrosion treatment approach.

Materials selection is a very expensive option to control or prevent corrosion (see Chapter 3 to understand the difference between corrosion control and corrosion prevention). In this option, the existing material of an asset is replaced with a more corrosion-resistant material so that the overall service life of the asset is increased. Figure 2.5 shows an example of materials selection for upstream oil and gas industry:

---

3 Engineering Encyclopedia, Saudi Aramco DeskTop Standards, entitled "MATERIALS OF CONSTRUCTION FOR PRESSURE VESSELS" considers corrosion allowance an important measure to control corrosion. On page 6 of this Desktop standard it reads verbatim "The most common method that is used to address corrosion in pressure vessels is to specify a corrosion allowance." While corrosion allowance is certainly not a corrosion treatment measure in the way that, for example, physical, chemical, or electrical treatment measure are, and despite that corrosion allowance is used in limited industries, it cannot be denied that these industries are themselves of appreciable importance to all industry sectors.

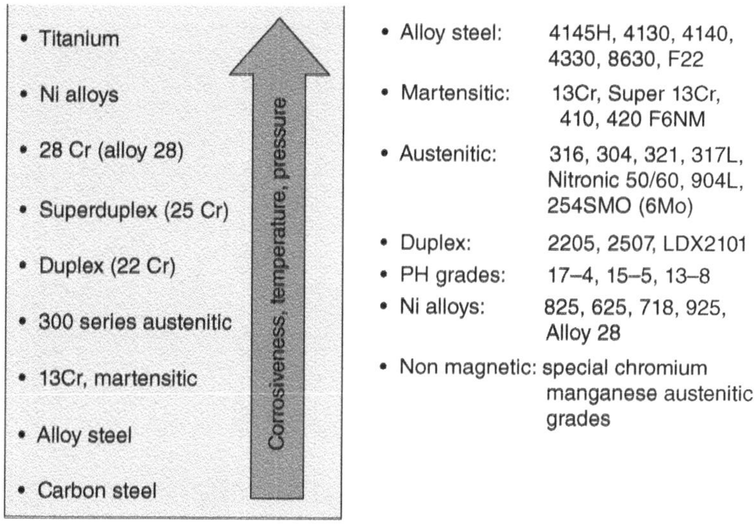

Figure 2.5 Materials selection chart for upstream exploration oil and gas industry.

While materials selection sounds like a viable option in treating corrosion by replacing less corrosion resistant materials with more resistant ones, one incredibly significant point to remember is its feasibility. In other words, while project managers may accept the superiority of the selected material over the existing material from a corrosion treatment strategy point of view, an important drawback could be the extra cost imposed by this option to the overall cost of the project. One has to bear in mind that the cost is not limited to the cost of the selected material per se, it is the person-hours needed to replace and operate, and the downtime between these steps. Say that we have decided to replace a carbon steel part with a stainless steel part. The price difference for January 2021 has been shown in Figure 2.6, where the overall difference is about 40% per mass of the steel selected. This can be translated to huge extra costs for shifting from carbon steel to stainless steel.

The issue will even become more considerable for projects that are being carried out in countries whose currency is not US dollars. When translated into their national currency, the imposed cost may become more eye catching, rendering it more likely to get negative responses to finance a materials selection option. Therefore, it is of vital importance to convince the management not only in terms of corrosion treatment-related terms but also in terms of the service life extension via an economic analysis such as life cycle cost analysis (we have briefly explained this and other models in Chapter 3 [Smart Corrosion Management Elements] and in more details related to economy in Chapter 4 [A Short Review of Some Fundamental Principles of Economics]).

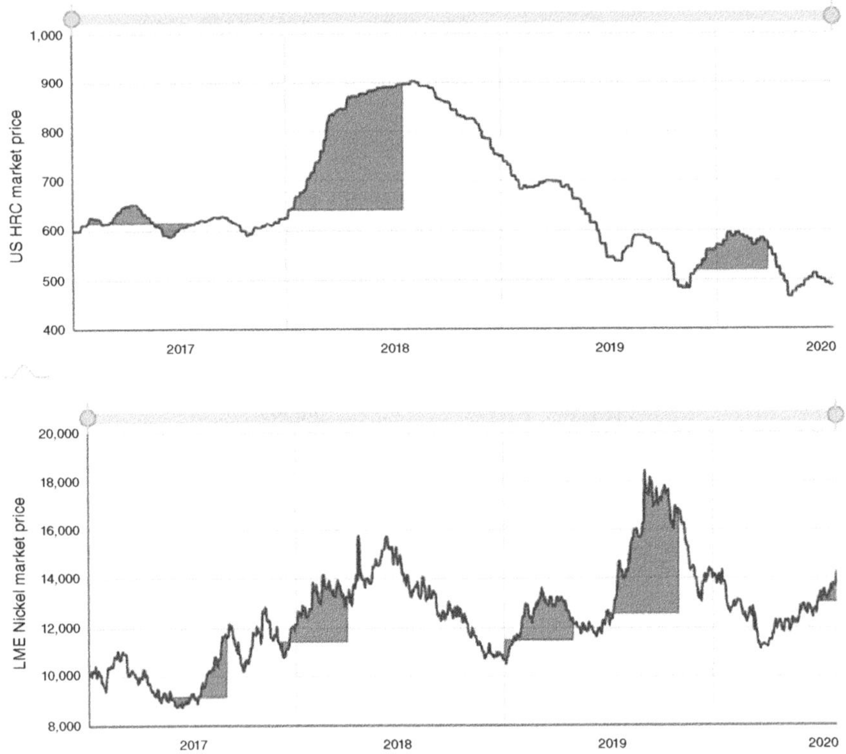

Figure 2.6 Carbon steel (hot rolled, cold rolled, and galvanized coils) price fluctuations with an average price of $0.5027/lb (top). Stainless steel (sheet) price fluctuations with an average price of $1.219/lb (bottom) [1] (all average prices based on prices given on 1 January 2021).

Change or modification of design can be performed without a need for materials selection if applied smartly. For instance, where possibility of galvanic corrosion exists, it is possible that by either changing the anode/cathode ratio, applying CP (to reverse the polarities), or coating either anode or cathode (or sometimes both) solve the corrosion treatment issue as a corrosion prevention/control approach.

Last but not least, the option for design modification-change/materials selection could be an option which is paradoxically both inexpensive and expensive; cost of design per se may not become too unbearable, whereas costs imposed by materials selection can become so exceedingly high that the project management will be pushed to consider other measures. Therefore, in dealing with corrosion treatment via this option, any proposal for doing so should come with a smartly planned economic analysis for whole process and not just relying on decreasing corrosion rates and severity alone.

2.2.2 Chemical Treatment

One the ways by which corrosion can be brought under control is application of corrosion inhibitors, and in case of MIC, biocides. The reason we emphasize mentioning biocides along with corrosion inhibitors is the incorrect belief by some corrosionists that by applying inhibitors, MIC can also be treated, or rather, MIC is less important than non-MIC corrosion problems. While MIC is also an electrochemical corrosion in essence, corrosion inhibitors (or nearly all of them) cannot treat MIC, as the contributing factor to corrosion is micro-organisms.

Corrosion inhibitors can be grouped by selecting various criteria, but we prefer to categorize them into main three groups; anodic, cathodic, and mixed effect inhibitors.

Anodic inhibitors form a protective oxide film on the surface of the metal. This film will be instrumental in causing a large anodic shift of the corrosion potential. This shift forces the metallic surface into the passivation region where the material can be assumed to become immune to corrosion. Due to this effect, anodic corrosion inhibitors are also sometimes referred to as passivators. Chromates, nitrates, tungstate, and molybdates are some examples of anodic inhibitors. On the other hand, cathodic inhibitors act mainly via two mechanisms; either by slowing the cathodic reaction itself, or selectively precipitating on cathodic areas to limit the diffusion of reducing species to the surface. The mechanism of mixed effect corrosion inhibitors can be summarized as the three points below:

- Mixed inhibitors work by reducing both the cathodic and anodic reactions. They are typically film-forming compounds that cause the formation of precipitates on the surface, blocking both anodic and cathodic sites indirectly.
- The most common inhibitors of this category are silicates and phosphates.
- Silicates and phosphates do not afford the degree of protection provided by anodic inhibitors such as chromates and nitrites, however, they are very useful in situations where non-toxic additives are required.

Biocides are chemicals that can serve to kill bacteria and this terminology is used particularly to address the bacteria contributing to MIC reactions. Biocides can be divided into oxidizing and non-oxidizing, depending on the detrimental mechanism and effects on the targeted bacteria. Important features of biocides [2] as well as pros and cons of frequently used biocides [3] have been reviewed elsewhere.

The main point about both corrosion inhibitors and biocides is that as they are both classified as chemicals, their entrance into the environment must be carefully assessed and monitored on a systematic basis [4].

2.2.3 Electrical Treatment

CP is based on reversing the anodic reaction(s) by returning the electrons liberated via anodic reactions back to the metal. This feature of CP can be schematically shown in Figure 2.7. As the figure shows, essential elements of CP are to control electron release via anodic reactions, that is via corrosion, are to be controlled by supplying electrons by either sacrificial anode CP technology or impressed current CP. It is possible to also mix the two methods to make the electron pool available to the object to be protected even more powerful. In addition, based on Pourbaix diagrams, CP "pulls down" the corrosion potential toward more negative potentials, thus bringing the whole reaction to where the atom of the metal remains an atom and not an ion. This way, CP also serves to protect the metal against corrosion.

CP, like any other technology, has its own advantages and disadvantages. The list of these advantages and disadvantages can be too lengthy and we see no need to mention them here. However it is important to note that installing CP systems near metallic objects (above ground or underground) without applying required

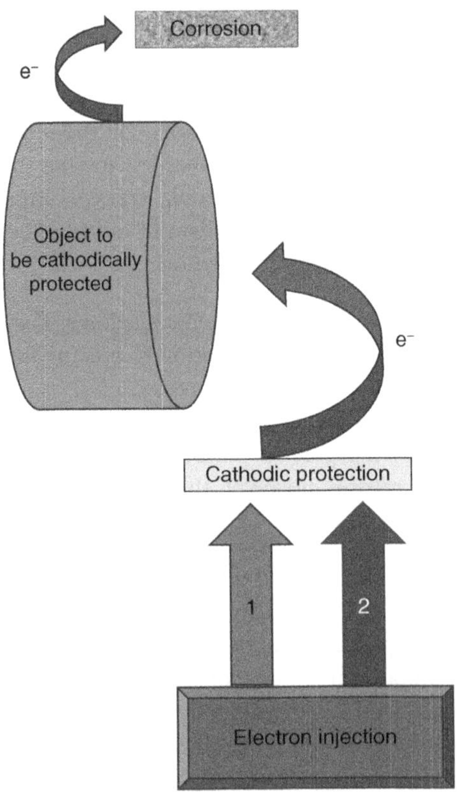

Figure 2.7 Conceptualization of the elements of cathodic protection (CP); 1: Sacrificial anode-galvanic CP, 2: Impressed current CP.

insulation as well not considering telluric effects on CP systems may actually decrease the workability of such systems. In addition, based on the locations these systems are installed, vandalization may also prove to be a serious matter.

2.2.4 Mechanical Treatment

PIG stands for Pipeline Inspection Gauge. Pigging is the act of running a PIG through a pipeline. The main aim of pigging from a CM point of view is not treating corrosion but somehow decreasing the likelihood of corrosion. Debris and scales formed within a pipeline are capable of facilitating the establishment of spots with differences in local partial oxygen pressure or concentration of chemical species. The former forms differential aeration electrochemical cells and the latter serve to form differential concentration electrochemical cells. In both cases, electrochemistry under the deposits will be highly likely to be leading to corrosion. Also, if temenos [5] (see footnote 1 in this chapter) forms, the corrosion products along with the temenos pieces can form deposits that would enhance the corrosive effects [6] of the deposits formed in this regards. This is when use of under-deposit corrosion will make sense. Pigging will damage and destroy these deposits and scales and thus will decrease the likelihood of under-deposit corrosion, by which these corrosion mechanisms may do detrimental effects on the mechanical integrity of the pipe.

Any action that will intervene with formation of deposits and establishing electrochemical cells by mechanical action can be grouped in this category. The deposits can be formed on the exterior of the asset, thus leading to external corrosion, or within the asset resulting in internal corrosion. Borrowing the terminology we have used in Chapter 3 to distinguish between corrosion control and corrosion prevention, mechanical measures can mostly be referred to as technologies to prevent corrosion by lowering the possibility of under-deposit corrosion in the way we previously addressed.

2.2.5 Physical Treatment

2.2.5.1 Paints, Coating Systems, and Premature Destruction in Industrial Facilities

Paints are very important in CM and its economy due to their high cost. Defects of paint and coating are inevitable. The users are trying to get optimal, long service time for paint systems.

During construction of refineries and industrial complexes, painting of structures and equipment programs is one of the last stages before getting started and commissioning. Managers and workers have a strong wish to get the job done, so painting teams, including contractors, inspectors, executors, and technicians, are under double pressure to complete the job.

Being in hurry leads to neglect of many standard considerations, and the results shorten the life of the paints and lead to their failure. The author and colleagues monitored seven petrochemical centers, two oil refineries, two oil offshore docks, and four gas refineries periodically during four years with emphasis on studying paint damages. Paint problems and solutions are reviewed in this chapter.

Type and size of the abrasive particle are important due to speed of work. Less attention to them may result low surface roughness, sand dirt, and surface contamination due to performing several operations simultaneously. In addition, many other factors contribute to the weakness of the surface preparation and that is one of the main causes of weakness in paints. Collecting atmospheric corrosion data and choosing the right coating system based on corrosion parameters is very important. It also helps in deciding the right time for the painting program.

Proper application of multi-layered paint systems, consideration the curing times and interlayer times, and observance of the weight balances of paint components and the skill of workers and technicians have led to successive results of long life paints and coating in many industrial projects.

Paint quality control tests almost always need a long duration time. However, managers and executives are in a hurry to start the operation of the units, and this sometimes leads to ignoring some necessary tests, and ultimately leads to using weak paints.

Defects of the paints often appear after the beginning of service. Several field variables play a role in coating defects, and these factors can work separately or intensify each other.

2.2.5.2 Features of Substrate

The material of substrate and shape of a surface are effective limits to useful life of coating and paint systems [7]. In industrial facilities, some base metals and subsurfaces include; steel, stainless steel, cast iron, galvanized steel, aluminum, zinc, copper, concrete, etc. The corrosion mechanisms of these substrates are various and each requires a different method of surface preparation before application of the coatings [7]. Stainless steel, zinc, and aluminum corrode slowly under the atmospheric condition [8].

Steel has a dense and smooth surface. It also has a high chemical reactivity with the atmosphere and pollutants. Therefore, complete de-rusting along with enough roughness are necessary in surface preparation.

Stainless steel 304 has 18–20% chromium by weight [8]. It has a much harder surface under normal atmospheric conditions than carbon steel, is smoother and denser, and is more resistant to corrosion under normal atmospheric conditions. It has many applications in atmospheric environments. Stainless steel strikes the rust point in an atmosphere containing chlorine and sulfur, however, it is more difficult for the coating to adhere to it. It is therefore recommended that you perform a weak and suitable acid wash on the surface of stainless steel before sand blasting, for proper adhesion of the primer [7].

Concrete is not dense and has a rough surface; its surface is porous and is vulnerable to the attack of undesirable chemicals such as acid, alkaline, and hydration of cement due to its surface pores. Therefore, it is necessary to create surface uniformity with the help of sealers which add consistency to the coatings. Concrete is also very sensitive to the application of acid base coatings.

Aluminum is a metal that has a dense and smooth surface covered with a layer of aluminum oxide. Alkaline materials degrade aluminum surfaces, and surface coatings on aluminum are less durable in alkaline environments. On the other hand, sulfur has many corrosive effects on iron, copper, silver, and lead substrates, but it is almost ineffective on aluminum; and coated aluminum is less problematic in a sulfur environment. One of the problems with aluminum substrate is the difficulty of paint adhesion on it. Anodizing is suitable for workshop treatments, but to overcoming this problem in the field, some chemical treatments and soft blasting are recommended.

Galvanized steel is prepared by hot or electrogalvanized zinc spray and provides better corrosion resistance in atmospheric environments. In the acidic environment, zinc reacts rapidly with acidic agents, and forms a white layer of zinc chloride on the surface in a chlorine-containing environment. If ions pass through the paint system layers on galvanized steel, whitening and surface porosity will appear.

A thin, pale green outer layer variously forms on the surface of copper by oxidation or other chemical processes, and is called a patina [9]. Although the patina can give a protective covering to the surface, it should be removed before painting.

Due to the high density and smoothness of copper surface, adhesion of the coatings to copper is weak. Copper has very good resistance in atmospheric conditions but is very sensitive in sulfur and acidic environments. It is recommended that the surface is prepared with a soft blasting and applying a very resistant coating for these environments.

Cast iron has from 1.8 to 6% carbon, and it provides a rather rough, porous surface, but it is much harder and brittle than the steel. Adhesion of coatings is weak, and the phenomenon of graphite corrosion is a major problem in immersed or buried conditions [7]. Sand blasting and using a suitable paint and coating system is essential.

Some examples of coating degradation on various substrates shown in the Figure 2.8.

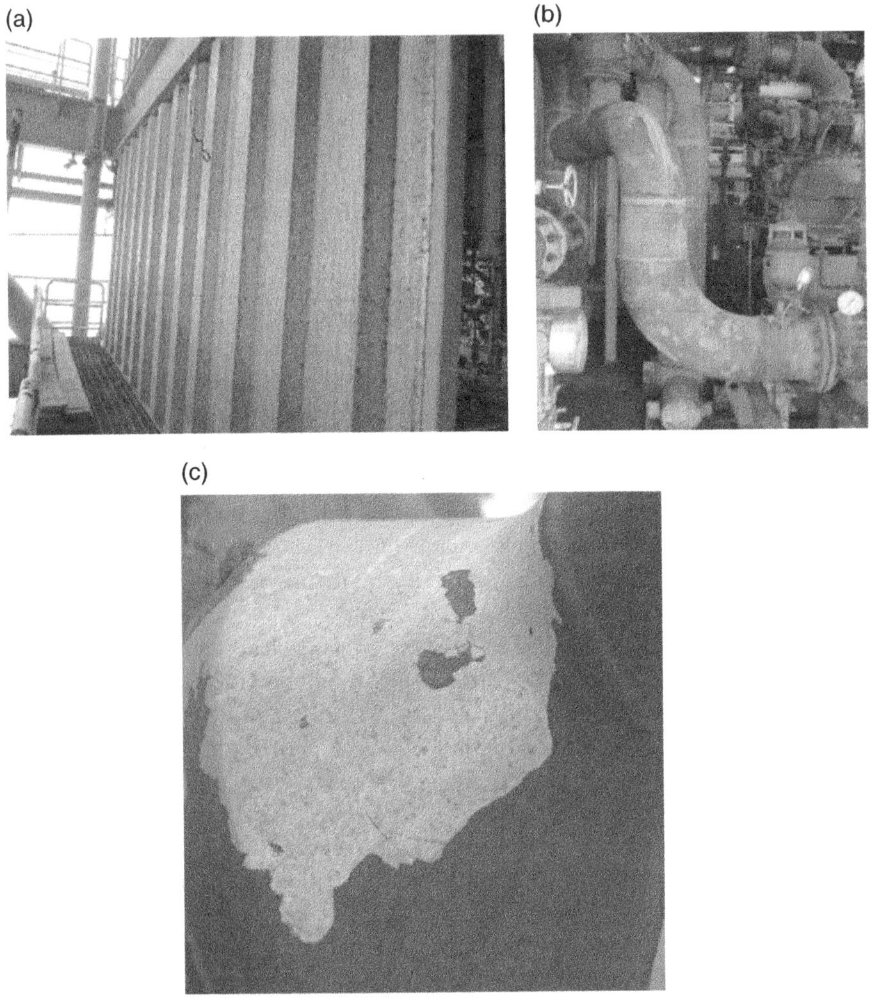

Figure 2.8 Paint degradation on various substrates. (a) Rust on stainless steel in an oil platform in the Persian Gulf. (b) Rust on water-line copper pipes in a gas processing unit. (c) Paint defect on cast iron.

2.2.5.3 Characteristics of the Environment and Local Features

The first cause of premature paint degradation is improper choice of paint for the intended application. Lack of exact knowledge on the type of paints, environmental properties, and operating conditions of the equipment leads to improper choice of paint. Figure 2.9 shows that the manufacturer of the device as well as the manufacturer of the paint did not predict the possibility of chemical leakage on the equipment. Obviously, the used paint system rapidly degraded by the spillage of this material.

Figure 2.9 Paint defects because of Amin leakage on the equipment.

Industrial paints are needed in a variety of environmental and climatic conditions. Different corrosion mechanisms of each region have led to the choice of specific paint systems. The atmosphere of offshore oilrigs and inlands is full of chloride ions. The inland petroleum and chemical industries are associated with factors such as NO_x and sulfur compounds. Carbon dioxide compounds are more common in large industrial cities. Large oil complexes often have sour oil-related mercaptans (thiol) and H_2S vapor. Various acid vapors such as sulfuric acid, phosphoric acid, urea, and ammonia are abundant in petrochemical plants. The passage of oil and gas transmission lines under the seabed is of special importance due to the high concentration of marine salts and the relatively high temperature of the substrate.

Environmental contaminants, especially chloride and sulfur, affect the atmospheric corrosion rate. Variations of the conditions in the area of oil and chemical facilities makes it difficult to choose a suitable paint system for each piece of equipment. Drawing an atmospheric corrosion map is very efficient for this propose at industrial complexes [10].

There is a lot of variety of liquids inside the tanks in a petrochemical complex. In addition, there is a huge difference in the internal ambient temperature of containers, vessels, tanks, etc. in industrial facilities. All of these environmental parameters must be considered when choosing paint and coatings for each application.

Recognition and classification of atmospheric and/or immersion environments, determination of their degree of corrosion, and choice of proper paint systems for each class is briefly explained in some standards [11, 12].

By measuring environmental characteristics and operating conditions such as substrate temperature, it is possible to extract the proper type and thickness of a suitable paint system and the number and thickness of layers based on standards [11].

Users must have knowledge of the standards for choosing the suitable paint systems for various industrial applications. They must be familiar with the environmental factors affecting the early degradation of paints.

Some of the factors needed to consider for choosing paint systems are as follows.

1) The substrate surface conditions;
2) environmental classification and determination of the environmental corrosion class;
3) the operating temperature of the equipment and the allowable service temperature;
4) the effect of UV rays;
5) the presence or absence of CP of substrate;
6) optimal life of the paint system;
7) possibility of workshop and/or field application;
8) the need for a special color of paint for luxury and/or identification;
9) health, safety, and environmental restrictions;
10) surface preparation and application conditions;
11) previous experiences;
12) possibility and cost of repairs; and
13) paint properties.

Lack of knowledge on the properties of paints and improper use of them is another major cause of paint degradation. If the inspectors and consultants of the industrial production of equipment do not have accurate information of the properties of ordered paints, unsuitable choices will cause premature destruction of the paints. For example, epoxy paint that is used in the open industrial environment of an oil refinery and exposed to direct sun radiation is rapidly damaged due to the weak resistance of the epoxy to the ultraviolet rays of sunlight (Figure 2.10).

In addition, due to the lack of knowledge about the method of curing ethyl silicate paint (requires high humidity for curing), its use in an industrial complex located in a dry environment has led to lack of primer curing, and ultimately the destruction of the paint system. This mistake was repeated for three times (Figure 2.11).

Therefore, in order to reduce or avoid paint problems, it is necessary for inspectors and users of industrial paints to have enough information about the properties of paints. Many documents have written with the knowledge of the type, classification, and use of paint [11, 13, 14].

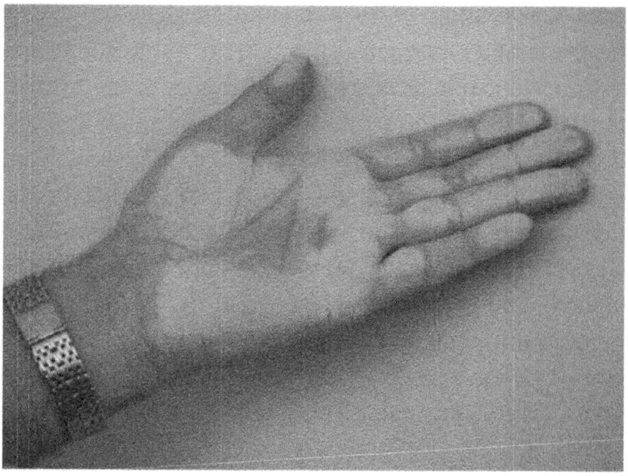

Figure 2.10 Paint checking due to weak resistance of the epoxy to the sunlight in an oil refinery.

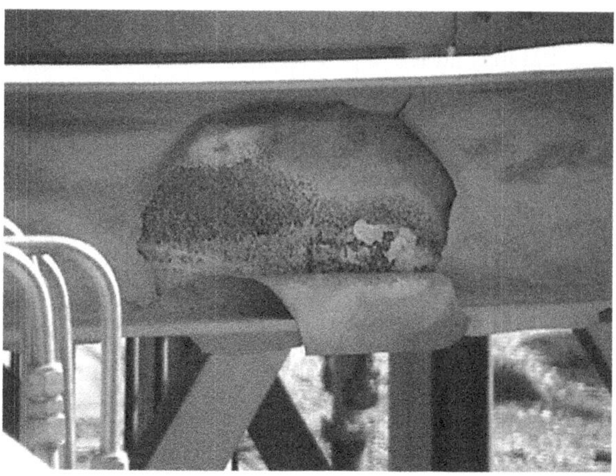

Figure 2.11 Lack of curing of ethyl silicate primer used in dry area.

The four main components of industrial paints, including pigments, binders, solvents, and additives play their own special role in paint performance. In addition, paint systems consist of appling industrial paints in continuous and sequential layers, each layer having certain properties.

Pigments play an important role in improving the quality of paint. Although some pigments have decorative properties, they also play an important role in optimizing the texotropic properties and rheology of the paints, such as reducing sagging and dripping problems. The strength of the paint and the gloss and hiding properties on the previous layer are other characteristics that are enhanced by the pigments. They also play an important role as corrosion inhibitors in the primer.

In the simplest classification, pigments are divided into two main groups; pigments and extenders [7]. Pigments include zinc powder and zinc oxide, iron oxides, titanium dioxide, lead chromate, basic lead-silico chromate, molybdate, zinc chromate, lead cyanimides, zinc powder, blue iron (Prussian blue), aluminum pigments (powder and paste), bronze, mercury oxide, etc. and their chemical properties are explained in handbooks and factory catalogs.

Zinc powder is one of the most common primer pigments. It is sometimes packaged as a third part for primers. Often, not observing the exact mixing ratio or insufficient stirring leads to poor wetting and causes a complication in the dry film (Figure 2.12).

Titanium dioxide is one of the most common pigments used in paints. Titanium is present in the structure of anatase and rutile. The rutile type consists of a denser crystal lattice than anatase, and it has a higher refractive index, and this type has good resistance to deterioration. The use of anatase as pigments of the top coat paints leads to severe chalking when the system is expose to sunlight (Figure 2.13).

Extenders are solid minerals that may added to paint formulation along with the pigments to modify some of the properties of the paints as well as to reduce the cost of production. They are also used to improve the following properties, and optimizing quality of paints.

1) Adjusting and optimizing the ratio of pigment to binder to meet suitable PVC.
2) Increasing the physical strength of the paint film.

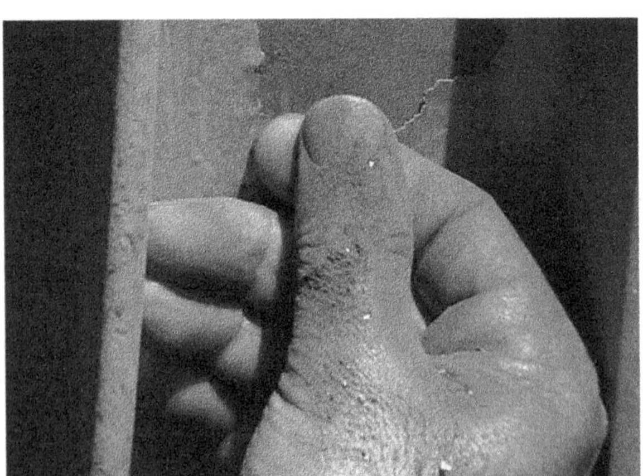

Figure 2.12 Poor wetting and undesired mixing of primer components.

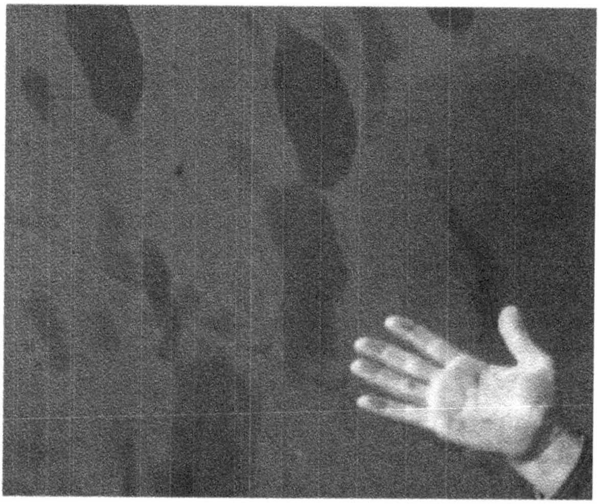

Figure 2.13 Chalking when the system exposes to sunlight because of the use of anatase TiO_2 pigments of the top coat.

3) Improving the rheological properties of paint.
4) Increasing the resistance to ion penetration.
5) Reducing the effect of microbial agents.
6) Optimization of optical properties.
7) Resistance or electrical conductivity.
8) Thermal resistance and fire retardant development.

Some extenders include calcium carbonate, white carbonate, aluminum silicate, magnesium silicate (talc), mica, silica, calcium sulfate, asbestos, barium sulphate, and zinc sulphate. Their chemical properties described in the reference [7].

Organic pigments and metal pigments also used in a few applications whose properties are described in some sources [7, 15]. For example, the use of the pigments of aluminum, copper, nickel, iron, stainless steel, zinc, tin, etc. in paints were described [7].

The resin is most important to form a uniform and continuous film on the metal surface. This component used in the paint formulation allows for adherance to the substrate and allows the continuity of paint components to bond in the dry film. It should also create a dry film that prevents the diffusion of water vapor and various corrosive contaminants from the environment to the subsurface.

Various polymers are used as binders in paints. Their properties, such as the molecular weight of the polymers, the mechanical strength of the bonds formed in film, the permeability of the film, thermal resistance, etc., are effective in the paint and coating capabilities. The paint curing mechanism is the most important method for classification of binders.

There is a group of binders in which the polymers react with air humidity or oxygen of the air for curing; they are known as air-dry binders. Drying oils and phenolic varnish castor oil (polished oil) are natural air-dry and alkyd binders; vinyl alkyd, epoxy ester, silicon alkyd, etc. are synthetic air-dry binders [7]. This group of binders is used in single-component paints. They begin to form a skin during shelf time, especially when the ambient temperature rises or the can or container lid is left open or the container damaged. It is also important to adjust the ratio of surface drying and deep drying additives in these paint formulations.

There is another group of binders where the polymerization reaction of the binder is completed during fabrication at the factory. When they apply to the substrate, the paint sticks to the substrate and forms a dry film only by the evaporation of solvent. Nitrocellulose binders, vinyl resins, and chlorinated rubbers are of this type. These paints are a single part similar to the previous group. Paint solvents play an essential role in the drying speed and only the use of suitable solvents made by the same factory is recommended for them to reduce the paint defects.

Another group of paints is known as co-reactive, and is packaged in two or three separate containers during storage. In order to start the polymerization reaction, they have to be mixed in right amounts and the mixture needs to rest for some minutes before application. The main binder is a pre-polymer and includes some other paint constituents packaged separately. Hardeners are other chemical compounds that are needed for initiation and to continue polymerization reactions and which are stored in separate container. The contents of the containers are added to each other before application. In this way, complementary polymerization reactions of the paint take place after mixing and application. Epoxy, polyurethane, polyester, and coal tar epoxy paints are of this type. To cut the defects in this group of binders, it is necessary to obey the exact ratio of mixing of the paint and hardener. It is often suggested to wait half an hour after mixing to start the pre-reactions and then start diluting and applying. It often takes more than a few days for full curing of these binders. Therefore, in the primer and middle layers, we have to wait at least 24 hours before the next layer and for the topcoat paint, and at least one week before service operation.

The polymerization of another group of binders can be completed by increasing the temperature to a certain extent; paints containing these binders are known as thermal curing. Pure phenolic paints and phenolic epoxy are two well-known categories of thermal curing [7]. It is possible to meet their last curing at the temperature of 230 °C.

Silicon binders are another heat-curing polymer. It is a mineral-based heat-bonding binder with the main constituents of silicon and oxygen. Polysiloxane-type silicone binders are also used in paints; this binder has a very good resistance to chemical agents due to having silica in the formula. Silicone binders are suitable for use at high temperatures (540–640 °C) and may be used to make paints resistant to the same temperature. For example, when using ceramic pigment particles, using them at up to 760 °C is recommended. This binder may also modify with alkyd for use at lower temperatures.

This group of paints has a very low mechanical strength before the final curing. Temperature of curing is mentioned in the manufacturer's catalog temperature. Therefore, special considerations are required when they are applied on large chemical equipment, especially before providing the curing temperature, which is to be provided at the start of activity of the factory in terms of the required chemical process (es).

Solvents may be used in paint formulation for two main purposes: first, to dilute the binders that allow the pigments to mix properly; and second, to balance the evaporation rate, which allows the wet film to dry over a desired time. They also play an essential role in promoting the rheological properties (fluidity) of paint.

Two categories of solvents can usually be used for each paint formulation, these are known as main solvents and auxiliary solvents. In terms of chemical structure, the main solvent has a special chemical group or band in its molecules that can better dissolve the binder. Auxiliary solvents can be used to optimize other properties, such as drying rate. Therefore, specific solvents for dilution of each paint may be used by the manufacturer, and using other solvents may be associated with complications.

Additives may also be added to formulation for optimizing some quality properties of paints. Complete information on specific applications for additives are discussed in scientific and commercial documents [16].

Sometimes there are some problems with the paint during the storage period, and they can only be recovered by returning it to the factory. The factory will improve the properties of the paint again by using additives, solvents, and homogenizaton.

In order to meet the proper performance of paints in industrial facilities, they apply a paint system in several layers. In general, the role of layers can summarized as follows.

2.2.5.3.1 Primer

The main role of the primer layer is to prevent corrosion reaction or to change reaction path on the metal surface to form a stable corrosion-resistant surface. For example, zinc content is very important for the CP of the steel substrates in zinc-rich primers. It is also essential that the primers have good adhesion to the metallic substrate and to the middle layer.

2.2.5.3.2 Intermediate Layers

Thick, continuous, and cross-linked intermediate layers will prevent the penetration of corrosive ions from the top surface to the substrate. This layer will also have proper adhesion to the primer and top layers.

2.2.5.3.3 Topcoat

Resistance to atmospheric factors, such as ultraviolet rays, is the most important role of the topcoat layer. Special properties such as abrasion, scratches, resistance to the growth of microbial agents, etc. are also dependent on this layer. The top layer should have good adhesion to the middle layer and have good decorative properties.

One of the causes of paint system degradation is the incomplete use or displacement of layers. In addition, in local repairs and touch-up paints, it is often observed that a new primer is applied on the middle layers or the old paint surface. This mistake can lead to premature damage to the paint. Removing the old topcoat layer, neglecting the new primer, and starting the implementation paint system from the new intermediate layer and then the new topcoat is a more proper method.

It is essential that paint users and paint inspectors heed the comprehensive scientific information of paint types and their potential capabilities on corrosion prevention.

2.2.5.4 Paints Quality Control

Sometimes, for economic reasons, the paint manufacturer substitutes the use of standard raw materials for some inexpensive ones. In addition, other reasons such as lack of proper formulation, exhaustion of equipment, use of raw materials in long storing periods, etc. lead to poor quality paint production.

Absent a valid, reliable QC laboratory, the risk of using weak paints is high. Therefore, sampling and sending for quality control tests is necessary by inspectors. In Figure 2.14 some typical complications of the paint system arising from the poor quality of the paints are seen.

It is recommended that you send samples of the paints to a reputable laboratory center for quality control to make sure their quality is in accordance to the standards at least three months before application.

In addition, before beginning the surface preparation in the field, check that the type and amount of paints, hardener, and solvents of each layer is proper and under the manufacturers recommendations.

Checking the color shade, density, and weight of the paint for each layer and reviewing its conformity according to the catalog and data sheets is also recommended. Other features to be careful about are appearance of the wet and the dry film (WFT and DFT) as well as possibility of complications for each layer preferred.

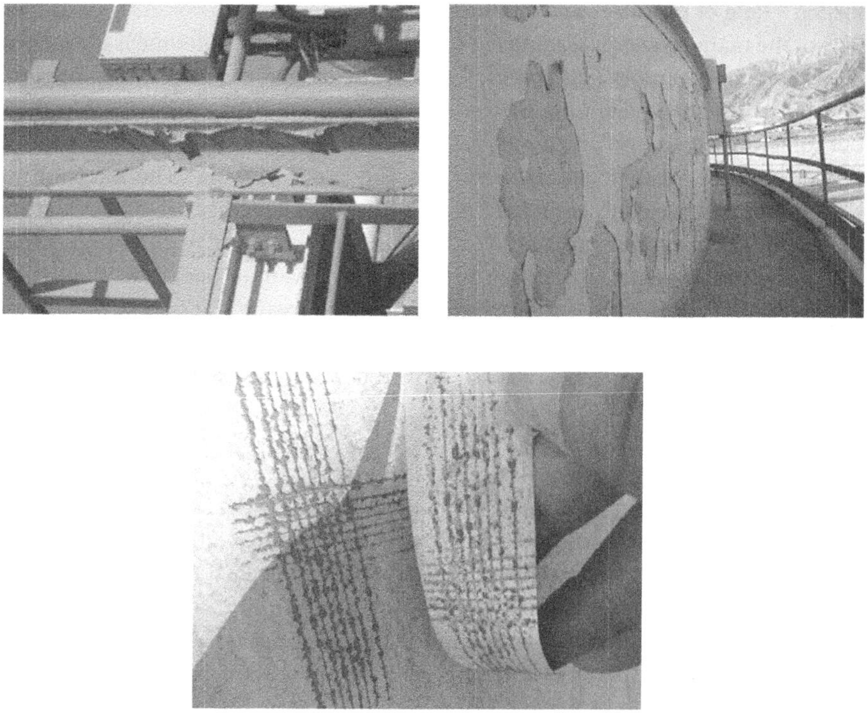

Figure 2.14 Some typical complications of the paints resulting from poor quality paint features. Weak adhesion of top coat by one manufacturer on the primer (left). Weak adhesion of topcoat to intermediate (right). Weak adhesion of intermediate to primer (down).

2.2.5.5 Paint Warehousing and Storage

One of the problems for paint application projects is the long-time duration between the paint production and its use in the field. Most industrial users buy paint a few months before using it due to the long QC test time, and storage of paints in most projects can be up to a year. In some countries, due to political and economic reasons early forecasting and purchasing are also needed.

In the executive projects, a separate warehouse is not provided for the paints and they may be stored with other equipment in a public warehouse. The standard conditions for paint storage, such as setting the right temperature, are not indicated most of the time; this is another problem that reduces the paint quality. The method of storing and using paints during storage is very important and stated in manufacturer catalog.

2.2.5.6 Role of Executors and Contractors

One of the main problems leading to the reduction of a paints lifetime is the lack of knowledge of executors and contractors and the lack of a precise working plan. The planning and scheduling of work in many cases is based on the weather, but sometimes it is very difficult to predict the field conditions and climate change can disrupt the work schedule, accuracy, and order of work. The selection of skilled and capable contractors and providing proper facilities for implementing the standards of surface preparation and paint application play an essential role in the life of the paint. In addition to being fully aware of operation standards, the contractor must have the proper equipment to assure the health and safety of the workers and comply with the local environment regulations. Examples of unqualified workers and contractors are seen at Figure 2.15.

2.2.5.7 Surface Preparation

One of the most important phases of a paint operation is the surface preparation, which sometimes costs quite a bit. Appropriate weather conditions, good blasting materials, equipment, and proper technical skills are needed to prepare the surface. Most of the paint problems of industrial equipment are due to the weakness of the surface preparation. Some examples of these problems are observed during the field operation inspections seen in Figure 2.16. In scientific documents, special attention is referred to this issue [11].

Dry abrasion blasting with silica sand, copper slag, steel grits, etc. is the most correct and common method of field and workshop surface preparation. In a scientific research, the author investigated the surface preparation factors affecting the primer adhesion to steel substrate; the primer was an epoxy zinc phosphate with a dry thickness of 60 μm. The surface was prepared by different methods. The results are according to Table 2.1. Primer adhesion on all samples was acceptable, however, there was a noticeable difference between the adhesions of the

(a) (b)

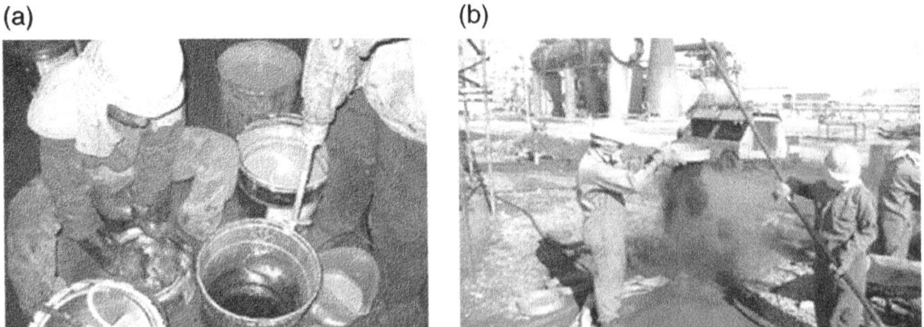

Figure 2.15 Examples of unqualified workers.

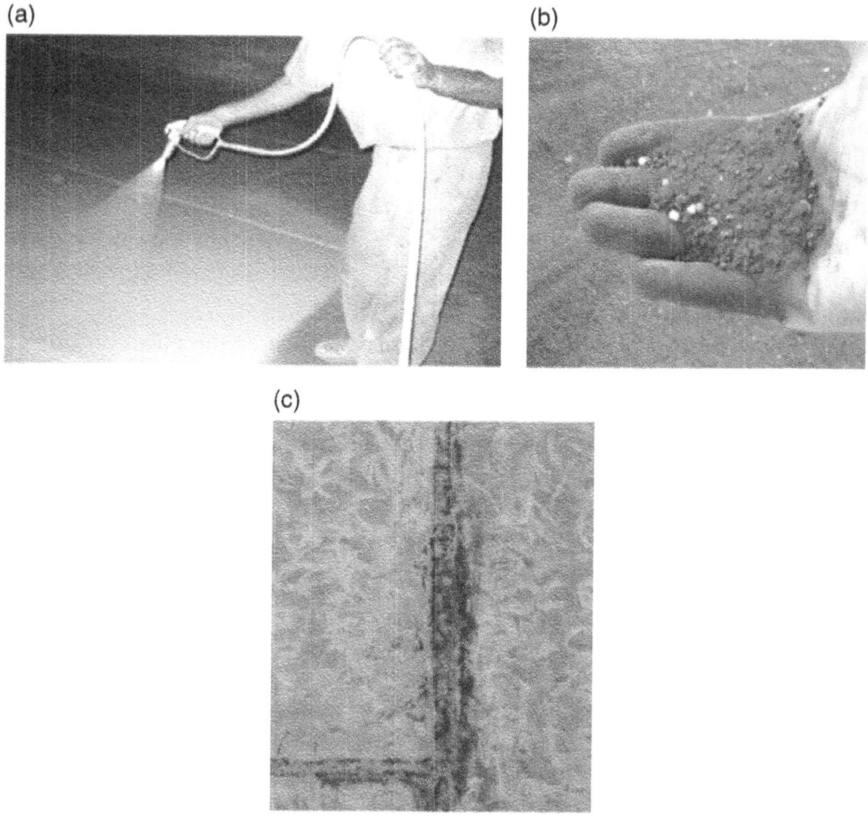

Figure 2.16 Some weakness of the surface preparation. Paint application on rusted surface (left). Unsuitable abrasives and reuse of them several times (right). Weak surface preparation on weld lines (down).

Table 2.1 Adhesion of a common zinc phosphate epoxy primer on the steel surfaces prepared with several methods.

Preparation methods	Adhesion (MPa) ASTM D4541	Disbanding condition
Washing with white sprite	1.73	Paint layer from metal surface
Wire brush ST3	3.19	Paint layer from metal surface
Weak acid washing	2.89	Paint layer from metal surface
Hard acid washing	4.48	Paint layer from metal surface
Phosphating	4.30	Paint layer from metal surface
Wet sand blast, water blast	8	Paint layer from metal surface
Dry sand blast Sa3	11.63	Paint layer from metal surface

primer to the steel surface prepared by different methods. Between some common methods of surface preparation, abrasive blasting especially by sand, to the degree of cleanness of Sa3, shows the highest quality on the paint adhesion.

In other research, the effect of different types of abrasives on steel preparation was investigated. Physical and chemical properties of tested abrasives are discussed in [17–19].

Properties of some abrasives are determined according to Table 2.2. In a series of tests, the author examined and compared the adhesion of a zinc phosphate epoxy primer with the same thickness on steel sheets that were prepared with these different abrasives. Results are summarized in Table 2.3.

Often, the functional performance of abrasives, such as the effect on primer adhesion is very important. It affects the mechanical stability of the film, as well as its chemical resistance, especially when the film is exposed to humid and saline environments during service time. The author investigated the effect of a number of abrasives on cured film performance by doing several tests. For that, some samples were blasted by different abrasives. They were coated with the same paint, having nearly dry film thickness. Then they were tested in a humidity chamber according to ASTM D2247, and in a salt spray chamber according to ASTM B117 [20, 21].

In these experiments, silica sand compared with different abrasives used for blasting surface. A noteworthy point about carbon steel samples prepared with copper slags was that their particles trapped in all samples. Slag particles were observed which could not washed out even by using air pressure (five bar pressure) and washing with solvent. However, for samples prepared with silica sand, leaving excess particles from the surface with compressed air were easily and completely removed from the surface. All samples were coated with the same paint layer with a similar dry film thickness.

Table 2.2 Properties of tested abrasives in blasting.

Test	Standard	Result for all samples
Hardness (mohz)	SSPC-AB-1	At least 6
Weight change on ignition	SSPC-AB-1	Max. 5%
Oil content	ASTM D-7393	No visible
Specific gravity	ASTM C-128	Min. 2.5
Moisture content	ASTM C-566	0.5%<
Water soluble contaminants	ASTM D-4940	<1000 μs/cm

Table 2.3 Effect of abrasives on adhesion of primer at humid and salty environments.

Abrasives	Adhesion test before exposure ASTM D4541 Psia	Adhesion test; 1500 hours salt spray and 24 hours drying ASTM D4541 Psia	Adhesion test; 1500 hours humidity and 24 hours drying ASTM D4541 Psia
Copper slag Sample A	1687	528	1603
Copper slag Sample B	1541	486	Test error
Copper slag Sample C	1569	388	1160
Copper slag Sample D	1446	516	1223
Cupper slag Sample E	1355	422	1306
Supper slag Sample F	1525	1329	1385
Silica sand Sample G	1687	1855	989
Silica sand Sample H	1900	1917	1898

As seen in the Table 2.3, after complete curing and before beginning the corrosive tests, the adhesion of all samples was in acceptable range. Separation of all samples were with an interlayer that did not return to the abrasive quality. The sheets were prepared with different abrasives, placed and tested for resistance in a humidity cabin environment at 100% humidity, and similarly in a salt spray chamber for 1500 hours, separately. The results were interesting; the samples prepared with abrasive silica sand showed a significantly better resistance than others. In addition, they had more strength adhesion at the end of the laboratory exposure tests. The results are shown according to Table 2.3. Therefore, type, shape, and size of the abrasive material have a significant effect on long life of paints in the abrasive blasting method.

2.2.5.8 Technical Painting Operations

Painting application requires special skills; the inexpert technician can lead to the destruction and complication of the film. Examples of paint application problems are shown in Figure 2.17. It is necessary for the executors be aware of the causes of

(a)

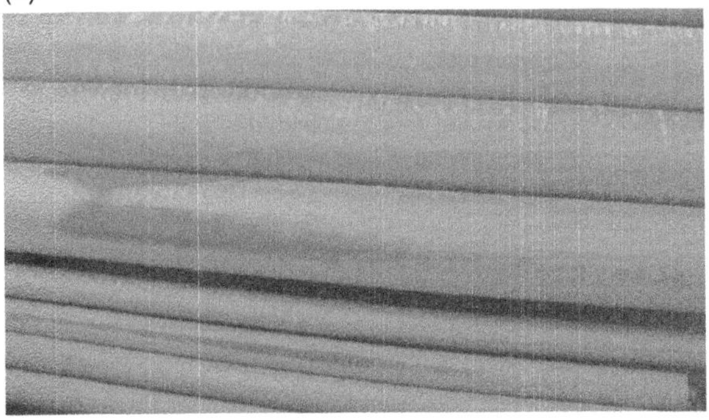

(b)

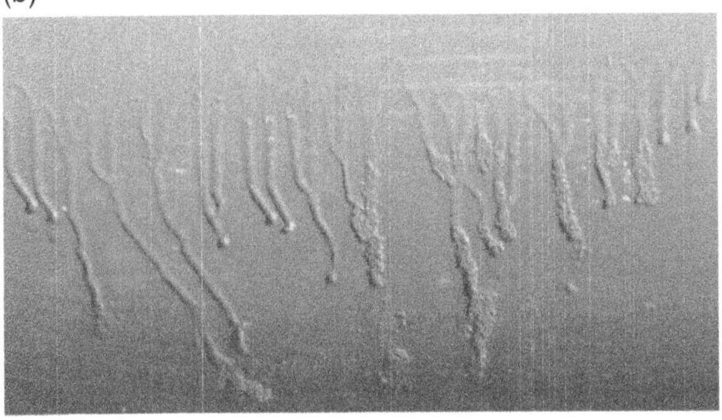

Figure 2.17 Some examples of paint application problems. Rusting on primer due to thin primer layer (a). Running of paint and settling of dirt due to adjacent blasting operation (b).

paint defects and the method of avoiding, eliminating, and correcting these complications.

Many paint defects are mainly related to the method of their application. The wrong choice of hardener and thinner for paints, incorrect mixing of paint components, and wrong choice of different layers of a paint system are the most common causes of complication in paint application. Accuracy at the important points in the catalog can cut problems. Lifting, orange peel, overspray, pinholes, spatter coat, etc. are some of the complications that may occur due to these problems [7].

Excessive hot or cold temperature of the substrate, high humidity, dust, and local pollution are other factors that may cause complications during paint application. Adherence to implementation instructions greatly reduces these problems. Complications such as amine blushing, blushing, holidays, rusting, etc. are observed in these conditions [7].

Inadequate regulation of paint viscosity and spraying pressure in airless spraying, as well as inadequate air-to-paint ratio in air spraying methods are other causes of complications during application. These factors lead to difficulty in achieving the right thickness as well as complications such as sagging and running, etc.

2.2.5.9 Inspection and Management

Industrial paint inspection plays an important role in the useful life of paints and paint systems. Unfortunately, in certain circumstances, surface preparation and paint application is performed without the necessary care by inspectors. For example, for problems referred to in Figure 2.18, the paint was applied only on visible sections. In addition, the primer was applied to the previous existing paint, which is another gross mistake of inspection.

On the other hand, a number of refinery project managers have no technical approach to the issue of paints and some of them consider it as one of the last priorities. Unfortunately, sometimes the project manager tries to compensate for the delay in completing the project by reducing the surface preparation and paint application time, and orders for increasing the speed of the paint operations.

Paint is very important as the most common way to prevent corrosion. It also gives beauty to the facilities and will give the workers a sense of security and safety while doing their work. Lack of attention to the problems of paint leads to the rapid development of damage, especially in chemical and marine facilities that are more

Figure 2.18 Weak inspection and management.

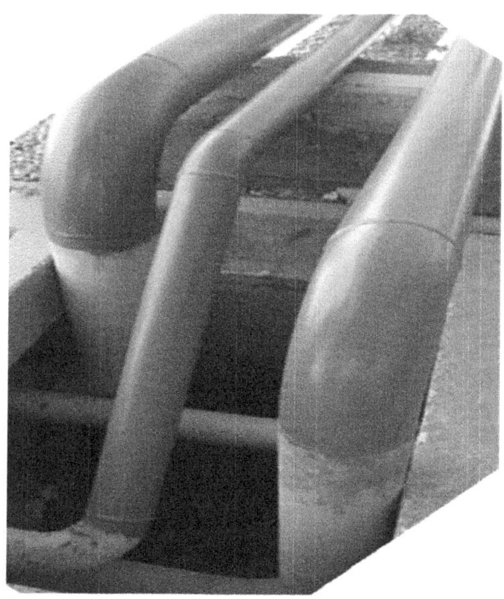

likely to exhibit paint problems. See some examples in Figure 2.19. In this comparison, the intervals of periodic visits were only two years.

Full documentation of the operation inspection can significantly reduce paint defects during the service life of the equipment. These documents show compliance with the standards. They show precision at choosing the right paint systems,

Figure 2.19 Rapid development of damage, especially in chemical and marine facilities that are more likely to exhibit paint problems. Defects developed during two years. Documented at May 2009 (left). Documented at October 2011 (right).

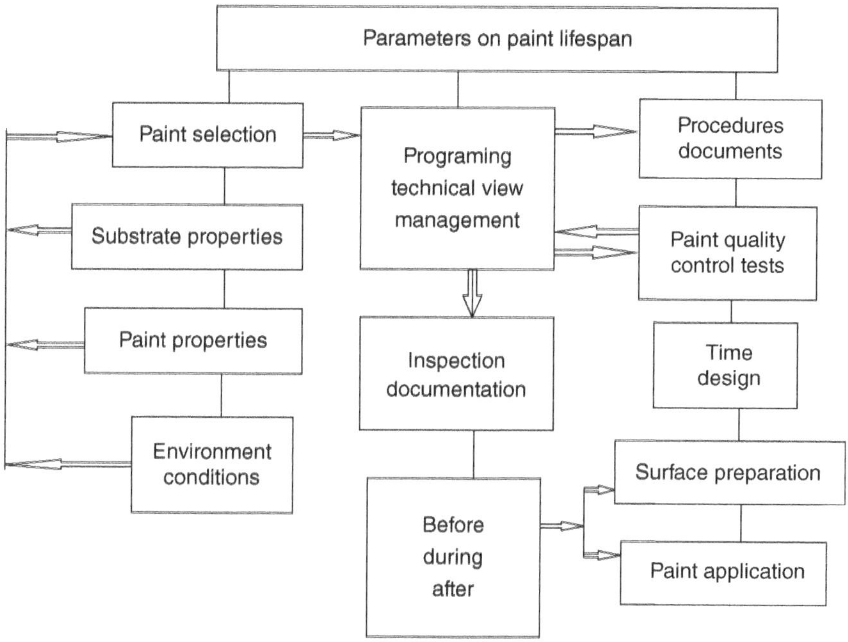

Figure 2.20 Effective parameters on paint useful lifespan.

performing the correct surface preparation, and good application. So, attention paid to proper instructions for the items listed in Figure 2.20 is essential.

2.3 Conclusion

Science of corrosion is based on understanding three main elements: what metal liberates electrons (thus becomes an anode and is consumed), which material takes electron (and becomes a cathode), and what fluid ions can move through easily. It follows, then, anything that can interrupt one or all of these elements is highly likely to affect corrosion and control it. If, for example, paint (more correctively, coating) is applied, then anode and cathode will not see each other or anode–cathode will be isolated from the external electrolyte; the former will not allow electron transfer and the latter will not allow ion exchange.

In this chapter, three main methods for corrosion prediction (SHE, Galvanic series, and Pourbaix doagrams) as well as categorizing five technical managements of corrosion have been briefly explained. In addition to a short description of each of the five categories, a rather detailed explanation about materials selection and paints has been given to allow the readers a feeling of some of the difficulties involved in technical treatment strategies to deal with corrosion.

References

1 https://agmetalminer.com/metal-prices, last visited 2 January 2021.

2 Javaherdashti, R. and Akvan, F. (2020). *Failure Modes, Effects and Causes of Microbiologically Influenced Corrosion: Advanced Perspectives and Analysis*. Elsevier.

3 Javaherdashti, R. and Akvan, F. (2017). *Hydrostatic Testing, Corrosion, and Microbiologically Influenced Corrosion: A Field Manual for Control and Prevention*. Boca Raton, Florida: CRC Press.

4 Javaherdashti, R. and Nikraz, H. (2010). *A Global Warning on Corrosions and Environment: A New Look at Existing Technical and Managerial Strategies and Tactics*. VDM Germany.

5 Javaherdashti, R. and Alasvand, K. (2019). *Biological Treatment of Microbial Corrosion*. Elsevier.

6 Javaherdashti, R. (2015). Corrosion and Biofilm. In: *Biofilm and Materials Science* (eds. H. Kanematsu and D. Barry), 79–83. Springer.

7 Munger, C.G. and Vincent, L.D. (2014). *Corrosion Protection by Protective Coatings*. Houston, TX: NACE International ISBN: 575902621 9781575902623.

8 Schweitzer, P.A. (2007). *Fundamental of metallic corrosion: Atmospheric media corrosion of metals*. CRC Press – Taylor & Francis.

9 Přikryl, R., Smith, B.J., and European Geosciences Union. General Assembly; Geological Society of London (2007). *Building Stone Decay: From Diagnosis to Conservation*, 295. ISBN: 78-1-86 239-218-2. Geological Society (accessed 26 September 2012).

10 Ghanbarzadeh, A., Neshati, J., Bagherzadeh, M.R. et al. (2013). Atmospheric corrosion map of an oil refinery. *Anti Corrosion Methods and Materials* 60: 106–114.

11 ISO 12944-Part1 to 8, Paints and varnishes – Corrosion protection of steel structures by protective paint systems. *ISI standard*, 2017.

12 BS 5493, Code of practice for Protective coating of iron and steel structures against corrosion (Formerly CP 2008), UDC 624.014.2:691.71:620.197.6, BS Standard, 1977.

13 Akbarinezhad, E., Faridi, H.R., and Ghanbarzadeh, A. (2009). Evaluation of zinc rich ethyl silicate primer by applying electrochemical methods. *Journal Surface Engineering* 25 (2): 163–166.

14 Hess, M., Mamburg, H.R., and Morgans, W.M. (1979). *Paint Film Defects, Their Courses and Cure*, 3e. London: Chapman and Hall ISBN: 47099018X 9780470990186.

15 Solverchem (2017). *Industrial Paints Formulation Encyclopedia 2*. Solverchem Publications Company, ISNN 9786056715730.

16 SSPC Painting Manual (2009). *Systems and Specifications, Society for Protective Coatings*, vol. 2. Pittsburgh, PA: SSPC.

17 Bagherzadeh, M.R., Jorsaraie, A., Ghanbarzadeh, A. et al. (2018). Preparation and investigation of anticorrosion performance of water-based epoxy coating containing nanohybrid of supercritical CO_2-synthesized self-doped polyaniline – expanded graphite. *Journal of Coatings Technology and Research* 15: 1–14.

18 Akbarinezhad, E., Ebrahimi, M., Sharif, F. et al. (2014). Evaluating protection performance of zinc rich epoxy paints modified with polyaniline and polyaniline-clay nanocomposite. *Progress in Organic Coatings* 77 (8): 1299–1308.

19 ASTM D4541 – 17 (2017). *Standard Test Method for Pull-Off Strength of Coatings Using Portable Adhesion Testes*. West Conshohocken, PA: *ASTM International*.

20 ASTM D2247 – 15 2020 (2020). *Standard Practice for Testing Water Resistance of Coatings in 100% Relative Humidity*. West Conshohocken, PA: *ASTM International*.

21 ASTM B117 – 19 (2019). *Standard Practice for Operating Salt Spray (Fog) Apparatus*. West Conshohocken, PA: *ASTM International*.

3

Smart Corrosion Management Elements

Reza Javaherdashti[1] *and Faranak Javaherdashti*[2]

[1] *Eninco Engineering B.V., Hengelo, The Netherlands*
[2] *IMI (Industrail Management Institute), Tehran, Iran*

3.1 Introduction

Management of corrosion differs from corrosion management (CM) in that it has a strong management component attached to it. Based on all definitions given so far for CM, it can be inferred that CM deals with the risk of corrosion and this is, at its best, just half of the truth about management of corrosion.

In this chapter, we want to look at a concept that so far has been known as CM from a different angle. We would like to address confusion that we have observed in our contacts with industries; and that is mixing up CM with corrosion engineering with a sugar coating of some cost analysis. Even in CM-related papers and documents published by reputable corrosion authorities, approaches toward CM seem to have been a mix of diagrams, procedures, and step-by-step documentation procedures that do not address anything but a technical issue related to corrosion; premature failure of a part in a system, coating disbondment, or concrete spalling.

However, as the very name implies, CM is composed of two parts: corrosion, which is a technical issue; and management, which is related to what humans do. In other words, any CM approach would have to include engineering/science of corrosion in addition to the human factor related to design and application of strategies, procedures, policies, and rules put forward by the management team or the manager.

To overcome this gap of understanding that exists about CM even among professionals, we have introduced a new concept that we call Corrosion Knowledge Management, or briefly CKM. In this chapter, we will concentrate on both CKM

Corrosion Policy Decision Making: Science, Engineering, Management, and Economy, First Edition. Edited by Reza Javaherdashti.
© 2022 John Wiley & Sons, Inc. Published 2022 by John Wiley & Sons, Inc.

and CM. The reason is that CM and CKM are not opposite to each other, but they must be regarded as completing each other.

In the following sections, we will also put emphasis on the concepts of Risk and Importance; these two terms are in fact the most significant sources of managers taking corrosion lightly. The reason, as we will discuss, is that risk of corrosion is not the same as its importance and vice versa.

We will order the topics to be discussed in this chapter by first touching the sensitive issue of risk and its vast difference from importance. We will then continue with CM and CKM.

3.1.1 Risk, Importance, and How They Are Interrelated?

Risk has different perceptions, although one can find meaning for it. Perception of risk is important to be defined clearly and within a certain frame so that it will be communicable to others, meaning that it will be possible to transfer to others the importance and gravity of the incoming danger. In everyday life, risk is usually associated with anything from working with kitchen shears to driving; there is always the risk of cutting your hand or hitting a car. It is in this context that risk and its significance must be of an essence that can be realized by others. Driving is an activity in which if the real value of risk of hitting each other is understood and appreciated by all drivers, so the risk of driving becomes controllable and manageable. The importance of risk from an engineering point of view is another story. That is because risk in engineering terms is much clearer and more attainable than the usual meaning of it in day-to-day activities.

For a corrosion engineer, risk of corrosion can be interpreted as how close the system of interest is to failure should corrosion be the only prevailed mechanism. In other words, the engineer measures the risk in terms of corrosion rate and the part's thickness. If, for instance, the thickness of a metallic part is 1 mm and the measured corrosion is 0.1 mm per year, it means that maximum 10 years can be taken as the safe margin of service life for that part (due to practical reasons, it is better, however, take that service life span as 8 or 9 years instead of 10). Figure 3.1 shows the categories by which corrosion severity can be addressed based on corrosion rates, as per NACE standard RP0775-2005.

However, risk has a much wider meaning. Risk is a function of likelihood of an event (hazard) and the consequences following it. In other words, if there is a small likelihood of a hazard occurring, but it does occur, then it will have rather serious consequences, its risk is high no matter what the likelihood could be.

Risk has a known definition: it is the by-product of likelihood of an event (here, corrosion) by its consequences. It is based on the level of risk thus defined that acceptable corrosion risk can be assessed. However, there are many uncertainties here; how, for example, we define consequences? In terms of the money to be lost or in terms of environmental impacts? If it is the former, is it the present value of

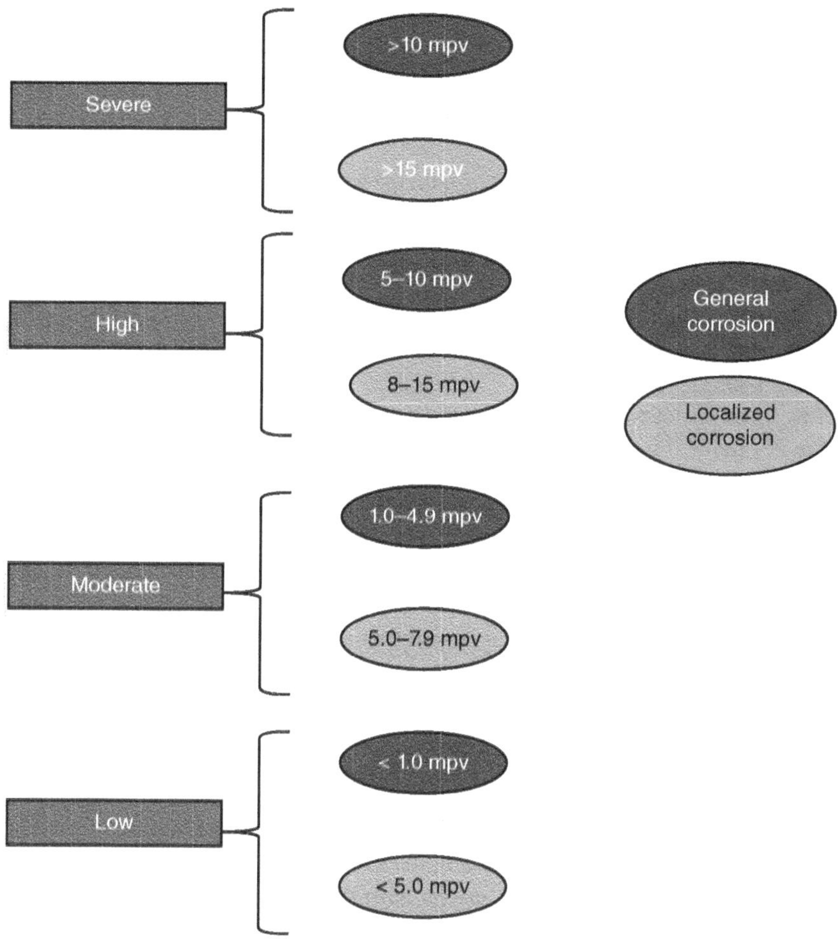

Figure 3.1 Categorizing corrosion severity based on corrosion rates (general and localized corrosion). *Source:* Mr. Ricardo Gonzalez.

the asset that concerns us, or its relative value with regards to where the asset fits in a chain of demands and purchases? In case of the latter, how it is measured? Or rather, how it can be proposed to be measured? Is there a systematic approach to estimate the indirect cost of corrosion? How can ecological impacts of corrosion be prevented, controlled, or assessed? What is the difference between 'corrosion control' and 'corrosion prevention'? Above all, how do we define importance in engineering terms? If we define engineering importance as a function of both risk and cost, meaning that if somethings risk and cost are high it is then important, how, then, can we connect engineering importance with CM?

First things first, how do we define consequences? API 580 part 4.1.5 defines consequence as *An outcome from an event. There may be one or more consequences from an event. Consequences may range from positive to negative. However, consequences are always negative for safety aspects. Consequences may be expressed qualitatively or quantitatively* [1].

Consequence could be interpreted as damage to the capital or environment – or both. The leakage of BP-owned pipeline (Prudhoe Bay Field event in 2006) was an example of a corrosion event whose consequences were financial as well as environmental.

In calculating the risk, one important factor is to know or at least estimate the consequences that could occur. The likelihood (probability) of the event is also of paramount significance.

Let us think about risk and the factors involved in it in a more detailed, somewhat mathematical way. There are four factors involved in calculating risk (R):

1) Life factor, shown with L. It is the number associated with the remaining useful life of a part/component of a system.
2) Consequence Factor, shown with C. It is quantitative assessment of consequences that can occur should corrosion becomes an event being worth notice, or being worth being branded as a serious cause. Some of the consequences resulting from a corrosion event are:

   a) Safety: it is possible that as a result of corrosion, safety become a serious matter of concern. The explosion of the natural gas pipeline Near Carlsbad, New Mexico on August 19, 2000 in which a young family dwelling nearby perished. There are both documented and witnessed cases where corrosion was the main culprit in sacrificing the safety of workplace [2].
   b) Production: corrosion can interrupt with the scheduled, normal regime of production. In 2010, mainly due to negligence of corrosion in a Petrochemical complex in Iran (Pardis Petrochemical Complex [3]), there was a disastrous fire that, reportedly, postponed production for nearly two months.
   c) Repair: the scheduled repair (overhaul) that is preferred to emergency repair (shut down). To show the adverse effect of unplanned shutdowns one can note that Indonesia experienced a reduction in its crude oil sales during the first half of the year 2014, that in part, was the result of 122 unplanned shutdowns from the operator BPMigas, which created 6860 bpd in lost production [4].
   d) Operability: meaning that when a part is corroded, it is not always easy to operate the equipment as easily as it used to be. An example of this condition can be seen in Figure 3.2.
   e) Environment: one of us (R. Javaherdashti) has written a book about a decade ago regarding environmental impacts of corrosion [5]. Some

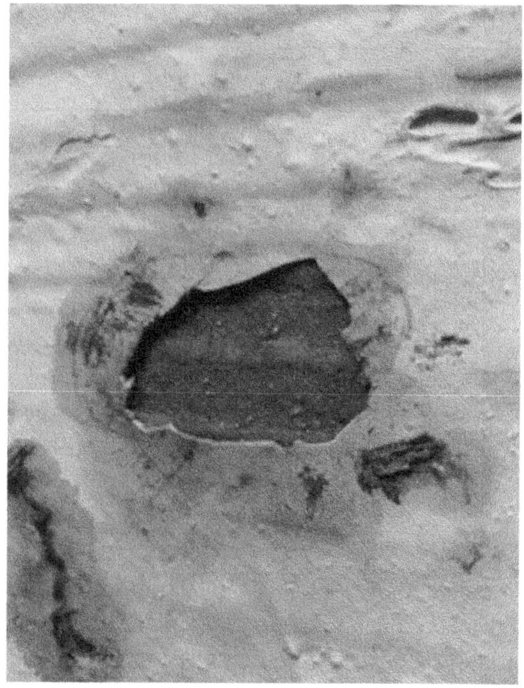

Figure 3.2 Damage of the inner layer of a protective coating applied inside an acid tank (biologically produced sulfuric acid) after being in service for about one year. The underlying substrate is the carbon steel body of the tank. A typical tank as such had a localized corrosion leading into severe leakage. (*Source:* Image courtesy of Dr. Reza Javaherdashti.)

environmental disasters and catastrophes related to corrosion have also been reviewed elsewhere [6, 7].

f) Third party interests: if corrosion happens on one's asset, its impact may not be limited to the asset owner only. As reported in ASM (American Society of Metals): *"the sudden collapse because of corrosion fatigue of the Silver Bridge over the Ohio River at Point Pleasant, OH in 1967 resulted in the loss of 46 lives and cost millions of dollars"* [8]. This and similar examples can clearly show how corrosion can affect several factors.
g) Customer and public perception: just imagine a car you purchased has a rusting exhaust or chassis. Defects as such, which are often not noticed by regular customers, are considered great losses in the life cycle of the car for anyone who buys that car. Factors affecting customer perception are typically "advertising, reviews, public relations, social media, personal experiences" [9]. In fact, it appears that corrosion scientists and engineers have taken this aspect very seriously, for example, for cases in which corrosion may lead to disasters such as water poisoning [10].

We have reviewed aspects of corrosion that are reflected onto our lives elsewhere [11].

Table 3.1 Consequence rating (F) of some consequence ratings.

Consequence	Rating (F)
1. Safety (to personnel and/or public)	10
2. Loss/delay of production	10
3. Environmental pollution	10
4. Loss of quality of the product	9
5. Loss of customer confidence	7
6. Interval between repairs	8

3) Consequence rating, F, is an arbitrarily selected ranking to designate what consequence is – relative to other consequences – of greater importance. On a range of 1 (the least important) to 10 (the most important) consequences, it is possible to arrange a ranking factor, or equivalently, a consequence rating as in Table 3.1, below:

While the figures given for F in Table 3.1 are chosen on an arbitrary basis (and obviously, open to discussion), the importance given to them is not that arbitrary. In fact, comparing the first three items show that for the operator/owner of the business, these three consequences are equally important, yet item 5 is of lesser significance. It is possible that based on the understanding of the business owner/operator, other F values are assigned, however, the relative values between the F values is of importance.

4) What is important is what to be considered as an event which is worth of being considered for its consequences. This will bring another rating factor that is called an Event Rating, designated by P. For example, within the safety category of consequences, can a pinhole in a line carrying toxic material be considered a serous event with serious consequences, what about a crack in the same line? Is a crack in a toxic fluid line a serious event, what about a crack in a line carrying flammable fluid? Within the category of pollution, is it the pollution induced in soil of more importance or in water? If we are talking about events in an offshore oil and gas platform, it is certainly the seawater pollution that deserves to be considered as an event to which an arbitrary value for P is assigned. For a solar panel farm, it is the atmospheric pollution that matters, and so on. Here, again, a ranking system can be constructed assigning a 1 (least important) to 10 (most important) value for the event factor. As with consequence factor, it is not the absolute values assumed for items that matter, but the relationship between them and how these relative figures can be interpreted with regards to each other.

Based on the above risk (R) can be taken as a function of life factor (L) and consequence factor (C). C itself is a function of F and P, and it will be the summation of each C factor (and related F and P factors) as mathematically shown below:

$$C = \sum C_x = \sum_{x=1}^{x=n} fn(F_x P_x)$$

The above equation relates the consequence factor with the ratings (F and P). It will enable us to construct a matrix that consists of probability (of corrosion) with related consequences, yielding risk categories, an example of which is given below, in Figure 3.3:

Obviously, it is necessary to avoid risk from zone 1 (high consequence and high probability) to zone 4 (low consequence and low probability). Figure 3.4 shows an example of equipment that more or less fits zone 1 conditions.

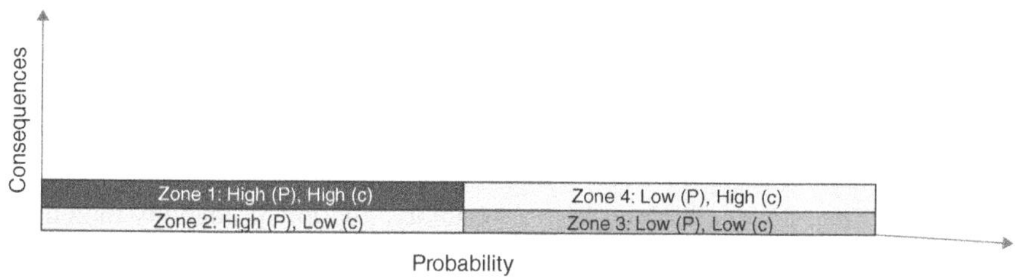

Figure 3.3 Risk categories and four zones created based on level of consequence and probability.

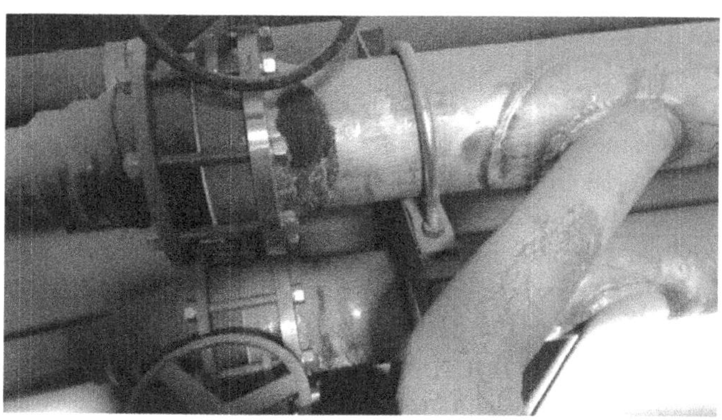

Figure 3.4 An example of CUI (corrosion under insulation) on a piping which is an essential part of a water treatment facility. Due to CUI, the (corrosion) probability is high and the equipment failure (consequence) is also high. (*Source:* Courtesy of Dr. Reza Javaherdashti.)

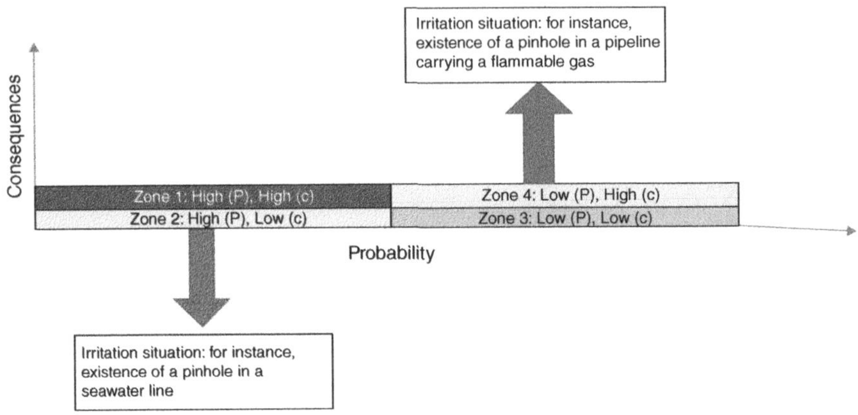

Figure 3.5 Irritation situation zones, where P (probability) and C (consequence) could yield various situations. Two examples for these zones are also given.

While zones 1 and 4 are serious and safe alerts, respectively, it is zones 2 and 3 that can be called irritation situations. Examples of these zones are given in Figure 3.5:

It is necessary for us to focus more on the meanings that can come to mind; when in Zone 2 we talk about high probability, and low probability in zone 4. However, before that we must answer a question whose answer may seem too obvious; how can probability about something be linked with the knowledge about it? In other words, why we call a case like that shown in zone 2 a high probability and a case like zone 4 a low probability case?

In fact, the more data we have about something (that is to occur), the more we will be able to assess if it is really going to occur with high or low probability. As an example, if we have data (prior knowledge) that a flammable material is placed near flame, then it is sensible to reason that the probability of fire is high. It follows with this assessment about probability that it is then possible to separate the flammable material from the flame, and therefore lower the consequences.

It can also follow, then, that risk is a very delicate issue often mixed up with likelihood by many, including professionals; likelihood shows the possibility of a hazard – here, corrosion – whereas risk measures the aftermath of the likelihood in terms of both the likelihood and the consequences that can occur after corrosion has taken place. Industries, not all of them of course, think that because they have taken all measures against corrosion, then they are safe. However, from experience, the bitter fact is that not all forms of corrosion may not be known. There are at least two reasons for that:

I) Different corrosion mechanisms as observed in different assets: basic assets in any plant consist of (i) utility, (ii) process, (iii) fire water ring/system, (iv) fuel lines, (v) feed lines, (vi) product lines, and (vii) waste lines [12]. Each of these assets has various types of corrosion processes already taking place in them that could be vastly different from each other. As an example, the corrosion problems in boiler, turbine, and hotbox-condenser in a power plant are different from each other. Also, problems associated with the waste material could be of another nature compared with corrosion problems in hot spots of the plant (for example, within water-wall tubes).

II) Different corrosion geometries [13]: even within one asset there can exist several corrosion mechanisms that are in competition with each other. These corrosion processes can be in series or parallel geometries with regards to each other, meaning that they can either enhance one another or occur independently from each other. This is a particularly important aspect of any problem-solving scheme. Figure 3.6 shows an example in which series and parallel corrosion reactions geometries can be observed. In the duct, examples of parallel reactions are (i) the damaging effect of atmospheric corrosion on the external coating, and (ii) internal corrosion of the fluid inside the pipe. This means that these corrosion reactions can take place independently from each other. However, when it comes to

Figure 3.6 An example of corrosion reactions geometries. (*Source:* Photo: Dr. R. Javaherdashti's collection.)

series corrosion reactions geometry, a likely corrosion scenario could be the detrimental, gradual corrosion of welds by the corrosive fluid inside the side-pipes.

Therefore, risk of corrosion is an important issue that must never be confused with the likelihood of corrosion, and being conscience about this difference must also be reflected in any CM scheme.

3.1.2 Corrosion Management: What It Is and What It Is Not

In both rather famous definitions of CM given in 2001 [14] and 2016 [15] by authorities,[1] CM is related to the management of risk of corrosion. In the 2001 definition of CM, there are two especially important concepts; risk of corrosion and corrosion control. In 2016, however, the pivotal concept of the definition is to manage the threat of corrosion. In a previous publication [12], we have differentiated the importance discrepancy between corrosion prevention and corrosion control and will not repeat it here. However, the most important aspect of CM, as both definitions also imply, is the risk of corrosion. The importance of highlighting risk is due to the role it plays in defining Engineering Importance.

Engineering importance is a function of both risk and cost. This means that from an engineering point of view, if somethings risk and cost are high, then its importance is high and must be regarded as something important. From the definitions given for CM it follows that CM is mainly concerned with the risk of corrosion. In other words, CM looks at those factors that are likely to create the risk of corrosion.

Being as such, CM will look at all the factors that will make a system vulnerable to corrosion. A part of these factors will be corrosion engineering/science, but what is not there is corrosion cost considerations. In other words, corrosion scientists and corrosion engineers try hard to control conditions that will lead to creating the risk of corrosion. It is in this regard that we believe corrosion professionals often use CM as a new way to look at their old problems; a recent example of such is the Bechtel CM model, presented at NACE in 2016 [16]. This model, as reported [17], is composed of some steps that are very briefly explained in the following:

1) list of all probable corrosion scenarios;
2) evaluation of each threat;

---

1 CM definition as per the reference belonging to 2001 is "Corrosion management is that part of the overall management system, which is concerned with the development, implementation, review, and *maintenance of the corrosion policy .... The corrosion policy provides a structured framework for identification of risks associated with corrosion, and the development and operation of suitable risk control measures.*" In 2016, IMPACT gives its own definition of CM; *A corrosion management system (CMS) is a documented set of processes and procedures required for planning, executing, and continually improving the ability of a company to manage the threat of corrosion for existing and future assets and asset systems.*

3) feasibility of using various remedies (such as using low-cost construction materials or corrosion-resistant alloys [CRA]), or non-metallic materials based on a life cycle cost (LCC) analysis;
4) determination of corrosion control (CC) technology;
5) selection of proper monitoring and inspection technique(s) to determine how well the remedies suggested and applied in step (3) work;
6) receiving feedback on the reporting and analysis of suggested monitoring and inspection methods in order to apply corrective measures, if necessary. This step would also work as a medium of interaction between corrosion the engineering team, operations team, and the facility owner.

In the same way that there is a Bechtel CM model, there are other CM models that are often referred to by the names of the companies in which these models have been developed and applied, such as BP model, Petronas model, Total CM model, etc.

Bechtel CM model is one of the most recent CM models developed in a reputable company, however there are points associated with this model that are more or less also true with other CM models. For instance, Bechtel model is too technical and rather an engineering model, meaning that it is only understandable and applicable by an engineer and not by, say, a financial manager at a company. Apart from the section that deals very briefly with economy in terms of LCC, this model has no room for a cost analysis in terms of what would happen without this model. In other words, Bechtel looks at the economy of using anti-corrosion measures from materials selection to selection of inspecting methods (visual, radiography, eddy current), but it assumes that anyone who is to use the model already knows how much damage it causes. This assumption is not necessarily true for all companies, whether in the oil and gas sector or elsewhere. We believe that any CM model must also be accompanied by yet another model that is specified for the cost of corrosion only. The Bechtel model lacks this economy and management side.

Each CM model must also come up with a management model that would mainly serve the higher management (the one with 'might and right'). Real life experience shows that while a CM model may be good for one company, another company may not want to use it or may want to have another model that is essentially the same as other models, but with a different manifestation.

Item (1) of this model is silent on corrosion geometry, in other words, it is not accompanied by a nomenclature that would define the following:

1) Based on the conditions, definition of corrosion scenarios may change. Note that there is a difference between knowing a corrosion process and knowing how it can be effective in an overall corrosion scheme. For example, one may be a very informative, skilled professional about microbial corrosion, microbiologically influenced corrosion (MIC), and in the corrosion system

(that is, a system whose corrosion is important for us), MIC does exist. The point is how MIC will become the main operative corrosion process when compared with other competing corrosion processes. This is what we call corrosion reactions geometry.

2) Failure in engineering terms is when a material is not functional any longer. In other words, a failed material/component is no longer being considered for the job it is supposed to do. However, failure has a much wider meaning that must be taken with some delicacy. In fact, before the material fully fails, its failure could have already started. This is a very important point often missed by CM model developers. They assume that their main tasks are definition of corrosion threats, then how to remedy and monitor it so that it will not recur. While this is certainly an especially important core of any CM model, it should not be taken as the main point; the main point in any CM model is not prevention of failure, but prevention of conditions that would *lead* to failure.

In fact, the main deficiency of all existing CM models is that they have no preparations for the persons who may have little or no corrosion knowledge, in addition to clear definitions of corrosion scenarios and failure. When we talk about our own approach toward CM, we will discuss the above with more details.

On the other hand, there is an incorrect perception even among professionals that CM just belongs to oil and gas industries. These authors believe the main source of this mistake is that corrosion specialists working in oil and gas have been much more active in constructing CM models compared with their counterparts in other industries. It seems that because of the prevailing nature of oil and gas industries and higher activities of corrosion professionals in this field, CM models are remembered with the names of famous oil and gas companies (BP, Petronas, Total, Bechtel).

In the next section, we will talk about our approach toward corrosion that we call Management of Corrosion.

3.1.3 Management of Corrosion

Management of corrosion is not just a simple play with words, it is an umbrella below which the following six concepts must be considered. These concepts are:

1) Geometry of corrosion processes;
2) Definition of failure within a new context;
3) Corrosion prevention (CP) and CC;
4) Javaherdashti CM model;
5) Corrosion cost estimation model;
6) CKM.

Based on our 20+ years of experience in various industries around the globe, that stretches from Papua New Guinea and Australia to the UK and the Middle East and South Asia, and being involved with CM in oil, gas, and petrochemical industries, along with automotive and mining, we strongly recommend that any approach toward management of corrosion must be taken as a package containing the four items we previously mentioned. We have talked about these four items in our numerous publications before, but it is the first time that all four are taken together as a package to address management of corrosion properly. Although we have explained some of the above in here and in other resources, it is worth of repeating them all here.

3.1.3.1 Corrosion Reactions Geometry

Corrosion reactions happen in one anodic reaction (losing an electron) and – based on several factors such as, but not limited to, oxygen availability, pH, electrochemical stability of chemicals in the environment, temperature, and the like – several cathodic reactions (gaining electrons) that if one is not happening, the other one can replace it. Therefore, right from the beginning, the electrochemistry of corrosion is based on not one but more than one reaction.

In addition to that, real-life working conditions are far different from controlled, design conditions of corrosion laboratories; under laboratory conditions corrosion reactions are mimicked, accelerated, or reconstructed. These three features which are essential to laboratory experiments, do show the deep gap between experimental and industrially experienced findings.

Under working circumstances, mostly several corrosion reactions are operative. Some examples could be:

a) Hydrotested pipelines; in these systems there are at least four corrosion processes that, based on the chemistry and microbiological as well as electrochemical conditions, could exist [18], the most important of which being microbiologically influenced corrosion (MIC). In a system as such, it is very important to know if the involved governing corrosion processes are acting in series or in parallel geometry to each other: if the MIC problem can initiate or launch galvanic corrosion between the weld and the parent material of the pipe, then by solving either end of the problem (treating MIC or resolving the existing corrosion potential between the weld and the parent material), the corrosion problem can be solved.
b) Bioleaching tanks; in these tanks the normal process for tank bioleaching is that bacterial culture containing Sulfur oxidizing bacteria (SOB) will dissolve certain mineral (for example copper) concentrates to yield an industrially pure metallic form of the required element. Due to the very nature of the reactions in the tank, both erosion (as induced by concentrate lumps accelerating by the

action of a mixer) and MIC, as induced by the action of the bacteria, can exist. It is then a matter of the electrochemistry, as well as the hydrodynamic of the concentrates inside the tank, that being in series or parallel geometry is determined.

Having a series or parallel corrosion reactions geometry has a very important impact on the way an overall corrosion problem is being treated; consulting efforts for a series corrosion geometry could be much easier and even more cost-effective than a parallel geometry, by picking up the main corrosion process and treating it, it is highly likely that the overall corrosion problem will be resolved. In case of a parallel geometry, however, such chances are very slim, as it is highly likely that the involved corrosion reactions are acting independently from each other so that by solving one, the others will not be resolved. In addition, it is also possible that one branch of an overall parallel corrosion geometry will itself consist of a series of parallel and series geometries, thus making the overall picture even more complicated. Figure 3.7 shows an example of such a complex system in which external corrosion of a buried pipeline has been illustrated. In this respect, overall corrosion geometry is in parallel geometry consisting of three rows. Each row consists of corrosion geometries which are in series form.

For instance, the first row discusses soil properties from soil particle size (I) to its microbiology (II), and its pH, moisture, and conductivity (III). Likewise, the other rows will be related to other components of the whole corrosion scheme.

The complexity observed in the two examples given above can also be seen in other equipment and assets. As can also be seen, it is not that easy to talk about all possible corrosion scenarios if the geometry of such reactions with regard to each other is not known. A skillful corrosion engineer may realize what the corrosion processes are in each asset, but it is equally important to realize how these corrosion processes are prevailing in the overall corrosion scheme in the first place.

3.1.3.2 Failure

Failure is normally regarded as a phenomenon where the extent of physical damage is so large that the equipment cannot be used any longer. Failure seems to have been understood as a self-evident issue so that by merely speaking about it, everyone would know what the topic is, but the fact is that it is not that straight forward to define failure.

In order to define failure, we must remind ourselves again about a principle that we mentioned earlier; our philosophy must not be treatment of failure but finding ways to avoid allowing failure to occur. In other words, our mission must be to prevent ways by which failure is highly likely to happen, not the failure per se.

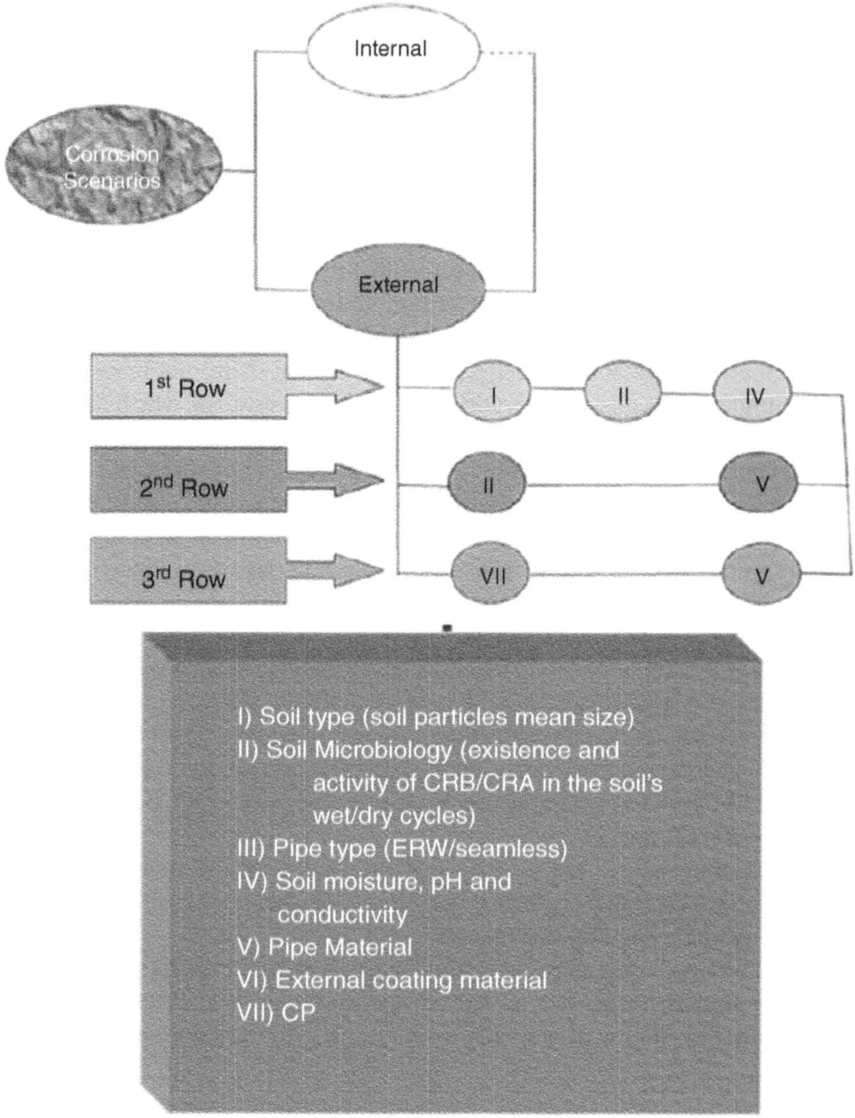

Figure 3.7 An example of a rather complex corrosion geometry related to external corrosion of a buried pipeline; overall corrosion geometry is in parallel form (geometry) where each branch itself consists of some series corrosion geometries. (*Source:* Figure taken from Javaherdashti and Akvan [12].)

In this regard, our job will be more or less like crisis management; crisis by definition has no management as it is unmanageable. It is in fact pre- and post-crisis conditions that can be managed. Pre-crisis conditions must be managed effectively not to give way to crisis, and post-crisis conditions must be managed to minimize the impact of crisis.

The thermodynamics of corrosion was emphasized to show that it is indeed a naturally occurring process. This means that corrosion cannot be stopped (the next section will discuss CP and CC and what the differences are between them). It can alternatively be said that corrosion is something that one has to accept, as long as pitting that would occur as a result of non-uniform, localized corrosion is not leading to leakage. Prior to that, we want uniform corrosion (that is easier to handle in both prediction of its effects and controlling them) not to be replaced by non-uniform, localized corrosion, an important result of which is pitting. While there have been many attempts to find an algorithm/pattern to predict pitting (even by using scholastic approaches), the unpredictable nature of pitting has always remained like a Damocles sword upon the head of corrosion technologists and engineers. Therefore, what is desired for us as operators or maintenance specialists with regards to CM can be summarized as in Figure 3.8 as an example of what we can call a 'corrosion safety procedure' for an asset such as a pipeline:

Taking the metaphor of a train reaching its last station, as far as corrosion safety procedure is concerned, leakage is the last station for the safety train. After the last stop, seen here as 'leakage' from the pipe, failure is inevitable. Therefore, failure in its classic meaning is where the mechanical integrity of the asset is lost and something like leakage from a pipe occurs. In pure engineering terms, failure is where the asset becomes useless.

The lack of usefulness and loss of mechanical integrity and strength has been introduced into corrosion engineers' books and nomenclature from a mechanical engineering understanding of materials.

For us, the stages through which failure is facilitated are important. In fact, it is only through these stages that a corrosion engineer can have the chance of controlling corrosion (in the next section we will explain the vast differences between

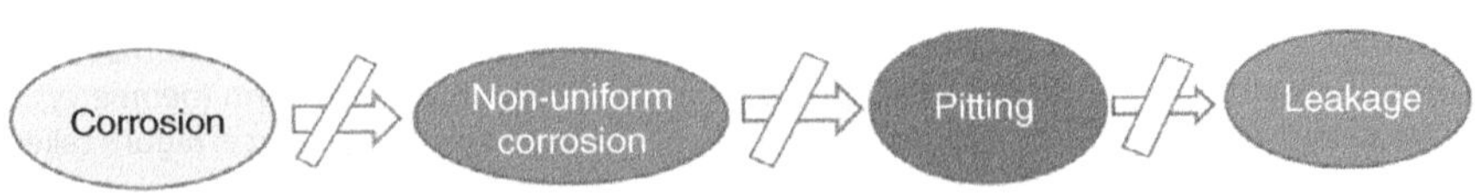

Figure 3.8 Corrosion safety procedure. The way corrosion must not proceed in order to make any damage; corrosion must not become non-uniform in its morphology (localized corrosion) to yield pitting. Even if pitting happens, leakage must not occur.

CC and CP). To explain these stages, we need to become familiar with three definitions that we will be using through this chapter:

- Fit-For-Service State (FFS): When an asset/equipment passes all the necessary tests as required by standards and professional approval of an inspector, it is said that the asset is ready to be put into service and it is fit for the intended service. In this state, it can safely be assumed that the asset is corrosion-free, or at least the extent/severity of corrosion is not significant enough to prevent the asset from being labeled as FFS.
- Pseudo-FFS State: As soon as the equipment is put into service, it will face conditions that will affect its corrosion resistance. In other words, the asset will experience corrosion immediately after being put into service. When the asset undergoes this status, it is still being used by the operator of the plant or asset manager simply because the measured corrosion rates are still within acceptable range or the corrosion is still within the limits of being controllable. This state is not the same as FFS state but is similar to it, thus the name pseudo-FFS.
- Zugzwang State: Zugzwang is a term in chess, simply meaning that whatever move one does, one will lose. No matter what your position is in the game, the game is doomed. In this context, we define Zugzwang state as per the below two definitions, and based on whichever would fit the situation of the asset, we may call it a Zugzwang effect status, Figure 3.9.

Figure 3.10 shows two examples of pseudo-FFS and Zugzwang effect state.

Zugzwang Effect in cases related to corrosion can be redefied with the following two criteria:

(1) For the structure/asset that has been exposed to corrosion, thinking of any repair is not economical any more. In other words, the extent of damage is so economically devastating that the management cannot/is not willing to handle any repairs as it will not make economic sense to them,

(2) Even if despite its lack of economic feasibility we decide to do it, corrosion case has become so widespread and hard to cope with that it will make it technically not feasible to even try to mitigate corrosion any more. This means that as the root cause of the problem is not known, even if it is tried to do some repairs, modifications or even part changing, an long as the root cause is not there, the results will always be the same.

Figure 3.9 Two alternative definitions of Zugzwang effect state. Based on the severity of the case and repeatability of repair costs, alternative definitions can apply.

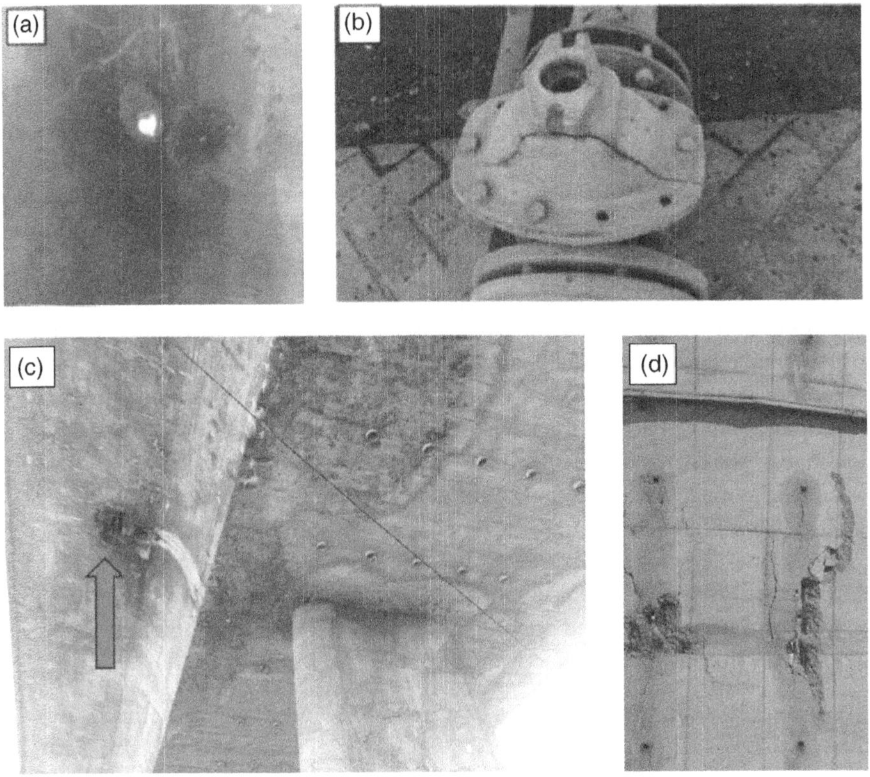

Figure 3.10 (a) A through-wall hole (at 6 o'clock position) of a stainless steel 316 pipe in which seawater remained stagnant for four months. (b) A cracked connection still used in a petrochemical complex. (c) Failure of reinforced concrete surface below a concrete highway bridge. (d) Spalling on the concrete wall of a cooling tower that looks toward sea. (*Source:* from Dr. Reza Javaherdashti collection [12].)

In Figure 3.10, figure A shows an example of an asset in its Zugzwang effect state, whereas figure B shows an asset in its pseudo-FFS state. In figure (A), the pipe is out of service completely due to the existence of the through-wall hole. However, in figure (B) the asset is still in use merely because its condition is not critical enough to push it out of service. In fact, a great number of assets already in use in all industries are in their pseudo-FFS status.

Based on the above, then, failure will be put into a new context which is simply this: *Failure is when an asset reaches its Zugzwang effect state*. Alternatively, the job of a corrosion consultant/engineer must be two-fold; increasing the time between FFS and pseudo-FFS states as well as the time between pseudo-FFS and Zugzwang effect states. The relationship between these status conditions through which the asset will travel is schematically shown in Figure 3.11.

As components in Figure 3.11 show, as the asset moves from FFS toward pseudo-FFS, and from pseudo-FFS to Zugzwang effect status, there is a need for both inspection and maintenance. As implied earlier, the shorter the distance between pseudo-FFS and Zugzwang effect status points, the more likely that failure will happen. The farther the distance between FFS and pseudo-FFS, the safer the asset is to operate. This is also applicable should the distance between pseudo-FFS and Zugzwang effect state be as far as possible, because that would mean that the equipment is in such good shape that failure is not likely to happen within a short time.

Therefore, by using the above concepts it is now possible to define what any useful CM must aim for. This can be stated as below diagram, Figure 3.12.

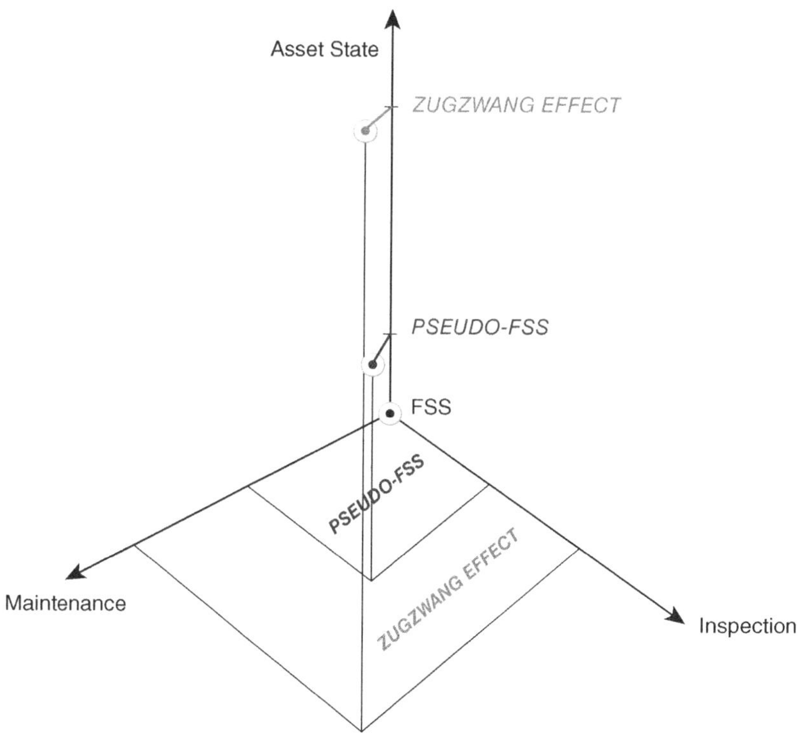

Figure 3.11 The relationship that exists between FFS, pseudo-FFS, and Zugzwang effect states by which the role and position of inspection and maintenance activities are also related.

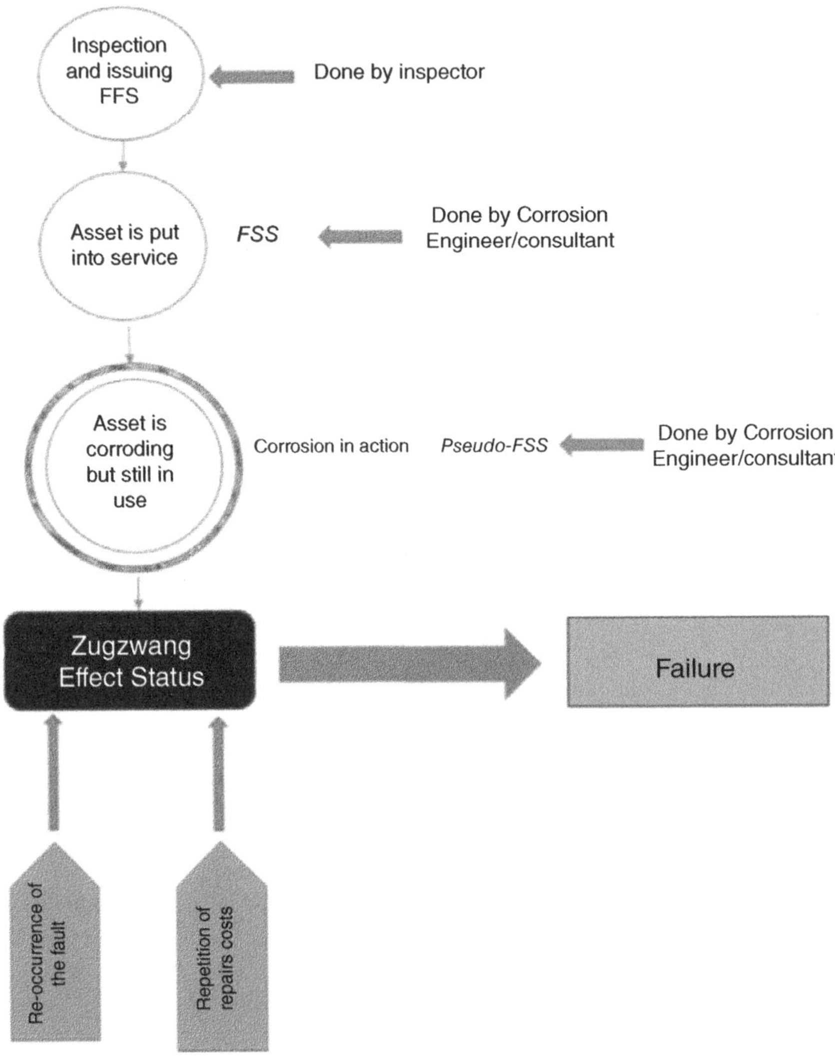

Figure 3.12 Milestones in the life of an asset from its fresh state of FFS to its failure.

As Figure 3.12 shows, there are milestones that can be assigned to the service life of an asset from the time it passes all the required inspection and fit-for service standards and codes, to the time it arrives at its Zugzwang effect state and it has failed. This pathway will also involve both inspectors and corrosion engineers/consultants, as well as integrity management technologists.

While the figure is self-descriptive, particularly regarding what was mentioned above, it is worth noticing that the expertise involved and highlighted in Figure 3.12 are not all that is required. In other words, there is certainly a blend of experts and expertise needed to gain a correct evaluation and assessment. This is particularly true when Figure 3.12 is looked at within the context of Figure 3.11. As it appears from Figure 3.11, inspection and maintenance experts are to accompany the CM consultant at all times. It is also obvious that to keep the asset at its FFS state for as long as possible, technical feedback from inspectors is vital; it is based on what they see and report that the CM engineer can decide how critical the situation is and how it must be handled.

It is also a very important to understand that any CM approach is a collective work and it flows from bottom to top, and not the other way. What is meant here is that top management will be provided the feedback from the CM team, and if the decision taken by the top management proves to be incorrect down the line, it is highly possible that the CM team have not communicated effectively enough with management. We have seen an example of such poor communication between the engineering team and the financial management so that just based on pure accounting (not economical) considerations, refusal to use a better inhibitor instead of a less effective but less expensive corrosion inhibitor cost the plant dearly [5].

No CM scheme would be complete without discussing the important difference between CP and CC. This is a concept that, even by corrosion professionals, is often neglected and used without paying attention within a CM scheme; if the differences between these two terminologies are not taken into consideration, several confusions may be created. We will explain these terms in the next section.

3.1.3.3 Corrosion Prevention and Corrosion Control

Thermodynamics of corrosion teaches us that corrosion cannot be stopped, nor can it be prevented. However, CP and CC are terms that are being used interchangeably even in many official documents related to CM; as reported by NACE [15], one of the objectives mentioned in the long-term corrosion strategy plan submitted by the US Department of Defense to the Congress in 2003, reads verbatim "Formation of a multi-service corrosion prevention and control ... team."

The fact, however, is that CP and CC are not the same at all. CP and CC differ from each other in features that can be summarized as in Table 3.2:

When we are in CP mode, based on the nomenclature we developed in the previous section, all efforts must be done to stay at FFS without letting the asset to move toward the pseudo-FFS state. When on CC, all efforts must be performed so as not to arrive at Zugzwang effect state. This is a very important matter, as by knowing where the asset is the required strategy will be prepared, and for that reason it will be possible to prepare the required strategy.

Table 3.2 Some important features of corrosion control (CC) and corrosion prevention (CP).

Corrosion control (CC)	Corrosion prevention (CP)
The equipment is in use (Brownfield projects).	The equipment is not in use or not extensively used (Greenfield project).
Corrosion Management is in the form of controlling corrosion and keeping it at less than 5 mpy, as per NACE corrosion rate requirements.	Using correct design and application parameters, corrosion can be kept at very low values so that no damage will be experienced.
OPEX (Operation Expenses) are high.	CAPEX (Capital Expenses) will be high but OPEX (Operation Expenses) will be low.

Distinguishing CP as described in Table 3.2 also reveals and explains the seemingly contradicting issues between thermodynamics of corrosion (that defies prevention of corrosion) and the way corrosion technologists normally refer to CP. As it appears, even when we are in CC mode, corrosion is not being prevented but it is kept within restrains to keep it manageable enough. On the other hand, CP in this context can only be mentioned if right from the beginning all possible conditions leading to corrosion are seen and addressed.

An example of CP could be by changing the design. An example for a design change that would eliminate the corrosion problem (appearing in the form of corrosion under deposition due to the establishment of differential aeration cells) is to solve a galvanic corrosion problem of a carbon steel pipe with a small total area, with respect to a stainless steel pipe with a larger area (small cathode versus a large cathode that will be leading into sever corrosion), by either using the same material as the stainless steel pipe or using stainless steel as the anode and the carbon steel pipe as the cathode (small cathode versus large anode). Yet another example could be putting a valve under a suitable place in a basin so that by discharging the fluid, no stagnant conditions for the fluid inside the basin will be created, leading to possible corrosion mechanisms under the deposit (such as but not limited to, MIC).

From another point of view, CC is dealing with 'as is' whereas CP is dealing with 'to be' conditions of the asset. As is, as the name implies, shows the present condition of the asset. In other words, it shows how tolerable corrosion conditions are when the asset is considered from a CM point of view. Taking CP as an ideal state of engineering CM measures, to be shows the ultimate solution or conditions that can be expected. It follows that if the current (as is) condition of flow rate in a pipeline is, for example, less than 1.5 m/s so that stagnant conditions are encouraged leading possibly into the likelihood for MIC to develop, the target (to be) must be increasing the flow rate to disturb conditions for biofilm formation.[2] When the flow rate is below the limit mentioned, the conditions will need to be controlled (CC), and when we increase the flow rate so that biofilm formation is prevented ([2]), it is then as if biofilm formation as a prerequisite for MIC has been prevented (CP).

3.1.3.4 CM Model

The corrosion model we are presenting here rests on more than two decades of personal experience in CM and troubleshooting in various industries around the globe. In the above sentence we have emphasized upon there important elements that serve to qualify us for presenting a model:

- Field as well as academic experience; CM models must not be developed based on either pure field or academic skills and must actually look into combining the two within a suitable context. While the framework and context of such a blend must rely on theoretical basis, the application must be sensible enough when applied,
- Different industries; so far all the CM models we have observed and to the best of our knowledge, have been proposed to deal with oil and gas industry, perhaps with the exception of one model suggested to deal with CM issues in mining industry in Western Australia [19]. The fact is that corrosion is not just an issue in oil and gas industry, but in all industries ranging from mining to power generation and others, therefore, the more diverse industries one has knowledge of, the better the engineering understanding of the systems and processes. In addition, in some industries it is the process-associated corrosion problems that are of interest. In fact, corrosion problems in any industry can be divided into three categories; (1) corrosion problems associated with the main process, (2) corrosion problems associated with auxiliary systems, and (3) corrosion problems associated with safety systems and equipment. In a power plant, for instance, category 1 problems are associated with corrosion problems for boiler, turbine, and condenser; whereas category 2 problems are those related to cooling tower, soot blowers, or boiler feed pump; and category 3 corrosion problems are mainly associated with underground metallic firewater rings. In a petrochemical plant, for example, category 1 and category 2 corrosion problems are of significance and most of the time category 3 problems are not taken seriously unless there has been a history related to it. Working as a corrosion consultant for a wide range of industries gives one the opportunity of getting familiar with the work culture of each, and therefore allows a much broader

---

2 We would like to focus the attention of the reader on the fact that having a flow rate over 1.5 m/s is a required condition, but not enough to avoid biofilm formation. For an explanation please see R. Javaherdashti and K. Alasvand, *Biological Treatment of Microbial Corrosion*, Elsevier, 2019, the term biofilm is not correct and we have suggested the term 'temenos' instead.

horizon for engineering disciplines than having worked in just one business in a given industry,

- Having a rather global culture about corrosion is a vital feature of any CM model that is to be proposed on an international level; the main defect of models proposed by mostly Western-minded professionals is that they look at CM through their own window. For an industry that exists in a country where the currency rate between internationally accepted monetary units such as dollars or euro versus its national monetary system is not stable, it is quite unrealistic to propose expensive means of corrosion monitoring, inspection, and treatment. In a culture where there is no previous history about cost of corrosion or even a record about parameters required to define a corrosion cost model (to be discussed in the next section), not giving an algorithm on how to obtain such data, any CM model is to be doomed with incompetency when faced with realities.

While the Javaherdashti CM model takes the essential elements of CM that, like any other CM model is more or less similar for all CM models, its successful application is only possible if all the six elements for a system of management of corrosion are also ready and applicable.

The Javaherdashti CM model has four phases and six steps, as we will now discuss.

3.1.4 Phase 1: Definition

In this phase the following steps are defined:

1) Define the corrosion system. Contrary to how it may seem, the meaning of a corrosion system is not too evident. In simple terms, a corrosion system is a part of a system that, due to a corrosion-related reason, is of interest to us. Therefore, if we are looking at corrosion problems related to the waterwalls in a boiler in a power plant, then the corrosion system is those waterwalls, but not other parts of the boiler. Another example is investigating MIC or CUI in a certain part of a given asset. Considering corrosion systems can be divided into two categories: CINS (corrosion in the system) and COFS (corrosion of the system). In a CINS approach, all possible corrosion scenarios for all parts of a given system are investigated, whereas in a CINS just a limited number of both the equipment and subsystems are of interest. More details have been given elsewhere [12].
2) Define the corrosion problem(s) and their geometry. This is a matter for which a strong scientific background is necessary. Obviously, to realize the corrosion reactions involved, a good knowledge of theoretical basis is required, not to address atmospheric corrosion, CUI, and MIC incorrectly, but because all of them are electrochemical in nature. As we mentioned earlier, in many real-life corrosion systems (hydrotested pipe, bioleaching tank, and the like) there is

more than one corrosion reaction involved, so that there is always a competition between these reactions. It is equally important to know the geometry of corrosion reactions to determine the one that is prevalent (in series corrosion reactions geometry) or those that can be expected (in parallel geometry).

3) Define control/prevention methods in terms of as-is and to-be. Alternatively, this means to distinguish if the model is to be defined for a Greenfield or a Brownfield project. Based on the corrosion reactions recognized as mentioned in the previous section, it is then possible to suggest control (CC) measures, or depending on the situation, preventing it (CP).

We believe that it may be worthwhile if we illustrate what has been mentioned so far (1.1–1.3) by giving an example; if the system of interest is the bioleaching tank we mentioned earlier. In this tank, copper concentrate particles are mixed via a mixer with a biotic medium containing SOB that can produce sulfuric acid with a pH down to 2. The volume of the fluid inside the tank is about 3 million liters and the mixer is made up of carbon steel with its propellers coated with rubber. As the bacteria are thermophile, the mean temperature of the fluid must be kept 50 °C by the stainless-steel heaters installed inside the tank and always in contact with the moving fluid. This overall, rough description is necessary because it will assist us in defining expected corrosion mechanisms and possible treatment recipes.

Therefore, the corrosion system is the bioleaching tank itself. We will adopt a CINS approach so that the tank will be broken into its subsystems. Based on this approach, we can think of the following hierarchy in sorting out the corrosion subsystems:

1) Internal equipment
   a) Internal coating of the tank
   b) Heaters
   c) Metallic body of the shaft of the mixer
   d) Metallic body of the mixer's blades
   e) Coating over the blades

2) External equipment
   a) External tank body
   b) External protective metallic layer

3) Reinforced concrete pedestal
4) Tank foundation
5) Fluid

Based on the CINS approach above, possible corrosive scenarios can be defined for each item. For example, the concrete pedestal could be vulnerable to moisture ingress from the atmosphere as well as from the ground. In either case, the steel bar inside the concrete will undergo corrosion and as per Tutti model for concrete corrosion, it is at risk to undergo spalling/failure. Tutti model and concrete corrosion along with its theoretical basis have been explained elsewhere [20]. Based on these corrosion reactions (that are in series geometry), the possible remedies could be using hydrophobic coating around the above-soil section of the tank and conditioning the concrete "root" in the soil with polymeric, anti-phobic wrapping materials or conditioning the soil around it with suitable anti-phobic materials.

In this regard, a "Decision Tree" is shown in Figure 3.13 that will serve to summarize the decisions (especially wrong decisions). One important feature about

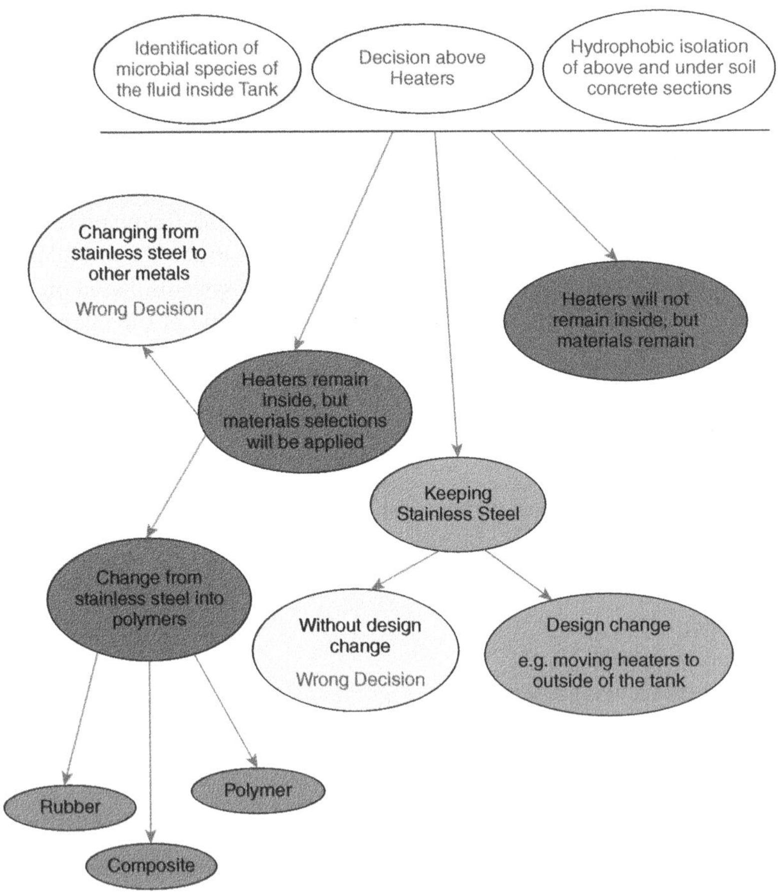

Figure 3.13 Decision tree for corrosion management of a bioleaching tank based on a CINS approach.

these decision trees is that they are fixed in time, as all these decisions are taken based on present conditions of the assets and according to precautions taken, can be changed. It is obvious that if conditions change, then the components of the decision tree will also change. For instance, if the decisions we have branded as wrong are to be applied, then the whole structure of the tree may change. In Figure 3.13, it is also evident that just one of the options has been discussed in detail and the other two (at the very top of the figure) have not been discussed. The main reason for this – apart from not adding too much detail into the figure – is that the decision tree cannot be confined just to these three issues, but must be based on the CINS approach and five subsystems we defined; it is required to define a decision tree for *each* subsystem. In addition, it is important to look for possible interactions between these decision tress and the possible synergistic/antagonistic effects they can have on each other.

3.1.5 Phase 2: Application

In this phase, the measures to treat corrosion as per the previous phase are applied. Remedies whose application as corrosion treatment measures has been approved will be executed to the corrosion system in such a way that can satisfy the minimum three requirements below:

- These measures are feasible enough to be applied
- These methods are eco-friendly
- They must be as easy as possible

The eco-friendly requirement is particularly important, and often the above three requirements may contradict each other under tough production conditions that are being imposed on the industry sector, in which the corrosion system is being investigated. Economy and ecology of anti-corrosion measures must be in place together to assure that the whole process of application meet the requirements of being called safe. A good example of such is replacing in-use synthetic biocides with their natural counterparts. Although research about some plants and their detrimental impact on several microorganisms has been known for some years now, they are not in use on a large scale by industries due to many reasons that we will not discuss here.

Application is also an important issue because there must be a convincing reason why a certain application method is preferred over others. While feasibility and eco-friendliness are both important, they are of no use if the application is not easy enough to be put into practice . Ease of application can quantitatively be measured with the time spent on it. If applying biocide takes less than applying coating, then biocide application is obviously preferred because of the economy involved in it.

Yet another issue that must be taken into consideration in application of anti-corrosion measures is how well they perform. There must be milestones in place that will show how positive or negative the measures we have taken are acting. This is the topic for the next phase; Monitoring.

3.1.6 Phase 3: Monitoring

This phase, in short, consists of two steps:

1) KPIs
2) Finding the best monitoring technologies

As briefly mentioned previously, it is very important to realize how well the applied anti-corrosion measure has performed. For this reason, it is important to define a measure against which the performance can be evaluated. This measure is what is often referred to as Key Performance Indicator (KPI). It is a measure that, when the obtained results are compared against it, it shows where the system is standing from a corrosion health point of view. It follows that to do the monitoring well, the first requirement is to have a set of KPIs. Certain standards can be used in this regard. For instance, to evaluate how severe corrosion is in terms of corrosion rates, or how serious corrosion is in terms of pitting, or the like there are many guidelines and standards that can be consulted with. KPIs show how and in which direction from as-is conditions of an asset the CM consultant must act to arrive in to-be conditions.

As there can exist parallel geometry for corrosion reactions already taking place in the system of corrosion, application methods could also be parallel and fit for each of the reactions. It is in this regards that possible antagonistic effects between these methods must be taken into consideration very seriously.

In addition to defining KPIs for the corrosion system for it's as-is and to-be conditions, the way by which corrosion must be monitored is also important. A web search for "Corrosion Monitoring" standards will certainly yield in what a corrosion technologist will be looking for.

It may be useful to advise that if one chooses to apply a CINS approach, monitoring must be defined as per subsystems whenever applicable. In the example of the bioleaching tank we gave earlier, it is possible to define the KPI for the outer metallic body of the tank in terms of allowable corrosion rate where no through-wall corrosion pitting will be likely to be experienced, nor will the entire mechanical integrity of the tank be sacrificed. For this, it is equally important to define KPI particularly for the internal coating. Having KPIs in place will also serve to get a sense about what to expect when the asset is being inspected during periodic overhauls.

Finally, we would like to once again remind our readers that all engineering assets are in their pseudo-FFS states and therefore it is very important to have KPIs in place, because in comparison with them the healthy condition of the asset will be evaluated.

3.1.7 Phase 4: Feedback

It is in this phase that the following must be carried out:

a) Feedback about the results of application phase
b) Re-application if the feedback is not desirable
c) Sharing the acquired knowledge and skills
d) Documenting the overall results in the form of recommended practice

Importance of Phase 4 is that it will allow others (next generations who will be recruited in the plant and current colleagues in your industry or similar industries who are sharing the same corrosion problem) to get a feeling about how to effectively rectify their problems.

Phase 4 is also a feature of Javaherdashti CM model that has not been addressed in many CM models. It seems that the ultimate goal of current CM models is to recognize the problem, find the best solution in terms of applicability and cost, and rectify the problem in as short a time and as cost effectively as possible. There is no plan B about what if the solutions provided did not work or what are the steps (to be documented for each plant/industry) that are necessary to share the information gained. Here we would like to mention that in our opinion, there is a vast difference between "data" and "information," as we can simply put as the following; raw information is data and processed data is information. It is not necessary to include all the figures and values one has obtained regarding the phases that have been applied, the overall picture would be enough. In addition, there must exist a procedure that what fraction of the gained information can be seen as lessons learned. This is a very important step, and if there is no legal framework for that it can lead into serious, unwanted results for the person(s) who are transferring the information in the form of a research paper, or as a part of a technical chat on social media.

It may be necessary to also explain what is meant by "re-application" within the context of this chapter. It is quite possible that the treatment measure(s) approved to be applied will not work when compared to the pre-determined KPIs. An example as such is if the wrong corrosion inhibitor[3] is applied, the wrong dosage of a

3 On one occasion, on a subsea pipeline that had been hydrotested with seawater and had developed microbial corrosion, a corrosion inhibitor was applied where the inhibitor itself acted like nutrient to sulphate-reducing bacteria (and possibly other microbial species present in the biofilm/planktonic forms) and increased the rate and severity of MIC in the pipe!

correct corrosion inhibitor is used, or the applied coating/material is not what it needs to be, and so forth. This will require going through previous phases, particularly Phase 1, to make sure that the corrosion processes and their geometry already active in the system have been picked up correctly. It is obvious that if the corrosion reactions and the way they interact with each other (series or parallel) are understood correctly, then chances are that the treatment strategies to be applied will work.

The Javaherdashti CM model may be stopped when it comes to Phase 2, particularly if the business owner has found out what his corrosion problem is and how to address it. An important issue for the business owner or the operator could even be how the corrosion problem can be addressed in a short time. This mentality is understandable because most of the time, the operator is under the pressure of production and commitments associated with it, and may not want to researchwhy the suggested remedies have worked and how they are going to publish a paper from the information (Phases 3 and 4). As long as they see that the suggested remedy worked well enough to allow them continue their production with a reasonable level of reliance, they (the business owner of plant management or operators) may not be interested in taking the CM business further unless the corrosion issues will have a vast impact on the ecology and/or economy of the production, or shorten the service life of the asset dramatically and bring it much closer to Zugzwang effect state should corrosion continue.

Therefore, it should not come as a surprise if any CM model is advised and even published, but remains half-executed due to many reasons. One of these reasons could be pouring the budget previously assigned for a CM project to an immediate case of COVID19 treatment or consequences associated with it (such as, but not limited to, having less workforce to carry out the required jobs).

3.1.7.1 Corrosion Cost Estimation Model

Thus far, we talked about technicalities that must be involved in any corrosion related matter for which a CM model is either to be developed or applied. However no CM model will be completed if there is no clear approach for considering the economy of the damage imposed by corrosion.

The importance of having a corrosion cost model in place is that it can largely serve to justify the costs associated with dealing with corrosion. These costs range from those are related to exploring corrosion and root cause analysis to finding the best treatment method. In all these scenarios, it is essential for the business owner to justify why these expenses are necessary.

While some economical aspects of corrosion will be discussed later in this book, it is essential to have an idea about what is currently available as the toolbox necessary for researchers and investigators, particularly concerning estimating the cost of corrosion. Details of such models have been given in part in several publications [15, 21–27].

Research about the cost of corrosion dates back more than 70 years; the cost must be expressed based on a measure that is a common denominator of all economic parameters in a country. It is obviously not reasonable to base the cost estimation model upon variable parameters that are not the same for developed and developing countries. However, the common outcome is that the cost of corrosion is expressed as a percentage of GDP (gross domestic product) or GNP (gross national product) of a given country. The cost of corrosion can be expressed regardless of the differences existing among countries; for example, if in country A the cost of corrosion is 3% of its GDP and in country B the cost of corrosion is 5% of its GDP, it does not mean that A and B have the same GDP, it simply shows how much of their wealth is being consumed by corrosion.

There are basically four models for estimation economic cost of corrosion as follows:

1) Uhlig Model, named after Dr. H.-H. Ublig, this model has features such as:
   a) It is rather a conservative model that gives minimum values for corrosion costs.
   b) Its basic assumption is that direct and indirect costs of corrosion are equal.
   c) Initial data to construct the model are taken from producers and not consumers.
   d) This model cannot express the corrosion cost distribution scheme within a given industry, nor can it calculate how much a certain industry can save on CM.
   e) This model does not present any particular method to calculate indirect costs of corrosion.
   f) Data collection is carried out by submitting questionnairs.
   g) This model has been used to calculate the cost of corrosion in the USA, India, Japan, and China.
2) The Hoar Method, named after efforts organized and conducted by Dr. T. P. Hoar has the following features:
   a) In this model, input and feedback on corrosion costs are inquired from both industry experts and academics.
   b) Hoar model is industry-based.
   c) Direct cost of corrosion as calculated by Hoar model is always higher than that calculated by Uhlig model.
   d) No clear methodology to calculate indirect cost of corrosion.
   e) Data collection is carried out by submitting questionnairs.
   f) This model has been used in the UK, Japan, and China.

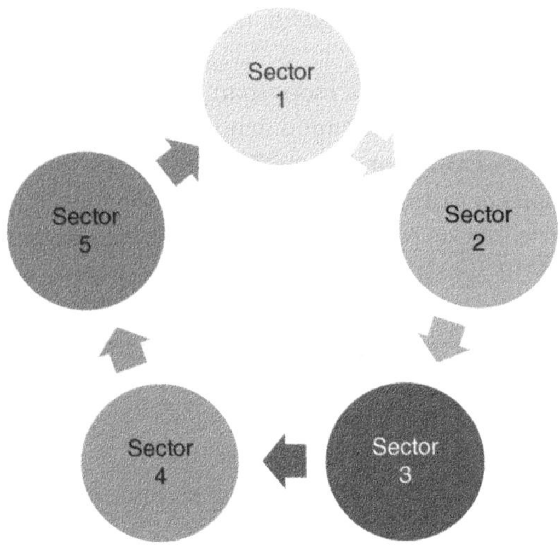

Figure 3.14 Simplified schematic presentation of I/O model, where the output of each sector is treated as the input of another sector. An example is that the output of mining industry as coke is treated as the input for steel making industry. The output of steel industry as galvanized steel then will be input to automotive industry.

3) Input/Output model, also known as I/O model can be schematically shown as in Figure 3.14.

   Some features of I/O model are as below:

   a) There are uncertainties about both the capital cost and intermediate outputs.
   b) For each industry sector, the tedious job of calculation of cost coefficients (the coefficients that show how different sectors are mathematically interrelated) is required.
   c) Despite the complexity of this model and difficulties associated with it, it is claimed that for a country like India, feeding the inputs for four decades into such a model has yielded a simple equation that can calculate the cost of corrosion in this country for each year required.
   d) There is no particular methodology to calculate indirect corrosion costs other than inter-sectors costs.
   e) This model has been used in the USA, Australia, India, Japan, and Turkey.

4) LCC analysis, has the following main features:

   a) Compared to the previous three models, this model appears as more realistic.
   b) LCC model is applicable to calculate cost of corrosion in equipment, structures, and factories.
   c) Applying this model to India has always demonstrated cost coefficients lower than those obtained by applying I/O model.

d) This model is not suitable for considering all factors and parameters involved in indirect costs of corrosion due to the ever-changing, dynamic nature of interactions among variables related to the indirect cost of corrosion.

All the models discussed above have their pros and cons and it would obviously be a more thorough approach if at least two of them are used together for one corrosion case and the results compared. In addition, it must be noted that these four models discussed earlier do lack a clear methodology to quantify the indirect costs of corrosion. This is a very significant aspect of these models that must be taken into consideration when these models are to be employed for any management of corrosion approach.

3.1.7.2 Corrosion Knowledge Management (CKM)

An integral part of a management of corrosion program is to have a plan ready to allow us to measure, either via quantity or quality, the impact of management parameters on the way corrosion is being treated.

We introduced CKM over two decades ago. CKM is a managerial tool specially design to advise managers with little or no technical knowledge of corrosion about how to deal with corrosion as an issue in both the present [28–31] and future [32] states of their industry.

While CM deals with the risk of corrosion, CKM deals with the management side of it, which also includes the economical/ecological impacts of corrosion. We have explained the similarities and differences between CM and CKM elsewhere [33].

The main algorithm to be followed regarding a CKM-based approach is as per the steps below:

Step 1: As per the industry and the plant to be investigated, at least two of the four internationally acceptable corrosion cost models are selected and applied;
Step 2: The outcomes of the models are compared, and the range of corrosion costs is determined;
Step 3: The findings are reported to the business owner/high ranking manager;
Step 4: The impact of cost of corrosion is measured against seven management parameters.

In applying the steps above as per a CKM scheme, a suitable CM approach must also be taken. This CM approach may opt to use any CM model available, from the CM model suggested by BP to that suggested by Total and the like. As we have mentioned earlier, a CM approach as such will be a technical matter and has been discussed in previous sections.

However, regarding the steps mentioned above, Step 4 needs to be explained in more detail; in fact, we need to specify what these seven parameters are and how they are (or can be) related to a CKM model.

The seven parameters of management are as follows.

1) Budget: When based on a combination of economy models related to the estimation of cost of corrosion, it was found how significant the cost of corrosion in the plant is, then it is decided how much it is logical to expect this cost to be lowered. It must be noted that as corrosion is thermodynamic, it is impossible to bring the corrosion cost down to zero. It is generally believed that up to 37% of the cost of corrosion can be recovered. Therefore, the remaining (or a percentage of the remaining) estimated recovery cost must be taken by the management as the budget necessary to manage corrosion. As an example, if in year zero, the economy models showed that for a particular plant the cost of corrosion is, say, $100m, then it must be planned to bring down this cost to, say, $70m. It is possible to take, say, $10m as the required budget to manage corrosion to save on later corrosion costs and have a net profit of at least $20m. We take it as self-evident that saving $30m cannot be expected to happen overnight and it requires careful planning (as per reference [16]; in Kuwait cost of corrosion in 1987 was 5.2% of the GDP, and was lowered to 1.7% of the GDP in 2011; it took 24 years for this country to lower cost of corrosion by 3.5%).
2) Human Resources (Expert Team): Based on management decision fighting against corrosion where management becomes serious enough, a team of experts will be necessary. There are three ways to handle this requirement; (i) employing corrosion experts (ii) employing an experienced consultant, or (iii) training the existing staff without going through the costs associated with previous options. If options (i) or (ii) are to be selected, for each or both options there may be a need to find the best possible solution as per existing regulations, based on the emergency of the case. In some countries, it may not be easy to create a position for a corrosion consultant to a refinery or a power plant simply because there is no position defined. It is then very important to bear in mind that any CM model that will neglect such cultural facts is to be doomed as a global recipe to treat corrosion.
3) Training Department: It is always a familiar policy by the industry to cut the expenses that may seem extra. It is quite expectable for a company not wanting to employ new person power or consultant(s) simply because it is seen as an extra cost. It is in this regards that training of existing staff may be put forward. The training department must then be very sensitive to the training programs it is planning, because most training departments have a passive rather than active role; they do not advise or publicize the available courses, but prefer to work based on the feedback they receive from the personnel. An example

is that the training department may never think that a course in galvanic corrosion (particularly galvanic corrosion, CUI, or atmospheric corrosion) is necessary for their colleagues. The most they think of is to offer a general course in corrosion. The duty of the training department is to continuously seek new topics of interest and use to the present and future operation scheme of the plant, not just wait until they have a request. To avoid and prevent any industrial nepotism, the required topic and industry skills both must be required. Our readers must note that the instructor is being called for an industry case which is far different from an academic research interest, therefore the invited trainer must ideally have both academic qualifications and trackable industry experience. One may not have very distinctive academic degrees (a BSc will suffice), but in return they must have at least 20 years of experience. Dealing with various corrosion processes does need a robust theoretical background, for example, not to call microbial corrosion as water corrosion, asbitterly reported by Tatnall [34].

4) Research (University-industry Gap): It is a bitter fact in many countries – perhaps with the exception of some Western countries – there is a huge gap between academia and industry, especially when it comes to corrosion. This author firmly believes that, at least in the field of corrosion, it is the industry that must be the leading force for research. Even very advanced research must come from industry needs and not just out of curiosity. Therefore, for a given plant it is necessary to be backed with a robust research scheme that will not only clarify corrosion reactions involved but the geometry among them. One very essential tool in conducting research is using TRIZ; the concept of TRIZ is discussed in Chapter 5 of this book, however what can be very briefly said about TRIZ is that it is an algorithm for innovation. By learning TRIZ and its principles, methods, and laws, the likelihood for achieving innovative results will be highly increased.
5) Information: There is a vast difference between data and information, as we said earlier. To process the data obtained from field measurements and convert it into information useful enough to create an idea about the current status of the corrosion system and what must be done about it in the future, both updated theoretical knowledge and field skills are needed. In addition to this aspect of information, it is vital to have a written procedure in place to lessen "dark information" and increase "white information" [5]. Dark information is the knowledge and skills hidden in people's minds and their own notes, and therefore so personalized that access to it is impossible (or too difficult) for others to understand. White information, on the other hand, is what is written and accessible by everyone. An example of dark information is verbal information, and examples of white information are standards, recommended practices, and the like. To deal with corrosion, the information must be shared

and easy to access so that no time will be wasted on activities that can be branded as re-invention.

6) Energy: In this context we use energy equivalent to motivation. Management has to deal with people and as such, they need to be motivated for the job they are going to do. It is necessary to motivate the people involved in the management of a corrosion scheme to feel that no matter their expertise – whether technical, such as engineering or non-technical, such as management – they are part of a project that is very significant. Perhaps one way of doing this is to ensure them that if next year there is a decrease in the cost of corrosion, they will all receive a bonus! Any measures that can be used to serve the target of getting motivation and so-called positive-energy is of extreme importance, as there are at least nine components of a working place that will render it stressful to the employees (Figure 3.15).

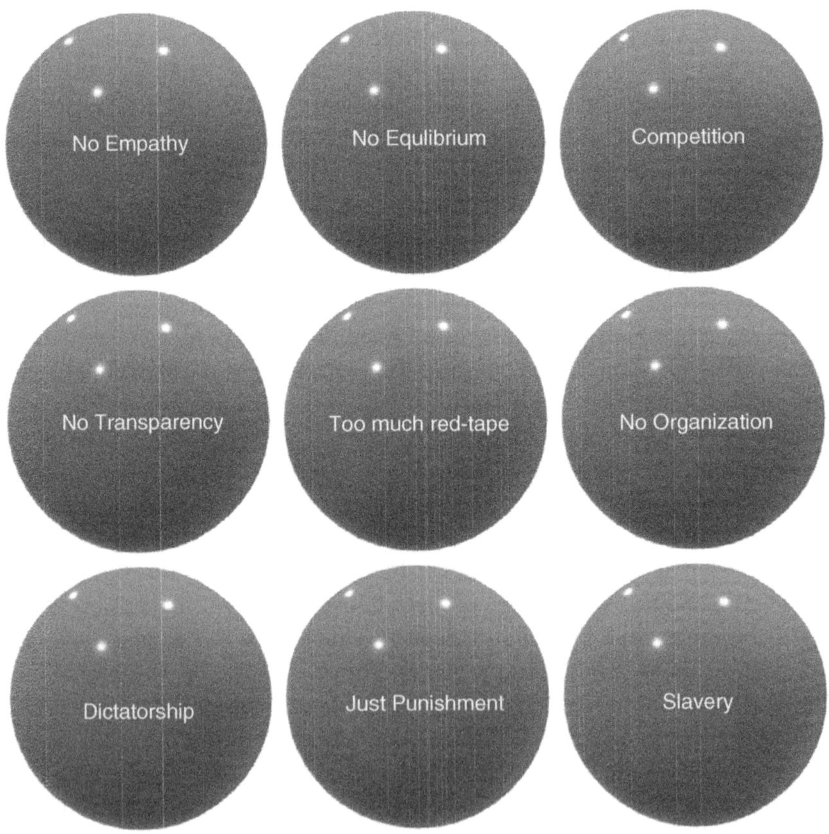

Figure 3.15 Components of an unsuitable workplace.

The nine components that can cause restlessness and stress in a workplace can be described as below:

1) No Empathy: Managers are disconnected from their employees in terms of not being able to relate to them. The wall around mangers will not allow their employees to feel attachment toward their managers.
2) No Equilibrium: Working hours and family hours are not recognized. There is no distinction regarding non-working hours for the employees, so their private lives can be severely affected by the long hours and what is imposed on them.
3) Competition: In this context, competition must not be viewed in its positive meaning. The positive meaning is when employees are in competition with each other for the good of the company. Here, however, competition is between employees themselves where methods used are not healthy. Competition in this regard can be lead to the collapse of the company, as does not aim toward advancing the company, but against other individuals.
4) No Transparency: There is no transparency about responsibilities and what is being applied. No one is aware who is doing what and who to go to for an explanation. As the responsibilities are not well defined, there is no way to distinguish where one's responsibility begins or ends.
5) Too much red tape: There are so many rules, regulations, and procedures that in reality what they do is lead to confusion. There is no opportunity to let people think and act freely, and the boundaries serve more to limit them than providing clear responsibilities.
6) No Organization: Everyone does his work in the way he thinks is correct. There are basically no measures that define where and when a job or a responsibility starts and how it must be carried out.
7) Dictatorship: In an unhealthy workplace, what the managers say is the law and everyone must follow. There is no room for critical thinking or constructive criticizing. The manager orders and then does not bear the responsibility for issues, and no one is allowed to question their decisions.
8) Just Punishment: One of the ways by which motivation can be constructed or re-shaped in accordance with the needs and aims of the company is by a reward-punishment system. Employees are promoted or receive a bonus for what they do that is favorable to the company, and are punished (such as fined, delaying promotions, or being fired) for what they do to harm the company, whether consciously or not. However, if instead of a dual reward-punishment system, we rely only on only one system, the employees will soon find their working conditions unbearable and will look for opportunities elsewhere and possibly sue their previous workplace.
9) Slavery: There is no defined relationship between the management and the employees so that management behaves with their employees in any way they

wish. More or less, the management assumes that their employees are like slaves to them; they may talk down to them, treat them poorly, or as lesser than themselves, whether skilled or not.

The nine components above may seem somewhat exaggerated, especially to Western readers, but this author can assure his readers that his presence in many industries around the globe have proved to him that these issues do exist.

While the nine components shown in Figure 3.15 can be related to any activity in which human beings are involved in a workplace, when it comes to CKM it becomes even much more important. One thing which often seems to be forgotten or overlooked by integrity managers and CM professionals when it comes to management of corrosion, and is a very important factor in this context, is how the members (both technical/engineers as well as technical/technicians and non-technical, administrative colleagues) of the plant and/or company at which CKM is to be carried out need to adjust themselves concerning having knowledge about corrosion, however small or limited this knowledge may be.

The issue of corrosion within a context related to CKM is like knowing first aid; you don't need to have in-depth knowledge of medical skills but you need to know just enough to see the essential needs. All personnel must be trained about corrosion and they must go through a carefully designed motivation system. While the level of education will be different for an engineer than an office clerk, it is something that must be determined by experts and fairly discussed with the training department of the plant/company, but the motivation program is something that would definitely require the help of an expert in this field. This expert most of the time is not an engineer but a psychologist with intense experience about people working in industries.

10) Time: What is meant by the context of time is not just completing jobs on schedule, but the entire plan. It is very important to know that a CKM scheme must start from year zero and go on for three years, at least in phases. The entire timing of the project should be like the one presented below, Figure 3.16.

The model suggested here consists of three territories:

- Blue territories: Phase 1 to Phase 4. In this territory, preparation for selection and application of corrosion cost models, as well as data acquisition and analysis to arrive at information are performed. What is meant by information here is a figure that shows direct cost of corrosion for the plant.
- While territories: In these territories, that are alternatively called waiting periods of 24 and 36 months, recommendations suggested by experts for lowering corrosion costs are applied.

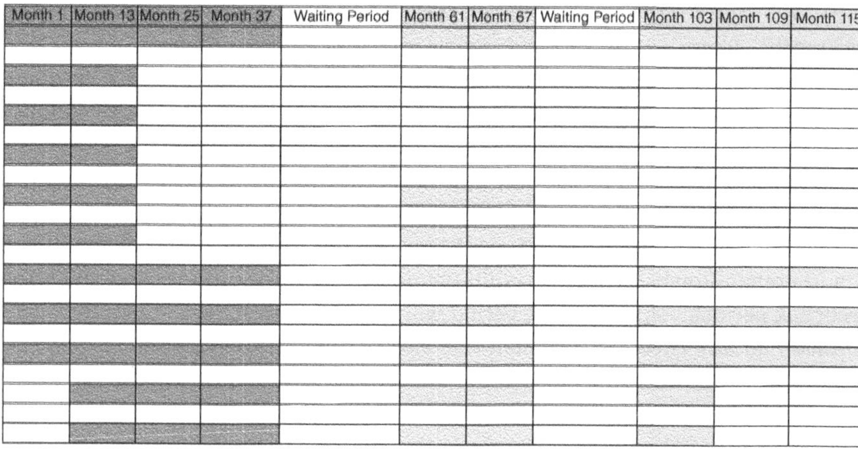

Figure 3.16 An example of a 115 month CKM scheme. This scheme is arbitrary and based on the CM culture that may exist in the plant, the overall duration may become longer or shorter than time milestones shown in the figure.

- Yellow territories: These are the periods during which the recommended CC strategies and tactics, and the results obtained from them, are examined and analyzed.

In more detail, the following can be said: phases of the CKM scheme given in Figure 3.15 are:

- Phase 1: Selection of the economic model(s) (1 month).
- Phase 2: Preparation and modeling based on provided data (12 months).
- Phase 3: Application of the model(s) in terms of data collection (12 months).
- Phase 4: Quantification of the data into information (12 months).

The above four phases (Blue territory) as mentioned above deal with selection, preparation, application, and results.

After phase 4, there must be a waiting period that we have chosen for our model to be two years (24 months). During the waiting period, within six months (month 61 to month 67), re-measuring corrosion cost model(s) is carried out. This is Phase 5 and is shown in yellow; Phase 5 can be called short-term feedback. It is during this yellow territory that the initial results of recommended measures against corrosion are taken into consideration and their effects on corrosion cost are being evaluated. In these periods, it is possible to re-use the models that were developed during the blue territory. To keep repeatability of the models and consistency of the results, it is advised to use the same models used for the blue territory for the yellow territories.

Phase 6 starts right after the quick feedback period and is a waiting time of 36 months. Phase 6 can alternatively be called mid-term feedback. Phase 6 is important in the sense that during this phase the assets are still in their pseudo-FFS state and the assumption is that during Phase 5 have all been applied. In other words, if the equipment has survived the short-term feedback period and Phase 5, then this implies that they must have been in conditions good enough to have arrived in the mid-term feedback phase.

Phase 7 is a 12-month period, dedicated mainly to documentation and final report preparation based on the results obtained from the short-term and mid-term feedback periods. Each feedback period will serve to show if the CM practices performed had any effect on corrosion cost as a part of a CKM scheme.

Although the two white time phases of 24 and 36 months are included in the entire scheme of a CKM project, they are excluded from it from a cost-management point of view. Therefore, the actual timing of the project in this example, not counting the waiting periods, will be 54 months or 4.5 years.

It is essential to note that changes to be made to seven management parameters must be done after short- and mid-term feedback periods.

For a typical project as such, the help of economists is necessary in addition to engineers. In addition, ecology experts will be needed if indirect cost of corrosion (for example limited study on the impact of corrosion on environment) is also to be regarded.

Obviously, the example given in Figure 3.15 is based on several assumptions that may not be true for a given plant. For example, based on the existing corrosion awareness culture, the time length of each phase can be shortened. However, it is strongly advised not to omit the waiting periods. These periods can be made shorter but they must included in such a way that they take at least six months.

CKM has some features that need to be addressed here:

1) CKM is to be useful mainly for Greenfield projects which are in their pseudo-FFS conditions and thus already experiencing corrosion.
2) CKM looks on both economic and ecologic sides of corrosion and therefore contrary to any CM scheme, a CKM scheme could take much longer. The main reason for this is in very few plants worldwide it is possible to find a corrosion cost scheme that has been advised by economists and executed by engineers. Equally rare is to find an ecological model for a given plant in which corrosion and its role has been highlighted.
3) CKM like CM is a part of a more encompassing management of corrosion program. The feedback taken from CKM studies, in both economy and ecology, must be carefully taken into consideration. Management of corrosion is a holistic approach and being as such, it requires that its framework also considers not only technical aspects of corrosion (such as research) but also non-engineering aspects as well, such as issues related to human resources and training.

CKM has four principles:

Principle 1: Corrosion system definition; this topic in terms of being able to define CINS and/or COFS has been covered previously in this chapter.

Principle 2: Transparency; this is a legal side of CKM. According to this principle, the business owner as well as the consultant or the project manager for a CKM project agree to reveal what aspects of finding to the professionals outside of the plant and the common people who have no technical information or background about the consequences of corrosion in terms of the costs it imposes. If the outcome of corrosion in terms of environmental impact is something that would affect people living nearby, there must exist regulations that would clearly determine how the obtained results from a CKM approach become known to others. In this regard, while CM could be an internal affair of the plant or company, CKM has both an internal affair side (rearranging of seven management principles in terms of making corrosion cost more affordable) and an external affairs side (how to share the findings with other colleagues working in the same industry as well as how to deal with the information gained in terms of legal and publicizing terms). It is a bitter fact that in some countries, speaking about the extent of corrosion-related catastrophes is taboo. This is even true for some Western reputable companies (the way some high-ranking managers reacted to the oil spill disaster in Prudhoe Bay, for instance). For example, water authorities are using many cast iron pipes for carrying drinking water, and this water is also to be disinfected by hypochlorous acid. Recently [35] it has been discovered that upon reaction with the disinfectant agent, cast iron releases toxic compounds of chromium that could lead to serious health problems such as cancer. What should the water authorities do with this finding? Can they share it with their colleagues? Can they share it with potentially affecting people? If the answers are yes, then to what extent must this sharing process continue? If this water authority company had been located in another geography, could we expect not to even hear about the problem and the possible underlying corrosion mechanism in the first place? It is because of these and similar questions that we believe having a CKM scheme along with a proactive CM is an essential part of any management of corrosion program that wishes to be taken seriously and become successful.

Principle 3: Use of Information; although we have previously discussed about dark and while information types, it is also worth repeating that a CKM scheme must also look at creating a systematic use of existing information (both dark and white) as well as creating information accessible to everyone, as per protocols and regulations.

Principle 4: Modeling; what is meant by modeling is that the CKM scheme must have the potential of using mathematics in order to construct mathematical models that later on could serve as the main platform to produce specialized software. While modeling certainly has limitations such as, but not limited to, the necessity of making assumptions to make the mathematics simple enough to become applicable to the system to be considered, it certainly has undeniable advantages in explaining the present state of a system and predicting its future. Assumptions, such as initial and boundary conditions, that are necessary to make a model may seem to simplify the system – and this is one of the defects always brought up by many – it is also an important feature of any mathematical modeling procedure.

Below we will introduce a mathematical procedure we have developed based on the mathematical modeling work by Schouten and Gellings [36]. This model introduces a mathematical tool that is called Corrosion Protection Factor, or c.p.f. The model can be applicable to any given CM approach to evaluate the usefulness of CM techniques frequently used. We have explained this model and how it can be applied elsewhere in one of our previous publications [12], but due to its very important features we will very briefly discuss the main features here.

Schouten and Gellings model, or as we have briefly named it, S-G model, looks at quantifying the best CM practice in terms of a mathematical algorithm that will combine frequency of use of certain CM measures (such as cathodic protection, coating, use of chemicals, etc.) with expert opinion of how useful these methods are when applied to a certain case of a corrosion system. In the example we have discussed in reference [12], we used a subsea pipeline example on which five CM measures, namely cathodic protection, coating, corrosion allowance (in terms of getting a "thicker" pipe), application of inhibitors, and application of biocides had been suggested as the main measures. Three experts are asked to evaluate both the usefulness and frequency of use of each method by answering a questionnaire. Based on overall opinions of the experts they are branded as optimistic, moderate, or pessimistic. For instance, the expert being branded as pessimist may have been of the opinion that most of the CM measurements are useless, less useful, or not frequently used, and the like. These opinions are quantified and certain c.p.f. percentages are calculated. The obtained c.p.f. values are matched with NACE references that define low, moderate, high, and severe corrosion categories, and based on that the average of c.p.f. values we calculated and compared with required ranges for which a given CM scheme can be called effective. In the example of the subsea pipeline and the CM measures taken against MIC management, the overall score of c.p.f. was 73.5% that corresponded to a "moderate corrosion management scheme."

Based on the c.p.f. model above, the CKM scheme can decide how well the applied corrosion measures are in inducing corrosion protection, and thus their effect on overall direct cost of corrosion. The costs imposed on the system by the available corrosion measurement methods and the level of protection they offer will then be an important decision-making factor in choosing the most feasible corrosion protection option. This, when joined by the impact of corrosion cost, can yield in a multilateral approach toward corrosion in a plant or in any given industry.

An incredibly significant side of any CKM is to have a say regarding indirect costs of corrosion, of which cost to environment could be the most significant. In our previous works [5, 12], we have tried to address an approach that will fill the existing gap between corrosion specialists and environment experts. Despite all these, there is no model that can be used to describe the present state of the environment when the assets are in their pseudo-FFS state, to the best of our knowledge.

However, we believe it is necessary to explain a point that can result in confusion; study of environmental factors on atmospheric corrosion, for example, is not what is meant in this context by the mutual impacts of corrosion and environment. As reported by Mattsson [37], ISO 8044 defines corrosion as *"Physicochemical interaction between a metal and its environment which results in changes in the properties of the metal and which may often lead to impairment of the function of the metal, the environment, or the technical system of which these form a part ... [where] this interaction is usually of an electrochemical nature."* Based on this definition, corrosion can also induce changes (impairment) to the environment. It is then this aspect of corrosion that has its impact on environment, that must be taken into consideration as a model capable of explaining the present and predicting the future of environment components (water, air, and soil) in terms of the effects corrosion can create on them.

While in our opinion a very important item in calculating the indirect cost of corrosion is its imposed costs on the environment, there are obviously other cost items that can be included under indirect cost title, such as but not limited to, overtime of production or overdesign [11]. There are also items whose costs must be included in any corrosion cost model. Normally, it is assumed that the indirect cost of corrosion is roughly equal to its direct cost; obviously, this is an incorrect assumption. To further illustrate an example of indirect cost of corrosion and how complicated it can be take the below example.

It is known that every 90 seconds one ton (1000 kgs) of steel corrodes, and to produce 1 ton of steel, between 6 to 7 tons of water is required [38]. It is then not too candid an assumption if we assume that every 90 seconds, 6000–7000 kgs of water is being wasted due to corrosion. This figure means that within one year more than 2 m liters of water per kgs of steel are lost due to corrosion. Obviously, the amount of water which is used for steel production, i.e. 6000–7000 kgs (liters) of water cannot be changed. However, the more we save on the water per kgs of steel corroded, by reducing the amount of corroded steel we will obviously save in wasted water. The cost thus saved (or rather, in case of inappropriate management of corrosion scheme, the cost thus imposed on the system) is also an indirect cost that corrosion can be lead to.

Any model that is to handle the indirect costs of corrosion must also take into consideration the very complicated nature of these costs. It may be wise to first simply choose one area in which an easily workable model can be developed. For instance, if the model is to quantify the corrosion impact on ecology, it can simply concentrate on its impact on only one component of environment; for example, water and as a subcategory, only seawater. The model may, for example, try to relate the number of failures in which corrosion has been documented, to be a major player with the existing models that consider the effect of such impacts on how seawater is being affected. An example of studying the impact of a regular CM measure (cathodic protection) on the marine environment [39] or similar studies, can be taking the required building materials to create an assessment model to consider indirect corrosion impacts (including costs) to be modeled and calculated.

3.2 Management of Corrosion and COVID19

These days our world is struggling with Corona virus pandemic. COVID19 has affected all aspects of life, including our private and professional lives. Meetings and seminars are now being conducted with the use of cyber tools, so that instead of having physical workshops and meetings, virtual meetings are becoming part of our schedules. Some of typical measures taken by industries against COVID pandemic are as follows.

1) Limiting contact with clients and suppliers and diverting such activities to digital tools wherever possible.
2) Observing social distance regulations wherever face-to-face meetings are inevitable.
3) Division of departments within a company to minimize the contact between the employees at inevitable occasions (such as breaks).
4) Minimizing the number of people simultaneously being present in each confined space.
5) Encouraging remote working when possible.
6) Increasing hygienic measures in terms of cleaning and disinfection of all general contact surfaces and areas.
7) Temporarily suspending all foreign trips.

Consequently, management of corrosion and its components in terms of CM and CKM are also being affected. The big challenge, particularly for corrosion industry, seems to be deciding which is more important; investing corrosion mitigation methods and technologies, or dealing with taking precautions against COVID19 as imposed by governing bodies.

The corrosion science community has already started academic work on the possible impacts of the COVID19 pandemic on the corrosion of materials in terms of accelerating corrosion rates due to frequent use of disinfecting chemicals containing chloride ions [40]. In a CM approach which is based on Javaherdashti CM model, a paper [40] is accommodated in Phase 1 which is the phase for definition. In this phase, possible expected corrosion scenarios are discussed for a case where anti-covid and anti-corrosion measures are to intersect, and having a clear Phase 1 is necessary as it will guide us on how to define the applicable, feasible, and available measure against corrosion. In our CKM approach, as one of the most important factors related to management of corrosion and being pro (not opposed to) CM, the important role of motivation, human resources, and training have been mentioned, and undoubtedly they are very significant parts of the larger issue of management of corrosion.

An integral part of management is dealing with human beings, and being as such it is important to also consider what can affect them. COVID19 is a crisis that has been affecting all aspects of life and activities including corrosion protection industry. It is in this regard, that human resource becomes of tremendous importance. As J. Eskenazi points out in his note "4 Things That Don't Change for Recruiting and Hiring During A Crisis" [41], these four essentials are:

- Communication
- Clear expectations
- Keeping interviews going
- Being ready to share

While the above points' main audiences are recruiters and hiring engines, they are also applicable to your company's present, existing, human resource, that is, your employees. The employees in this context are both corrosion-related professionals and, seemingly remotely corrosion related people who are supposed to have no ties with corrosion issues at all. The first includes engineers and technicians working at maintenance and/or any department related to corrosion protection (such as cathodic protection or water treatment). The second will contain professionals who work in relation to process, and the third will be related to people working in departments such as human resources and accounting, to name a few. These people need to become ready for the upcoming changes due to restrictions imposed by the pandemic. Currently what most companies are doing is encouraging their employees to work from home. While this may be a temporary remedy, it does call for having a suitable infrastructure that would support all the services required from webinars (on- or off-line) to virtual seminars and kick-off meetings. However, the obvious point is that the operation of the plant cannot be carried out virtually, and even if the operation would become as such, the corrosion problems that would arise do require to be addressed by physical presence. This will require having an interactive, viable, and robust network of all the people involved. To design this network, I believe the four essential elements put forward by J. Eskenazi will also be applicable.

If COVID19 is to be included in any CM of corrosion scheme, there are a few questions that the plant/company that wants to apply management of corrosion under COVID19 conditions need to answer, the most important of which are as follows [42].

1) What is the emergency plan for the plant for working remotely?
2) How can the plant management persuade the concerned experts (particularly regarding corrosion issue) to attend the virtual gathering sessions, including training?
3) Study the effect of COVID19 on financial issues of the plant (as per instructions proposed by, for instance, international management consultants "Deloitte UK").

It should be noted that the conditions imposed on working conditions by COVID19 can affect factors such as human resources, training, research, and energy, and when applied to CKM principles, all activities that are required to be controlled by experts attending the duty (for instance, preparing and application of procedures related to establishment of transparency) can be affected.

Getting back to Figure 3.12 that shows and summarizes steps to be taken during the life of an asset from its fresh state of FFS to its failure, it may be of interest to see how COVID19 or any other stressful situation of that magnitude that has a biological component attached to it can affect the life an asset.

The first step, as also stated in Figure 3.12, is inspection and issuing FFS by the inspector. The impact of pandemic on the inspector is a risk that cannot be belittled. Regardless of how small the likelihood, due to its serious consequences, the inspector must take the risk very seriously. In this regard, the inspector must observe all required conditions and take every precaution possible. The inspector must note that they will be in contact with many parts that could have the potential of already being contaminated; Corona virus has been reported to survive on metallic surface from two hours to a few days (on aluminum surfaces from two to eight hours and on stainless steel surfaces two to three days) [43]. Therefore, inspection to assess for FFS is a crucial step in which precautions against COVID19 is of vital importance for the inspector. When the asset is put into service and it is assumed to be at FFS stage, then it is corrosion engineer(s) and consultants who must monitor the condition of the asset. This will of course require many interactions that may take place via meetings, and it is during these physical meetings that the likelihood of contracting Corona could become high. It is then advised at this stage, to have an IT infrastructure updated and powerful enough to support virtual meetings as video conferences and live webinars, if required. It is also important to note that issues such as training will be accommodated in this step. The same precautions as just mentioned can also be applicable to pseudo-FFS stage; in this stage, corrosion engineers and corrosion consultants will be engaged again. This, although not happening too frequently, is still of significant importance and must be taken into consideration seriously. Therefore, from a COVID19 consideration point of view, there is extraordinarily little difference between FFS and pseudo-FFS stages.

CM can be affected by COVID19 too. If our CM model is such that involvement of human beings is necessary, then it is also of paramount importance to consider controlling measures for COVID19. CKM is an approach that is directly related to the involvement of humans. An example of such is anything that is related to gatherings (technical meetings) that will require serious precautions against COVID19.

The most important aspect of any strategy that aims at management of corrosion is to first have thought enough about the environmental factors that can affect the health of humans working in this project. If anything we do to rectify corrosion can have one atom of effect of improving the way humans are interacting with environment in the context of corrosion, it is worth it!

3.3 Environment

Corrosion has an important aspect attached to it that has not been studied in the way it deserves. While in one of our previous works [5], we tried to touch on this subject, it was too brief and like any pioneering work, there is much that needs to be done. Most corrosion related environmental impacts are expressed in a way that can be shown to be, at the best, the cause of the disaster. It is quite natural to express that such and such disaster happened due to leaking of such and such flammable gas or liquid, but how many wonder if the leak itself was indeed a result of poor management of corrosion (that obviously consists of both CM and CKM)? In addition, a very fundamental question in this regard is if enhanced corrosion in the form of leakage (through wall pitting) is the only way by which corrosion can adversely affect environment. There is currently no model or methodology that can be used to assess the environmental impact of corrosion on the three components of environment (water, soil, and air). Just to give an example, take the two examples below that give us a glimpse into how corrosion can have an environmental impact even without causing a leak.

For the first example, suppose that we have a buried pipeline that requires 400 A/y (Amperes per year) – to be supplied by impressed current cathodic protection – to be protected against corrosion. If we use the so-called lead/silver anodes for this purpose with a consumption rate of 0.1 kg/Ay for a design life of 20 years, this will give an average mass of 941.0 kg of Pb-Ag anode to be required for such an current output over the expected service life. Alternatively, this means an annual anode consumption rate of 47 kg (these anodes typically can have a mass balance of lead of about 93%). All this will mean that for this particular case of an underground pipe to be catholically protected, about 42 kg lead each year (or more than 115 000 mg per day) will be released to a component of the environment, which is soil in this case. This figure becomes more important when we notice that natural soils usually contain less than 50 ppm lead (about 50 mg lead per kg of soil). Therefore, by applying cathodic protection via impressed current for this pipe, each year the average amount of lead which enters into the soil is about 800 000 times its natural lead content. Multiply this by the number of pipelines that have (more or less) the same requirements, and it can be seen that huge amounts of lead are being pushed into the soil each year and thus poisoning it.

Regarding the second example, take an offshore oil platform that has been estimated to require 400 A/y for protection. For a useful design life of 15 years, and provided the use of aluminum anodes, during the design life of this platform more than 3800 g of aluminum will enter into the seawater on a daily basis. In our calculations, we have not considered other ways of metallic pollution (such as industrial wastes). If just one platform with the above working conditions can contribute to contaminating seawater about four tons per day, it is not farfetched how other platforms can do the same harm to the marine environment, especially knowing that there are currently more than 1400 offshore platforms around the globe. Therefore, environmental impacts of corrosion cannot be limited to issues regarding unexpected leaks; corrosion can slowly but surely impose its impact on pollution via quite normal CM practices as well.

How can these impacts be addressed in the most appropriate way? Apart from an urgent need for an environmental model of corrosion, perhaps a very effective way to address this issue will be looking at it from an organizational point of view.

Due to mutual links that exists in many safety/HSE accidents and near-miss incidents with corrosion being the main cause, it is suggested to establish a Corrosion-Safety Management (CSIM) Department with any organizational unit that is given the job of looking at CKM within a plant or a company that operates a company or a series of companies.

What can be shown as the necessity for establishing a CSIM Department? In any plant; regardless of whether the plant of interest is a gas processing refinery, a power plant, a steel complex, or anything of this sort in which corrosion can play a very significant role, corrosion can be a major role player in four issues.

1) Corrosion in the main process equipment;
2) Corrosion in the auxiliary equipment;

3) Corrosion in the process safety equipment;
4) Corrosion in the physical space in which the equipment (main and auxiliary) occurs.

To clarify the above four corrosion issues, let us take the example of a refinery:

1) Example of corrosion in the main process equipment. The equipment in which various processes, such as but not limited to, distillation, refining, converting, and the like take place, are themselves vulnerable to corrosion. For example, corrosion could be a serious hazard in a furnace, in an Isomax unit, distillation tower, and other main process equipment.
2) Example of corrosion in the auxiliary equipment. If this refinery has a base-load power plant to maintain the required power to run the motors needed for the refinery, the power plant can be regarded as a very important element without which main processes can undergo difficulties in operation. Corrosion in these auxiliary units and equipment, then, is an incredibly significant issue. An example as such could be MIC issues in the cooling system.
3) Example of corrosion in the process safety equipment. To manage the risk of fire, for example, fire water rings must be available to service and operate. One must make sure that these rings have not been clogged with corrosion deposits or there is no chance for pitting corrosion being developed under these deposits leading into the loss of the fire water. Therefore, the equipment that is to be used to maintain safety must also be considered, and suitable CM must be applied to them as well.
4) Example of corrosion in the physical space in which the equipment (main and auxiliary) occurs. The concrete slabs, structural steel pipes, and the walls of the units in which the main and auxiliary equipment are sitting may also be vulnerable to various corrosion scenarios.

The above could also have safety and HSE aspects. A pipeline in which leaking has occurred because of development of a crack in the body of the pipe while within the pipe flammable /toxic fluid is being carried can create very serious safety and HSE threats to both the workplace and the environment. The root cause of the crack in this pipeline could be due to pre-commissioning practices such as hydrotesting, or post-commission operation-related causes. In any case, while the root cause is related to CM, the environmental impact is to be handled by HSE and safety personnel.

All the above calls for establishing a department that will look after both corrosion and HSE issues and will study the risk, not only in terms of HSE risk, but corrosion risk too. This interdisciplinary department can be called as a Corrosion-Safety Issues Management (CSIM) Department.

Who will benefit from a CSIM Department?

1) The owner/management of the plant, whether from private sector or government sector. This is because by establishing a CSIM Department, the economic–ecologic costs to be imposed on the owner will be removed. The reason for this is that CSM Department will assess both corrosion and safety risks (likelihood and consequences).
2) The workers of the plant. In addition to humanitarian reasons, it is necessary to save them from any loss they may experience such as losing their jobs because of injury or imposing costs on the health system.
3) The environment around the plant. Soil, water, and air are the main elements of any environmental risk assessment. These three elements are also the main elements that corrode any physical structure; atmospheric corrosion can be linked to air, microbiological corrosion can be linked to both soil and water environments, cavitation could be a result of water hydrodynamics, and so on. The three elements mentioned above must be addressed to both control corrosion and contain safety hazards.

The required expertise for this department would be a team made up of both HSE and corrosion experts. The cause-effect relationship between safety issues and corrosion causes must be studied and categorized. A risk matrix must be developed for all observable corrosion processes in any given plant.

3.4 Application of Management of Corrosion Scheme to Underground Fire Water Ring[4]

The case we will explain below is an underground fire water ring as shown in Figure 3.17

The alarming symptoms for the possibility of corrosion in this system are as follows:

- Observation of corrosion in external coupling system as suggested by high concentration of aluminum oxides, that upon visual inspection, appeared as white stains very hard to clean chemically.
- Observation of colored water upon discharge from hydrants.
- There are neither chemical nor microbial analyses about the water stagnant within the ring, or the soil surrounding the buried ring.

4 Both corrosion management schemes for the bioleaching tank and the underground fire water ring are examples of corrosion management consultations done by Dr. Reza Javaherdashti and used with consent.

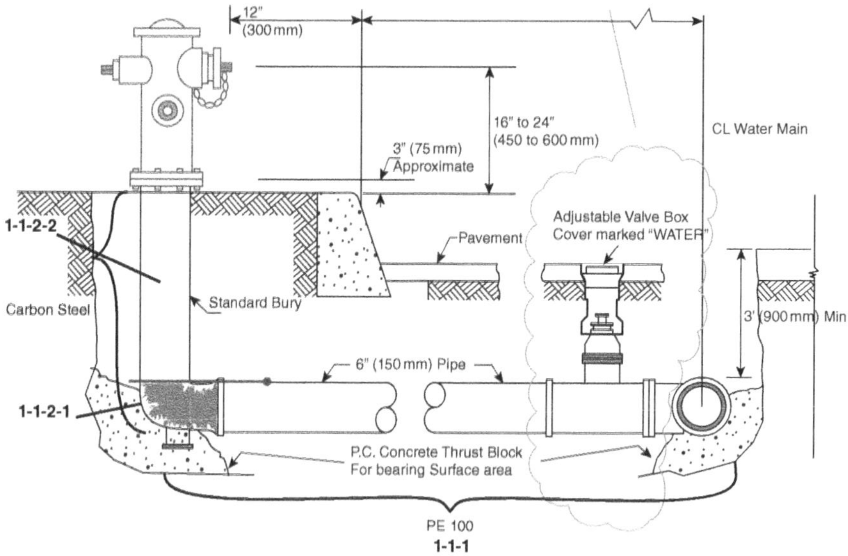

Figure 3.17 General schematic of a fire water ring as the corrosion system and the accompanying corrosion subsystems.

- No inspection of any parts of the ring since it started operation.
- As the water used in the ring is coming from a deep water well and no chemical treatment has been carried out (application of either corrosion inhibitor or biocides), the lack of data on present chemical and microbiological state of the water may both suggest that there is a possibility for corrosion.

Figure 3.17 is an example of CP and not CC. The reason is that although the asset (an underground fire water ring) has been in use for over one year and can still be considered as a newly installed and operated asset, it has not shown severe signs of corrosion, although some alarming signs as briefly addressed above. It is because of these corrosion signs that one could argue that corrosion was taking place, therefore setting the whole ring system vulnerable to corrosion. In other words, the asset is in its pseudo-FFS state (it is being used despite showing some signs of corrosion), and we are going to apply a CP strategy so as not to allow the asset to approach the Zugzwang effect state.

The first step is that we apply a COFS/CINS approach to highlight the corrosion subsystems. A COFS approach will yield the classification below (Only important subdivisions are shown in the Figure):

1.1 The ring

1.1.1. Polyethylene (Pe) section

1.1.2. Carbon steel section

1.1.2.1. The elbow section

1.1.2.2. The steel section above the elbow and below the hydrant

2.1. The hydrant

Based on a color coding system, green showing the least expectable risk, amber showing medium risk, and red showing high risk levels, we will color code the corrosion subsystems identified above as the following:

1.1.1. Pe section (orange arrow)

A very good reason to replace metallic parts with non-metallics, and particularly composites, is the high corrosion resistance that the latter shows. However, polymers may also be vulnerable to microbiological attack. The polymer used for this section is Pe 100. Pe 100 is a high-density polyethylene (HDPE). To make it UV resistant, carbon black is used in its structure (for the particular case of the Pe used in this ring, the carbon black content is 2%). It must be noted that if polymers are treated to become UV resistant, they can also become susceptible to microbial attack [44]. While Pe may not be consumable by microorganisms, it must be noted that this polymer also contains carbon black, which can be used by microorganisms [45, 46] and it could be a vulnerable material by the action of bacteria existing both in the water and in the surrounding. Therefore, the main (possible) degradation[5] mechanisms for this section could be microbial degradation from inside as imposed by the fire water existing in the ring and microbial degradation from outside as imposed from the surrounding soil.

1.1.2.1. (Carbon steel) elbow section (r).(Red Arrow)

It is in this section that stagnation of fire water mostly happens. Under the conditions existing in this section, the dissolved oxygen in water starts to scavenge over time [13]. This will allow anaerobic (oxygen-free) conditions in which – based on the chemistry and temperature of water – growth and activity of corrosion-related bacteria and archaea will be possible. This will prepare the grounds for inducing microbiologically influenced corrosion (MIC). It is also possible that anaerobic, non-MIC reactions will occur. It is possible that non-MIC corrosion can lead into enhanced MIC [47], therefore a series corrosion geometry may be the prevailing corrosion geometry.

It is of great importance to consider the issue of MIC in fire water ring system as it may lead to catastrophic situations if not properly taken care of [48].

2.1. The hydrant (yellow arrow)

Although this section is not in contact with the water, due to its very physical layout which is immediately above section 1.1.2.1 in Figure 3.17, its surfaces are being covered with water vapor coming from the stagnant water inside the elbow.

5 "Degradation" is used for non-metals whereas "corrosion" is sued only for metals. "Rusting" is corrosion of ferrous materials.

This can increase the likelihood of corrosion in the form of electrochemical, uniform corrosion.

Looking at the possible corrosion scenarios in three main sections of the fire water system – the Pe pipe, the carbon steel elbow, and the carbon steel pipe below the hydrant – it appears that the main likely corrosion reaction will be MIC. This will finalize Phase 1 of Javaherdashti CM model, which is defining possible corrosion scenarios and their geometry.

Phase 2 will be highlighting what might be available as countermeasures. Normally and under these circumstances, there are some anti-corrosion measures that can be thought of, for example, one may add suitable additives to Pe to make it immune to MIC, along with the use of biocides.

However, to do that, it is necessary to know the chemical and microbial composition of the water as well as the surrounding soil. It must be noted that while a material like Pe could be degradable by many microbes, these microbes may have not corrosive effects on metallic substrates. This is not a significant issue as long as the corrosion of the elbow is not important; in other words, there bacteria can exist in either the water or the surrounding soil that can serve to degrade and lose Pe. If they can do so, that means they can impose destruction to the fabric of the polymer and thus cause water leakage. It is also possible that corrosion-related bacteria or archaea will not have any destructive impact on Pe, but be capable of inducing corrosion to the metallic elbow. The result will be the same, which is leakage of the water and depletion of the line from the precious water that would have been otherwise used for fire extinguishing purposes when needed.

The best that can be done is to find any bacterial/archaea species that can have any effect on accelerating the destruction of either metal (elbow) or polymer (line) so that the main result will be water leakage.

3.5 Damage Management

In the fall of 2020, at least three accidents occurred that did not find the place they deserved in the news due to the frenzy created by COVID19. A gas explosion and fire at a gas plant (December 2020, El Salvador); a massive fire at a refinery (December 2020, South Africa); release of poisonous material, and a massive explosion at gas station (August 2020 Russia). Of course, the list does not end here.

In a previous work [48] that focused on the role of MIC on fire management, we mentioned three stages:

1) Management of the cause of the fire, (pre-damage)
2) Management of controlling/extinguishing the fire, (in-damage)
3) Post-disaster management, (pro-damage)

Obviously, damage is not limited to fire only. Damage can have a much wider meaning than fire, including leakages, mechanical failures, fractures, etc. Relating this to the concepts we discussed earlier, the most important is CP/CC. It means that it will make sense if at the pre-damage stage any plans for CP or CC are executed. Where will an activity like failure analysis be accommodated? It will be in the post-damage step whereby applying conventional standard methods of failure analysis, a root cause analysis report is made. This report will then help to construct the requirements that need to be applied to avoid damage in the future. If, for example, failure analysis showed that overheating was the cause of the damage, then our pre-damage task needs to concentrate on measures that would either prevent overheating or control it as a potential corrosion-inducing damage mechanism.

The main structure for a damage management scheme can be within the boundaries by which applying management of corrosion will make sense. In other words, CM and CKM components of a management of corrosion scheme both need to be defined within the executable frame of a pre-damage step. It can also be safely said that the engineering involved in the pre-damage step will be much less than that required in the post-damage step; the amount of engineering work needed to assess and analyze the root cause of the damage is certainly greater when the damage has already occurred. This is an especially important factor that can be used to distinguish the division of labors between pre- and post-damage steps.

The place pre-damage approach has within a management of corrosion scheme will be explained in more details in the next section. However, it is significant to know that any cost that may seem to be imposed during the pre-damage step will be highly paid off in the next two stages should the next stages (in-damage and post-damage) ever occur in the first place.

3.6 Algorithm

As we hope our readers appreciate, we have introduced in this chapter all the elements necessary to manage corrosion in terms of both technical and non-technical managerial terms. To be able to transmit our approach more efficiently, we had to invent a new nomenclature of terms that, to the best knowledge of this author, are being introduced for the first time. Some of these terms are related to a new definition of failure not as a physical form of deterioration, but as an equivalent to what we call the Zugzwang effect status of the asset in its service life. We defined failure in this context not by the mechanical inability of the asset/equipment to do its job, but in terms of (1) repeating damages and (2) lack of economic justification for repairs. In doing so, we stated that each asset has to pass through three major steps; The first step is the Fit-For-Service (FFS) where the asset is inspected and guaranteed by standards, and the skillful technical persons that are appropriate to be used in the production or safety process. When the asset is put into service from the time FFS certificate has been issued, due to both thermodynamic, non-stoppable nature of corrosion as well as the working conditions, the asset starts to degrade/corrode. This is the second step that we call pseudo-FFS, meaning that although corrosion is already taking place on the asset, the asset is still in a condition that can be used. This can be due to the asset undergoing regular overhauls and not unexpected shutdowns, which is itself a result of strong inspection/monitoring techniques and skills. The third step is what is called "The Zugzwang Effect," where the asset essentially becomes out of service and cannot be repaired any more.

We also defined CC and CP being different from each other; while CP could be possible via measures such as design and material selection, CC may not be achievable that easily. Besides, we distinguished between CM and CKM and everything related to these two concepts that are in fact complementary to each other.

All the above-mentioned definitions, concepts, and approaches toward management of corrosion can be referred to under an umbrella collectively called "Smart Corrosion Management." Essential elements and relationships dealt with as a Smart CM approach are explained in Figure 3.18.

Based on what has been shown in Figure 3.18, we define the algorithm that can be used to effectively and more efficiently treat corrosion in terms of management of corrosion. However complicated it may seem at first glance, Figure 3.18 is actually a summary of what has been discussed thus far and contains all the necessary elements for a smart CM system. To understand this figure, one must understand the consciousness levels hidden in it.

Figure 3.18 contains three main levels; level 1 is appreciation of necessity of a pre-damage stage. As our universe of discussion and context of this book revolves mainly around corrosion, level 1 consciousness can be interpreted as a serious appreciation that corrosion is indeed damaging in its broadest possible sense, and must be included in any pre-damage strategic plan and design. However evident and easy-to-understand it may sound, it is in fact one of the biggest challenges one can ever have:

The first step is the philosophy that we adapt about a pre-damage decision-requiring challenging situation; in 2015, this author published a short article on LinkedIn titled "Why Bill Gates Doesn't Pay Attention to Corrosion" [49]. In that article, we reasoned that why Bill Gates–and in fact, the public – was not paying the attention that an issue like corrosion deserved was "*because corrosionists all around the world prefer to yell at their own counter-parts but not the public and the decision-makers.*" As long as we are sure that the first required level of

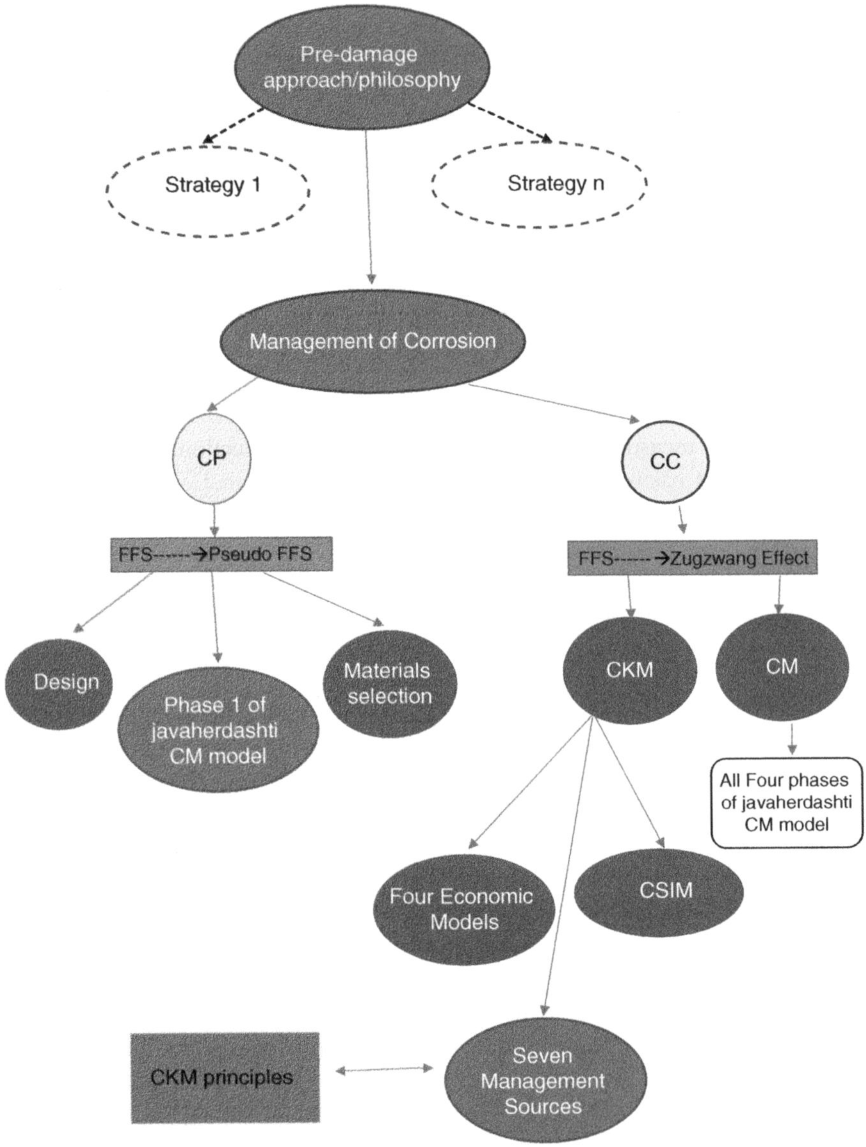

Figure 3.18 Overall view of management of corrosion defined as per smart corrosion management elements.

consciousness has been taken, it means that the necessary background for execution of any scheme of management of corrosion is ready.

The second level of consciousness is to have a specific strategy for the management of corrosion. Obviously, damage is not confined to corrosion but if the context of our activities is corrosion, then it is of significant importance to define our strategy for dealing with a corrosion-related pre-damage step to be very relatable to a management of corrosion scheme.

It is only after securing these two levels that the third level of consciousness may be executed. Based on where the asset or the set of assets of interest are in their line of life – whether they are at FFS, pseudo-FFS, or Zugzwang effect step – we need to carefully decide if we want a CP or a CC approach to be taken. It is actually based on this choice that the entire set of activities will commence.

For a CP approach, not only is it essential to consider options such as design and material selection into serious consideration, but also first phase of Javaherdashti model that essentially concentrates on definition of corrosion system, corrosion process(s) to be expected, and how they are going to be positioned with regards to each other, along with what counter-measures could be applied. By design option, we can modify the existing or future design of the equipment in such a way that a certain corrosion process will not become available at all, and if that certain corrosion process was going to ignite a cascade of corrosion processes (series corrosion geometry), then by a smart design, those corrosion processes would have also been prevented. An example of such smart design to bypass corrosion under insulation (CUI) has been addressed elsewhere [50] as well as in Figure 3.4. When it comes to materials selection, this author would say that in comparison with the two other measures just mentioned, that is design and execution of Phase 1 of Javaherdashti CM model, materials selection can be a costly option. Normally and under real-life working conditions, what happens is not "selection" of the fittest material but "upgrading" it, for example, from a carbon steel material, which is rather inexpensive, to a more expensive material such as a member of the stainless steel class. Whatever it would mean, material selection will always remain a viable option to prevent corrosion in the sense we have defined in this chapter.

The real challenge, however, starts with CP; we call it as such because CC can be applied to equipment and assets that are already in use, and therefore the challenge needs to be two-sided. By using CM methods combined with corrosion engineering techniques, one has to control corrosion so as not to lead to failures (both mechanical failures and failures addressed by Zugzwang effect definition) to satisfy the client, and on other side one has to overcome the 'why's' and 'but's' by justifying that the expenses to be undertaken to control corrosion are to control both the direct (economic) as well as indirect (environment) costs of corrosion. For this, the corrosion specialist will need to not only apply the Javaherdashti

CM model, and within that model, also use corrosion science and engineering principles, but also through application of one or a combination of the four economic models for estimation cost of corrosion, as well as using information received from CSIM, and developing ecological models to estimate environmental costs. In doing so, the corrosion specialist will also need to apply CKM models as well contemplating over the seven management sources to coordinate them toward a better management of the corrosion scheme. For example, the research tool as well as the researchers must not confine themselves to ordinary, conventional routes and paradigms in CM and research, but by using new, much more effective methods such as using TRIZ, try to make their research goals much more reachable and practical.[6]

3.7 Final Remarks

For the time period between 1979 and 2016, more than eight billions Euros have been lost and 586 lives have been affected by various corrosion-related accidents; that is almost 15 people per year or one person per month!

Corrosion is important in terms of anything one may assume. By a quick look at the overall costs of all natural disasters in a given year and comparing it with the global cost of corrosion, one can easily see that cost of corrosion exceeds that of all natural disasters combined by a significant amount. To give a better idea about the importance of corrosion in terms of lives (and money) that are lost because of it, Figure 3.19 can present a good example.

Corrosion is a multi-disciplinary area of expertise; this is contrary to the notion some industrialists may have of the notion that corrosion should (must) be the concern of certain people; corrosion engineers, forensic engineers, integrity management professionals, and certain departments, inspection and NDT or the equivalent. As we have briefly discussed earlier, anything related to forensic investigation of corrosion including Root Cause Analysis and the accompanying failure analyses are not all particularly useful but *post-mortem* practices; these practices as per their own definitions try to locate a corrosion scenario to explain the cause of the failure, and are especially useful. The point is that failure has already occurred, and findings of these disciplines may become useful only in modifying the working conditions or application of materials selection or

6 It is based on this understanding about the vital importance of TRIZ and what corrosion research lacks that we have dedicated an entire chapter of this book to TRIZ (Chapter 6). In this author's opinion, no corrosion management scheme in which corrosion research is also playing a role will deserve to be called completed if TRIZ and its effect on algorithmization research have not been considered.

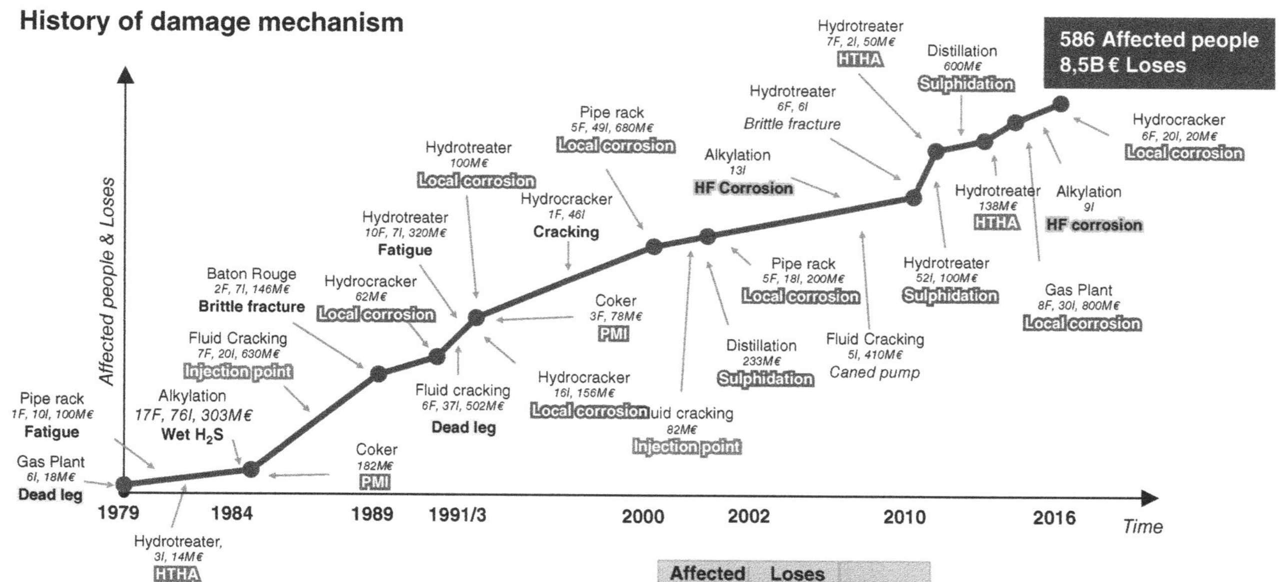

	Affected People	Loses M€	Incidents
Local corrosion	143	2970	14
Sulphidation	56	1023	6
Dead leg	52	525	3
Injection point	34	714	4
HF Corrosion	26	59	5
HTHA	18	291	5
PMI	3	307	4

References
On top: *Refinery unit*
#F:*Number of fatalities*
#I:*Number of injured*
Loses:*in M€*
HTHA:*High temperature hydrogen attack*
PMI: *Positive material identification*
HF: *Hydrofluoric acid*

TOTAL Classification: Restricted Distribution
TOTAL -All rights reserved

Figure 3.19 History of corrosion damage. (*Source:* used with permission from Mr. Ricardo Gonzalez who originally constructed it.)

application of anti-corrosion measures. There are no guarantees that the same corrosion process will not repeat itself in the future (as is well-known to many operators), or other corrosion processes that for many reasons had either not been active at the time when the failure was investigated for the first time or that by applying new measures to prevent failure, new corrosion-related problems occurred. An example of the former is be parallel corrosion reactions, and an example of the latter could be introduction of nitric acid corrosion due to application of anti-microbial measures such as bio-competitive exclusion and/or bio-augmentation [12]. However, nobody can deny that practices such as failure analysis and Root Cause analyses are very instructive, they can assist a great deal in understanding the situations by which a failure has occurred for a large number of people involved who may or may not include operators as well as opthers. A true story of such can be found by what a principle Engineer put as a Linkedin note that was a real eye-opener with regards to potential advantages a failure analysis investigation can have on top of just technical merits [51].

Elements of any approach toward corrosion need to identify it as a multilateral issue with not only technical failure consequences, but also with economic and ecologic consequences. Any system that wants to handle corrosion and manage it must look at all these elements as much as possible. In other words, such CM systems must also have the capability and programs to explain the non-technical aspects of corrosion.

Before the outbreak of COVID19, we corrosion experts were candidly thinking that the causes of corrosion might only be found within some limited factors such as [28] human error, lack of awareness and knowledge, and lack of supervision and control. However, now that the impact of the COVID19 pandemic and the lockdown resulting from it have imposed on us additional incidents, we have a better understanding of the factors involved. An example of this is a chemical leak from large tanks unattended due to restriction executed during COVID19 lockdown at an LG plant in India for six weeks, resulting in 11 people being killed and 1000 people becoming infected and ill [52]. This alone may serve to show that no CM models thus far put forward and executed may be taken as being the last word on corrosion management. These models largely depend on corrosion engineering with a glimpse of environment and management information. The point is that a good deal of all of these models suggested basically consist of flow charts that are guidelines to look for recognition and possible prevention and control of corrosion (Corrosion Control).

One important point that must not be overlooked is that all processes which can be referred to as corrosion processes are known today. There are still some gaps to be filled with regard to either their exact working mechanism (such mechanism[s] involved in MIC) or the ways by which these corrosion processes can be treated (such as pros and cons of CUI). However, the main types of corrosion are known and the best that can be done concerning their treatment and underlying mechanisms are the main topics for research. So the core problem seems to be static, while the environmental factors encircling it are subjected to continuous changes. Any method that aims at treating corrosion not under laboratory, controllable conditions but under ever-changing real-life working conditions must embrace the change and must have the capacity for it. One must always bear in mind that a man cannot swim in a river twice, as neither the river nor the man is the same.

We are not claiming that the Smart Management of Corrosion approach we have proposed and explained in full in this chapter is the only answer available to the old problem of corrosion and its efficient management, but it is a new perspective and a new avenue opened in front of those who want to embrace the change. We have coined many terms and have been careful in re-defining concepts that either seemed too obvious (such as definition of failure) or borrowed (such as Zugzwang), or even those that have not been taken into consideration before and have been used interchangeably (such as CP and CC). In my opinion as a professional with more than a quarter of century experience in more than one industry and more than one geographical site, this author firmly believes Smart Management of Corrosion scheme could be one of the best answers to the problem of corrosion and its efficient and feasible management and control in industry.

References

1 (2009). *API Recommended Practice 580*, 2e. USA: American Petroleum Institute.

2 https://corrosion-doctors.org/Forms/Accidents.htm, last visited 13 September 2020.

3 https://news.kodoom.com/iran-economy/%D8%A2%D8%AA%D8%B4-%D8%B3%D9%88%D8%B2%DB%8C-%D8%AF%D8%B1-%D9%85%D8%AC%D8%AA%D9%85%D8%B9-%D9%BE%D8%AA%D8%B1%D9%88%D8%B4%DB%8C%D9%85%DB%8C-%D9%BE%D8%B1%D8%AF%DB%8C%D8%B3/story/1068411, last visited 14 September 2020.

4 https://www.oilandgasiq.com/integrity-hse-maintenance/articles/shutdowns-and-turnarounds-in-the-oil-and-gas-indus, last visited 14 September 2020.

5 Javaherdashti, R. and Nikraz, H. (2010). *A Global Warning on Corrosions and Environment: A New Look at Existing Technical and Managerial Strategies and Tactics*. VDM Germany.

6 Singh, N., Case study on some of the major corrosion catastrophes in the history. A report submitted at completion of the course on corrosion & environmental degradation of Materials (MT41013), https://www.academia.edu/2233096/Case_study_on_some_of_the_major_corrosion_catastrophes_in_the_history, last visited 18 September 2020.

7 Javaherdashti, R., Nwaoha, C., and Tan, H. (2013). *Corrosion and Materials in Oil and Gas Industries*. Florida: CRC Press/Taylor & Francis.

8 Corrosion: Understanding the Basics (#06691G), Chapter 1, The Effects and Economic Impact of Corrosion. *ASM International*, www.asminternational.org, 2000.

9 Kiruba, J.E. and Nivasini, K.D. (2018). A study on customer perception towards Nescafe products with special reference to Coimbatore city. *International Journal of Scientific & Engineering Research* 9 (2): 337–339.

10 A better informed society can prevent lead poisoning disasters. https://www.sciencedaily.com/releases/2020/09/200918154522.htm, last visited 12 October 2020.

11 Javaherdashti, R. (2000). How corrosion affects industry and life. *Anti-Corrosion Methods and Materials* 47 (1): 5.

12 Javaherdashti, R. and Akvan, F. (2020). *Failure Modes, Effects and Causes of Microbiologically Influenced Corrosion: Advanced Perspectives and Analysis*. Elsevier.

13 Javaherdashti, R. and Akvan, F. (2017). *Hydrostatic Testing, Corrosion, and Microbiologically Influenced Corrosion: A Field Manual for Control and Prevention*. Boca Raton, Florida: CRC Press.

14 Review of corrosion management for offshore oil and gas processing. Capcis Limited, Offshore Technology Report 2001/044, 2001, UK.

15 Koch, G., Varney, J., Thompson, N. et al. (2016). *International Measures of Prevention, Application, and Economics of Corrosion Technologies Studies (IMPACT)*. USA: NACE International.

16 Ghalsasi, S., Fultz, B., and Colwell, R. (2016). Primary Constituents of a Successful Corrosion Management Program. NACE Corrosion Risk Management Conference, paper no. RISK16-8734, Houston, TX: NACE International.

17 Essential Elements of a Successful Corrosion Management Program. https://www.materialsperformance.com/articles/material-selection-design/2016/07/essential-elements-of-a-successful-corrosion-management-program, last visited 16 October 2020.

18 Darwin, A., Annadorai, K., and Heidersbach, K. (2010). Prevention of corrosion in carbon steel pipelines containing hydrotest water – An overview, Paper No. 10401, NACE 2010, USA.

19 Harrision, G. (2013). Setting up a corrosion management system: 3 points to consider. *Corrosion & Materials* 38 (3): 44–45.

20 Olivia, M., Moheimani, N., Javaherdashti, R. et al. (2013). The influence of micro algae on corrosion of steel in fly ash geopolymer concrete: a preliminary study. *Advanced Materials Research* 626: 861–866.

21 Payer, J.H., Dippold, D.G., Boyd, W.K. et al. (1978). Economic Effects of Mettalic Corrosion in The United States: Appendix B, Part II, A Report to NBS by Battelle Columbus Laboratories, U.S. Department of Commerce, USA.

22 Hou, B., Li, X., Ma, X. et al. The cost of corrosion in China. http://www.nature.com/npjmatdeg, published on-line, 25 July 2017.

23 Koch, G.H., Brongers, M.P.H., Thompson, N.G. et al. Corrosion Costs and Preventive Strategies in the United States. *NACE International*, Publication No. FHWA-D-01-156, USA.

24 Akintemi, O.O., Nwaokocha, C.N., Adesanya, A.O. et al. (2012). Evaluation of corrosion cost of crude oil processing industry. *Journal of Engineering Science and Technology* 7 (4): 517–528.

25 Knight, S., Palanisamy, S., and Colwell, J. (2019). Corrosion costs of the Royal Australian Navy Frigate Fleet. *Materials Performance* 58 (8): 52–57.

26 Moshaweh, A.R., Al-Shahrani, M.F., and Teh, L.S. (2020). Cost of corrosion: Key performance indicator, pp. 50–53.

27 Bayatmakoo, A., Venturini, P., Rispoli, F., et al. (2019). Corrosion Cost Prediction of the Tube Coils used in Oil & Gas Industrial Fired Heaters by Using Artificial Neural Network. M. Sc. Thesis, University of Rome; La Sapienza (Italy), July.

28 Javaherdashti, R. (2016). Corrosion knowledge management for managers. *Materials Performance* 55 (9): 58–61.

29 Javaherdashti, R. (2002). How to manage corrosion control without a corrosion background. *Materials Performance* 41 (3): 30–32.

30 Javaherdashti, R. (2003). Corrosion management: a guide for industry managers. *Corrosion Reviews* 21 (4): 311–326.

31 Javaherdashti, R. (2006). Corrosion knowledge management: an example of an interface for more efficient managerial communication. *Materials and Corrosion* 57 (12): 945–950.

32 Javaherdashti, R. and Akvan, A. (2015). On the link between future studies and necessity of including corrosion in a "Desired Future" scenario: Presenting a model. *International Journal of Engineering Technologies and Management Research* 2 (4): 1–8.

33 Javaherdashti, R. and Akvan, F. (2016). Corrosion Knowledge Management Versus Corrosion Management: An Essential Tool for Assets Integrity Management. *EuroCoRR 2016*, France, 11–15 September.

34 Tatnall, R.E. (1999). Case histories: biocorrosion. In: *Biofouling and Biocorrosion in Industrial Water Systems* (eds. H.C. Flemming and G.G. Geesey). Heidelberg, Germany: Springer-Verlag Berlin.

35 https://cen.acs.org/environment/water/Pipes-release-toxic-chromium-tap/98/web/2020/10, last visited 31 October 2020.

36 Schouten, J.C. and Gellings, P.J. (1987). Quantitative measures of corrosion and prevention: application to corrosion in agriculture. *Journal of Agricultural Engineering Research* 36: 217–231.

37 Mattsson, E. (1989). *Basic Corrosion Technology for Scientists and Engineers*, Chapter 3. Ellis Horwood Limited.

38 Costs and Importance of Corrosion. https://membership.corrosion.com.au/training/topic/costs-and-importance-of-corrosion, last visited 4 November 2020.

39 Gabelle, C., Baraud, F., Bire, L. et al. (2012). The impact of aluminium sacrificial anodes on the marine environment: a case study. *Applied Geochemistry* 27 (10).

40 Bonin, L., Vitry, V., Olivier, M.-G. et al. (2020). COVID-19: effect of disinfection on corrosion of surfaces. *Corrosion Engineering, Science and Technology* 55 (8): 693–695.

41 Eskenazi, J. 4 Things That Don't Change for Recruiting and Hiring During A Crisis. http://www.coatingspromag.com/articles/business-tips/2020/11/4-things-that-dont-change-for-recruiting-and-hiring-during-a-crisis, last visited 22/11/2020.

42 KIanpour, M. (2020). *Living with Corona*. Persian. Shaparak Associates.

43 How Long Does the Coronavirus Live on Surfaces? https://www.webmd.com/lung/how-long-covid-19-lives-on-surfaces, last visited 26 November 2020.

44 Pathak, V.M. and Navneet (2017). Review on the current status of polymer degradation: a microbial approach. *Bioresources and Bioprocessing* 4 (15) https://doi.org/10.1186/s40643-017-0145-9.

45 Malits, A., Cattaneo, R., Sintes, E. et al. (2015). Potential impacts of black carbon on the marine microbial community. *Aquatic Microbial Ecology* 75: 27–42.

46 Black carbon in air pollution aids bacteria which cause respiratory disease. www.belfasttelegraph.co.uk/news/uk/black-carbon-in-air-pollution-aids-bacteria-which-cause-respiratory-disease-35498399.html, last visited 11 October 2020.

47 Javaherdashti, R. (2011). Behaviour of stainless steel 316L in a marine mixed culture containing sulphate reducing and iron reducing bacteria. *Corrosion & Materials* 36 (1): 52–57.

48 Javaherdashti, R. (2019). A quick review of the role that microbial corrosion can play in the engineering and management of fire. In: *Phenomena and Theories in Corrosion Science, Methods of Prevention* (ed. A. Gergely). USA: Nova Science Publishers.

49 https://www.linkedin.com/pulse/why-bill-gates-doesnt-pay-attention-corrosion-reza-javaherdashti, last visit 14 December 2020.

50 Javaherdashti, R. (2014). Corrosion under insulation (CUI): a review of essential knowledge and practice. *Journal of Materials Science and Surface Engineering* 1 (2): 36–43.

51 Turcott, S. (2020). Learning From Failure; Leaking Sewage. 15 December 2020, https://www.linkedin.com/pulse/learning-from-failure-speed-record-shane-turcott/?trackingId=qQq39JTI1Fp9cKXmnl6SZQ%3D%3D, last visited 16 December 2020.

52 https://www.notion.so/Chemical-leak-at-LG-plant-in-India-kills-11-about-1-000-ill-091f4944e939453ea1a22a211587311c, last visited 17 December 2020.

4

Economics and Corrosion

*Mahsa Mostashar-Nezami**

CEO and Founder of ATAVART, Privately Held co., Tehran, Iran

4.1 Introduction

I try to consider corrosion as a factor which impacts an economy because it contains a vast amount of product. The NACE report in 2013 said 3.4% of the world's products deplete in the corrosion processes [1], and a considerable percentage of this cost is avoidable.

In this chapter, first, you will be familiarized with some of the main principles of Economics. Economics is the knowledge of big pictures. To show the essential trace of every engineer's work in an economy, it is important that they can imagine how their work and their occupation impacts the economy. To address this idea in this chapter, there are two main sections: Economics and Corrosion Economics.

In the first section, I try to teach the basic economic principles and show this complex view, and then bring some of the most useful economics approaches to estimating a productive society.

In the second section, two methods for estimating corrosion cost must be considered. The first one, the I/O, is a time-consuming and difficult method for estimating corrosion costs. The results of the I/O method are accurate and reliable but I suggest this method use in nation-wide or heterogeneous projects. The second method is about the life cycle cost (LCC) method and there are many different ways to calculate cost with this method. It is easier to to measure than the I/O method and is suggested to use with small or medium sized projects.

In the last section, try to show the impact of corrosion on sustainability, as it is important to show the readers how their job affects lives for many years; this allows them keep a closer watch on their actions. I choose sustainability rather

*https://www.linkedin.com/in/mahsa-mostashar-nezami-a539b578.

Corrosion Policy Decision Making: Science, Engineering, Management, and Economy, First Edition.
Edited by Reza Javaherdashti.
© 2022 John Wiley & Sons, Inc. Published 2022 by John Wiley & Sons, Inc.

than socio-economic or environmental impact because sustainability contains both the social and environmental impact of corrosion, and also takes economics into account.

In this chapter, the words shown in italic emphasize the importance of understanding the meaning of the content.

4.2 Economics

This section aims to familiarize engineers with some basic economics definitions such as gross domestic product (GDP), what it is, and its different measurement methods; and present value (PV), and so on.

In this section, I try to be as brief with these principles as I can; for this reason, reading the charts and tables carefully is important for understanding the content.

4.2.1 What Is Economics

Economics is the knowledge of coping with *scarcity*. It means that economics follow our unlimited needs within limited resources. There are some conflicts to classify economics as a science or art. In the Avestan and Pahlavi languages, which were the ancestors of the today's Persian Language and the decedent of the Indo-European languages, the word art can be used with two parts "hu" and "nara." The part "hu" means good, well, beautiful, very, sufficient, expressive, strong, and steadfast; and "nara" refers to the human [2]. With this means is economics is an art because it tries to structure human behavior in a sensible and sufficient way.

Economics is almost a young major in contrast with, for instance, Mathematics or chemistry. In the past 250 years, philosophers had tended to this knowledge. Gradually, this major will segregate from philosophy. Macroeconomics appears as a separate subject within economics in the 1930s, when the great depression occurred [3]. Today, we have many economics subjects such as macroeconomics, microeconomics, international economics, and economic development. There are also some interdisciplinary subjects such as environmental economics, socio-economics, and so on. In recent years, an important major has emerged; sustainable development, which considers the unity between economics, the environment, and aspects of society. In this section, I will speak about macroeconomics first, then and try to explain the corrosion impact on sustainability in the last section.

The above mentioned subjects are about mainstream economics, but we have two kinds of *economists*; the Heterodox Economists and Orthodox Economists. The heterodox economists are those who do not accept the basic underlying models used by a majority of economists as the most useful model for analyzing the economy [3], while the orthodox economists accept the mainstream principles of economics.

There are six heterodox approaches; Austrian, Post-Keynesian, Institutionalist, Radical, Feminist, and Religious.[1] One of most the famous approaches in the Radical view is Marxian economics. Radical economists are much more prepared to discuss social conflict and stresses in our society than are mainstream economics. The characteristic of many heterodox economists is their analysis tends to be less formal than the mainstream. The less formal is not about whether right or wrong, or better or worse, because in the actual situation it is difficult to say precisely what the logic is [3]. On another side of less informal economists are mainstream economists trying to explain people's behavior with many strict models.

Macroeconomics focuses on the big picture of a society or the whole of the world. The specific focus of macroeconomics are the *big four* issues. These four issues are inflation, unemployment, the rate of Economic growth, and business cycles (fluctuations in the economy) [4].

Two main economic *theories* are Classical and Keynesian. The most challenging issue between these two theories is the size of government and its interference. In 1776 when Adam Smith wrote his well-known book "The Wealth of Nations" the world population had half a million people [5]. It is understandable with this population that Adam Smith believed in an endogenous adjusted system he introduced as his famous idea, "invisible hand." Thus, he believed in the minimum government interference, and a system that corrects itself automatically in the long run and provides full employment. In the 1930s during the great depression, Professor John Maynard Keynes challenged this condition in the neoclassical idea and said his famous quote, "In the long run we are all dead." He brings the short and medium term correction by the interference of the government. Although his idea was challenged in the 1970s by these years' stagflation, his recommendations could not solve this problem, and the Monterey economics birthed. However, today's governmental policies are almost a mix of these views.

Today's mainstream economics are based on Alfred Marshal's theory on supply and demand, and he is the founder of neoclassical economics. In addition, he brings the ideas of supply and demand in the book "Principles of Economics" in 1890 [6]. His theory is the fundamental theory of today's economics; the supply and demand are about the aggregate supply and demand of the goods and services in an economy.

There are vast principles in economics, but in this chapter, I prefer to introduce the most useful principles will need to understand in the model of corrosion estimation.

---

1 You can read a brief well-writing of these heterodox approaches in the book of Macroeconomics by Professor Colander.

4.2.2 Gross Domestic Product

Understanding the operation of the real world is the aim of macroeconomics [7]. To measure an economy, we focus on what happens in the entire economy and its relationship with other countries. In an economy, we have two main parts which allocate its resources; the *government* and the *private sector*. The private sector is divided into two parts: *households* and *firms*. Thus, households, firms, the government, and the rest of the world are the four main parts of an economy.

We need some definitions to understand how economists try to measure the GDPs of an economy.

First, they know two kinds of users for goods and services; final users and intermediate users. Second, they separate the goods and services produced in the entire economy in three ways; final goods, intermediate goods, and value-added goods.

Final *users* are the ones who buy goods and services for their *daily* life consumption. Households purchasing furniture for their home, or an administration purchasing a laser printer for their daily work are final users. However, the intermediate users are the one buy the goods for their production; for instance, a furniture manufacturer buys leather for their goods.

Final *goods* and *services* are the final users who buy and use these goods in their *daily* life. For example, we buy dairy goods for our meals; these dairy goods are the final goods. Intermediate goods are from the manufacturers that buy or produce their goods and sell them to other intermediate users or final users. For example, packets a dairy manufacture purchases for their goods are the intermediate goods. Some of the products are intermediate goods sold to other intermediate users; for example, some petrochemical products are examples of these intermediate goods.

Familiarizing ourselves with these kinds of users, goods, and services is essential; because when measuring the GDP we use the final goods for final users.

GDP calculation: Three different methods of estimating the GDP are (i) Expenditure, (ii) Income, and (iii) Product or value-added approaches.

The *expenditure* approach added all of the final goods and service values for an economy within a given time.

The *income* approach added all of the income – wages, rents, interest, and profits – that all production factors earn to produce final goods and services in a given time [8].

The *product* or *value-added* approach adds the value of primary goods with the value of the products to produce the final goods. For example, if a firm purchased wood, leather, glue, and screws for their production with a sum up to \$100, and sold their products to the final customer for \$250, this method added \$100 for primary goods and \$150 for a value that add to these materials, which may contain some other intermediate materials and services.

These three types of estimation have an equal value to each other. In the following tables which introduce these three measurements of GDP, compare the amount of GDP to each other.

4.2.2.1 The Expenditure Approach

The most useful method which economists use for measuring the GDP is the expenditure approach. The GDP is equal to the total expenditures for all final goods and services produced within the country in a stipulated period [9].

$$Y = C + I + G + NX \tag{4.1}$$

Y is GDP; C is personal consumption expenditure (PCE); I is Investment, gross private domestic investment; G is Government consumption expenditure and gross investment; NX is Net export, Export minus Import. Table 4.1 shows this method for USA GDP in 2019.

Personal Consumption expenditures (C) consist of expenditure people pay for goods and services. *Goods* includes durable goods and nondurable goods. *Durable* goods are something that lasts for a long time (usually for more than one year); motor vehicles and parts, furnishings and durable household equipment, and recreational. *Nondurable* goods are those that last for less than one year; basically a citizen's daily life consumptions. Food and beverages purchased for off-premise consumption, clothing and footwear, gasoline or other energy goods, and other items. *Services* consist of household consumption expenditures (for services) and final consumption expenditures of nonprofit institutions serving households. Household consumption expenditures contain housing and utilities, health care, transportation services, recreation services, food services, insurance, and other services. Final consumption expenditures of nonprofit institutions serving households contain the gross output of nonprofit institutions minus receipts from sales of goods and services by nonprofit institutions.

PCEs are the highest part of the GDP for 2019; reaching 67.9% of the GDP of the United States.

Gross private investment (I) consists of fixed investment and change in private inventories. *Fixed investment* has two parts; nonresidential and residential investment. Nonresidential investment contains structures, equipment, and intellectual property products. Change in private inventories includes farm and nonfarm investment. Gross private investment contains 17.5% of the GDP. Almost all of these investments are about fixed and nonresidential investments.

Net exports and imports (NX) include exports and imports of goods and services. A trade deficit occurs when the number of goods and services a country imports is more than goods and services export. The United States mostly has a trade deficit, and it was −2.8% of GDP in 2019.

Government consumption expenditures and gross investment (G) have two parts; federal, and state and local for the US. Federal consists of national

Table 4.1 GDP, expenditure approach (billions of dollars).

Line		2019	Share of GDP
1	**Gross domestic product (GDP)**	**21 747.4**	**100.0**
2	**Personal consumption expenditures**	**14 759.2**	**67.9**
3	**Goods**	**4562.4**	**21.0**
4	Durable goods	1554.1	7.1
9	Nondurable goods	3008.2	13.8
14	**Services**	**10 196.8**	**46.9**
26	**Gross private domestic investment**	**3732.6**	**17.2**
27	**Fixed investment**	**3729.2**	**17.1**
28	Nonresidential	2902.3	13.3
29	Structures	652.3	3.0
30	Equipment	1229.3	5.7
35	Intellectual property products	1020.7	4.7
39	Residential	827.0	3.8
40	**Change in private inventories**	**3.4**	**0.0**
43	**Net exports of goods and services**	**−549.8**	**−2.5**
44	**Exports**	**2515.7**	**11.6**
45	Goods	1629.1	7.5
46	Services	886.5	4.1
47	**Imports**	**3065.4**	14.1
48	Goods	2460.7	11.3
49	Services	604.8	2.8
50	**Government consumption expenditures and gross investment**	**3805.3**	**17.5**
51	**Federal**	**1447.9**	**6.7**
52	National defense	874.7	4.0
55	Nondefense	573.3	2.6
58	**State and local**	**2357.4**	**10.8**

Source: U.S. Bureau of Economic Analysis. [Online]; 2020 [cited 10 19 2020. Available from: https://apps.bea.gov/scb/2020/08-august/0820-nipa-update.htm#pce.

defense and nondefense, and state and local consists of government consumption expenditures and gross investment. The government consumption expenditure and gross investment was 17.5% of the GDP in 2019, and most of it was from consumption expenditures of states and local governments.

4.2.2.2 The Income Approach

Another approach for measuring the nation's economy is the income approach. This approach counts the income earned and costs spent during production. The Gross domestic income (GDI) should equal the GDP because of the different sources used to measure it. You can see this method in Table 4.2 for USA income in 2019. There are some differences between these two kinds of measurements and is known as the "statistical discrepancy."[2]

The name of the accounts in this table can show which activity counts; but an account that is less speaks about is the consumption of fixed capital (CFC).

CFC is an account considered to degrade and corrode the economy's capital. This item was $3474.4 billion in 2019, and most of this includes the private sector. In the Bureau of Economic Analysis, this account is defined as, "The charge for the using up of private and government fixed capital located in the United States. It is the decline in the value of the stock of fixed assets due to wear and tear, obsolescence, accidental damage, and aging. For general government and for nonprofit institutions that primarily serve individuals, CFC serves as a measure of the value of the current services of the fixed assets owned and used by these entities" [10]. The System of National Accounts (SNA) defined this as, "Consumption of fixed capital (CFC); the decline in the value, in current prices, of the stock of capital goods held by producers resulting from physical deterioration, normal obsolescence, or normal accidental damage" [11].

In the later sections we speak about this account in more depth.

4.2.2.3 The Value-Added Approach

The value-added approach is another source for measuring economic production. It concerns the values that each institutional unit adds to the components which produces a good or service. This approach estimates the sum of employee compensation, taxes on production, and imports fewer subsidies and gross operating surplus. Table 4.3 explains this approach. Each item in the following table includes one of these three components. In the following table, you can see the share of industries in domestic production.

As you can see, these three types of measuring the GDP are almost equal, but the points of view are different.

4.2.2.4 Income, Consumption, Saving, and Investment

Saving and *investment* are two factors related to the main topic of this section. To explore these two factors, we must familiarize ourselves with the relation between income, investment, consumption, and saving. All of the consumption, investment, or saving in the private and governmental sectors include these items.

In an economy, income is distributed between savings or consumption. Thus, this relation is between these three items:

---

2 BEA's definition of GDI.

$$Y = C + S \tag{4.2}$$

where Y is the income of the economy, C is consumption, and S is savings. Savings is the part of income that society prefers to put away without spending, for example, in a bank account.

On the other side, this relation is between consumption and income:

$$C = C_0 + cY \tag{4.3}$$

C_0 is the undebatable consumption in an economy and is called autonomous expenditure.

There is another definition for C and Y:

$$\frac{\Delta C}{\Delta y} = MPC \tag{4.4}$$

Where MPC means marginal propensity consumption.

Furthermore, this relation is between saving and income.

$$S = sY \tag{4.5}$$

Also, marginal saving is:

$$\frac{\Delta S}{\Delta y} = MPS \tag{4.6}$$

The relation below is between marginal consumption and marginal savings:

$$MPS + MPC = 1 \tag{4.7}$$

In sum,

$$S = Y - C \tag{4.8}$$

$$S = -C_0 + (1 - c)Y \tag{4.9}$$

This equation shows the relation between *savings* and *consumption* in an economy. As you can see, if autonomous consumption increased, the level of savings would decrease.

4.2.2.4.1 Investment

Investment is an expenditure society spends on capital stock, such as renewing factory machinery, infrastructure repair and maintenance, or construction of a building. In economics, saving and investment must be equal, which means all of the income savings must be spent for investment in an economy; however, in the real world, these are not the same. Most of the world's countries' investments are more than their savings because they have foreign investments. In the expenditure approach, this item was 17.2% of US GDP in 2019.

From corrosion points of view, while the amount of money spent on corrosion issues such as repairing and maintenance belong to the consumption items, the amount of money spent on investment to capital stocks decreases. Spending money on investment helps society in growth and development. In Table 4.2 this amount shows in the CFC account.

Table 4.2 Gross domestic income by type of income (billions of dollars).

Gross domestic income (GDI)	**21 721.2**
Compensation of employees, paid	**11 577.4**
Wages and salaries	9435.1
To persons	9415.7
To the rest of the world	19.4
Supplements to wages and salaries	2142.4
Taxes on production and imports	**1509.6**
Less: Subsidies 1	**81.1**
Net operating surplus	**5240.8**
Private enterprises	5248.7
Net interest and miscellaneous payments, domestic industries	808.9
Business current transfer payments (net)	152.0
Proprietors' income with inventory valuation and capital consumption adjustments	1697.7
Rental income of persons with capital consumption adjustment	795.5
Corporate profits with inventory valuation and capital consumption adjustments, domestic industries	1794.6
Taxes on corporate income	312.3
Profits after tax with inventory valuation and capital consumption adjustments	1482.3
Net dividends	968.7
Undistributed corporate profits with inventory valuation and capital consumption adjustments	513.6
Current surplus of government enterprises1	−7.9
Consumption of fixed capital	**3474.4**
Private	2878.9
Government	595.6
Addendum:	
Statistical discrepancy	26.2

Source: U.S. Bureau of Economic Analysis, September 30, 2020 – Next Release Date October 29, 2020.

Table 4.3 Value added by industry (billions of dollars).

Line		2019
1	**Gross domestic product**	21 747.4
2	**Private industries**	19 075.1
3	**Agriculture, forestry, fishing, and hunting**	180.4
4	Farms	140.1
5	Forestry, fishing, and related activities	40.3
6	**Mining**	300.6
7	Oil and gas extraction	188.8
8	Mining, except oil and gas	60.9
9	Support activities for mining	50.8
10	**Utilities**	340.6
11	**Construction**	905.5
12	**Manufacturing**	2370.1
13	Durable goods	1328.0
14	Wood products	42.1
15	Nonmetallic mineral products	65.9
16	Primary metals	63.1
17	Fabricated metal products	161.7
18	Machinery	162.5
19	Computer and electronic products	308.3
20	Electrical equipment, appliances, and components	66.2
21	Motor vehicles, bodies and trailers, and parts	164.7
22	Other transportation equipment	164.3
23	Furniture and related products	31.4
24	Miscellaneous manufacturing	97.7
25	Nondurable goods	1042.1
26	Food and beverage and tobacco products	270.1
27	Textile mills and textile product mills	17.0
28	Apparel and leather and allied products	9.2
29	Paper products	61.2
30	Printing and related support activities	40.5
31	Petroleum and coal products	161.7
32	Chemical products	398.0
33	Plastics and rubber products	84.4

Table 4.3 (Continued)

Line		2019
34	**Wholesale trade**	1270.4
35	**Retail trade**	1180.3
36	Motor vehicle and parts dealers	228.5
37	Food and beverage stores	163.0
38	General merchandise stores	154.1
39	Other retail	634.7
40	**Transportation and warehousing**	710.0
41	Air transportation	150.0
42	Rail transportation	43.6
43	Water transportation	15.0
44	Truck transportation	176.8
45	Transit and ground passenger transportation	57.1
46	Pipeline transportation	51.8
47	Other transportation and support activities	136.7
48	Warehousing and storage	79.0
49	**Information**	1156.1
50	Publishing industries, except internet (includes software)	313.3
51	Motion picture and sound recording industries	90.0
52	Broadcasting and telecommunications	471.9
53	Data processing, internet publishing, and other information services	280.9
54	**Finance, insurance, real estate, rental, and leasing**	4615.5
55	**Finance and insurance**	1693.2
56	Federal Reserve banks, credit intermediation, and related activities	721.5
57	Securities, commodity contracts, and investments	309.2
58	Insurance carriers and related activities	632.4
59	Funds, trusts, and other financial vehicles	30.2
60	**Real estate and rental and leasing**	2922.3
61	Real estate	2658.3
62	Housing	2122.9
63	Other real estate	535.4
64	Rental and leasing services and lessors of intangible assets	264.0
65	**Professional and business services**	2753.9

(Continued)

Table 4.3 (Continued)

Line		2019
66	**Professional, scientific, and technical services**	1668.0
67	Legal services	285.0
68	Computer systems design and related services	377.4
69	Miscellaneous professional, scientific, and technical services	1005.6
70	**Management of companies and enterprises**	413.6
71	**Administrative and waste management services**	672.3
72	Administrative and support services	609.1
73	Waste management and remediation services	63.2
74	**Educational services, health care, and social assistance**	1898.4
75	**Educational services**	275.3
76	**Health care and social assistance**	1623.1
77	Ambulatory health care services	792.8
78	Hospitals	522.6
79	Nursing and residential care facilities	160.1
80	Social assistance	147.6
81	**Arts, entertainment, recreation, accommodation, and food services**	923.4
82	**Arts, entertainment, and recreation**	243.5
83	Performing arts, spectator sports, museums, and related activities	146.4
84	Amusements, gambling, and recreation industries	97.1
85	**Accommodation and food services**	680.0
86	Accommodation	190.0
87	Food services and drinking places	490.0
88	**Other services, except government**	469.7
89	**Government**	2672.3
90	**Federal**	822.2
91	General government	759.5
92	National defense	438.7
93	Nondefense	320.7
94	Government enterprises	62.7
95	**State and local**	1850.1
96	General government	1687.7
97	Government enterprises	162.5

Table 4.3 (Continued)

Line		2019
	Addenda:	
98	Private goods-producing industries[a]	3756.6
99	Private services-producing industries[b]	15 318.4
100	Information-communications-technology-producing industries[c]	---

Note: Detail may not add to total due to rounding.
[a] Consists of agriculture, forestry, fishing and hunting; mining; construction; and manufacturing.
[b] Consists of utilities; wholesale trade; retail trade; transportation and warehousing; information; finance, insurance, real estate, rental, and leasing; professional and business services; educational services, health care, and social assistance; arts, entertainment, recreation, accommodation, and food services; and other services, except government.
[c] Consists of computer and electronic product manufacturing (excluding navigational, measuring, electromedical, and control instruments manufacturing); software publishers; broadcasting and telecommunications; data processing, hosting and related services; internet publishing and broadcasting and web search portals; and computer systems design and related services.
Source: U.S. Bureau of Economic Analysis, Last Revised on: September 30, 2020.

4.2.2.5 Gross National Product

Production of final goods and services by *a country's production factors* are the GNP, regardless of where they are produced. The three primary production factors are land, labor, and capital [8]. Everywhere these factors generate an output, it measures as that country's GNP.

The differences between GDP and GNP are about the *border* of a country; all of the goods and services produce within a country's borders, regardless of its citizenship, are the GDP.

An example of these two types of estimations is the Honda plant in Marysville, Ohio. This is a Japanese firm owned by the Honda Corporation. The workers of this company are US workers, so the output of this plant is included in the US GDP. The wages paid to US workers are US GNP; Honda's profits count as Japanese GNP but not as Japanese GDP because the products are created within the other country's borders [8].

4.2.3 Introduction to National Account

This section is designed to show the *CFC* account of an economy in a country's national account. It is essential to follow the pattern of corrosion in the entirety of an economy in order to reduce its avoidable cost.

The SNA is an internationally agreed standard[3] recommending how to measure the economic activity by strict accounting conventions such as GDP based on economic principles [11]. These recommendations are designed in terms of a set of concepts, definitions, classifications, and accounting rules. This SNA accounting framework helps with economic analysis, decision-making, and policymaking, and they record the complex economic activities in an economy. These records have these elements:

- Comprehensive: it is designated for covering all activities and the consequences of all agents;
- Consistent: using the same accounting rules for identical values of the side effect of a single action;
- Integrated: resulting in all the consequences of a single action, it reflects the impact of the accounts on the wealth on balance sheet measurements.

This standard is vast and includes all information about the economy. Thus, one specific part cannot provide this enormous standard calculation; the Table 4.4 shows a snapshot of the items covered by this standard. Then I introduce information a corrosion engineer must be familiar with to comprehend the level of corrosion action.

4.2.3.1 Production Account, the Intermediate Consumption, and the Consumption of Fixed Capital

The most crucial point in this standard relating to this book's purpose is the production account. This account displays the account sequence for institutional units, and sectors show how income is generated, distributed, and used in an economy. The economy-wide production account includes the account aggregation for an institute or a sector and as a whole for the entire economy. The production account allows the study of industrial activity and permits the compilation of supply and use tables, and inputs-output tables, extending this table in the other sections of this chapter [11].

The production account shows *the value-added,* and it is one of the main balancing items in the SNA. Table 4.5 shows the production account in uses view. However, this account does not cover all transactions within production processes or show the final production (output); it is sample information for corrosion engineers [11].

3 The system of national account is provided by the United Nations, European Commission, International Monetary Fund, Organization for Economic Cooperation and Development (OECD), and World Bank Group. The latest book published in 2008, and this information is from there. I highly recommend reading this book, you can download it for free at this URL: https://unstats.un.org/unsd/nationalaccount/sna.asp.

Table 4.4 The Standard of National Account usage.

A) The SNA as a system	
B) Accounts and their corresponding economic activities	
1) The sequence of accounts	
- Current account	
- Production account	
- Distribution and use of income account	
- Accumulation accounts	
- balance sheet	
2) Activates and transactions	
C) The institutional sectors of the economy	
- Non-financial corporation	
- Financial corporation	
- Government units, including social security funds	
- NPIs serving households (NPISHs)	
- Households	
D) Other features of the system	
1) Supply and use table	
2) Price and volume measures	
E) Concepts	
1) The production boundary	
- Household production	Such as grocery growing by a household.
- Other production boundary problem	Some interchange between firms without exchange of money
1) The consumption boundary	Households consume their production
2) The asset boundary	Natural resources
3) The national boundary	Determining taxes the resident of a country does not pay when work in the foreign country
F) Uses of the SNA	
1) Specific use	
Monitoring the behavior of the economy	

(*Continued*)

Table 4.4 (Continued)

Economic policy-making and decision-taking
International comparisons
2) Flexibility of implementation and use
3) The SNA as a coordinating framework for statistic
G) Harmonization between different statistical system
H) Implementation of concepts
1) Final consumption, intermediate consumption, and gross fixed formation
Training, research, and development Education Repairs, maintenance, and gross fixed capital formation
2) Interpretation of the distinction between consumption and gross fixed capital formation

This arrangement from the latest this global report under the name: National Accounts Statistics: Main Aggregates and Detailed Tables, 2019.

The intermediate consumption consists of inputs of goods and services in the production process. It neither covers the cost of the fixed assets use nor any expenditures on the acquisition of fixed assets. Measuring the intermediate consumption is troublesome because the borderline between consumption and gross fixed capital formation is not always easy to determine in practice. One example of this dichotomy appears when repairs and maintenance occur; ordinary maintenance and repair for fixed assets constitute intermediate consumption, these actions improve their performance and increase their capacities. Drawing a line between common repairs and necessary improvements is problematic, although SNA provided enough recommendations for this issue [11].

CFC is a flow of physical deterioration, normal obsolescence, or accidental damage such as storms, fires, or human errors that decline the current value of fixed assets. This must measure accidental damage, normal obsolescence, and physical deterioration, so we calculate it because of changing value-added estimation. The CFC is one of the most important elements of the SNA. This account calculated fixed assets owned by producers. It declines over time and use for output in the production process. In addition, capital consumption measures all types of

Table 4.5 The production account – uses.

Transactions and balancing items	Non-financial corporations	Financial corporations	General Government	Households	NPISHs	Total economy	Goods and services	Total
Output	xx	xx	xx	xx	xx	xx	xx	xx
Market output	xx	xx	xx	xx	xx	xx	xx	xx
Output for own final use	xx	xx	xx	xx	xx	xx	xx	xx
Non-market output	xx	xx	xx	xx	xx	xx	xx	xx
Intermediate consumption	xx	xx	xx	xx	xx	xx	xx	xx
Taxes on products	xx	xx	xx	xx	xx	xx	xx	xx
Subsidies on products (–)	xx	xx	xx	xx	xx	xx	xx	xx
Value added, gross/ Gross domestic product	xx	xx	xx	xx	xx	xx	xx	xx
Consumption of fixed capital	xx	xx	xx	xx	xx	xx	xx	xx
Value added, net/ Net domestic product	xx	xx	xx	xx	xx	xx	xx	xx

structures, such as railway tracks or roads that are scrapped or demolished. These types of assets are owned or maintained by government units. This account does not consist of precious metals or stones, or *natural* asset degradations such as oil, coal, etc. Some of the losses infrequently occur as major earthquakes, volcanic eruptions, tidal waves, or hurricanes. These items do not include accidental damage, and they consist of cost of production. CFC is one of the most troublsome accounts to define conceptually and estimate in practice [11].

There is an important reason for calculating the CFC account. The assets may have been purchased in the past, the consumption occurs in the future years, and its prices are different from the original supply. The CFC account reflects underlying resource costs and relative demands at times of production, but, the cost value with a set of current prices is used for output and intermediate consumption. The historic costs[4] of fixed assets become irrelevant for this account calculation, as a result, a reduction in the PV derives from the CFC [11].

The final item is *the gross* or *net* value added in this section. The gross term is used before you deduct the amount of the CFC. When it is deducted, the value-added becomes the net. This term applies to all the major balancing items in the accounts from value-added to savings [11].

4.2.4 Net Present Value (NPV) and Net Future Value (NFV)

The core concept of Net Present Value (NPV) is about the worth of money in the present time. It may refer to monetary spending in the past or future, and with the NPV method, we bring this amount to the present and calculate how much it worth.

The largest issue affecting in changing the worth of money is *inflation*. For example, if you have $1000 money in your home and save it for one year, with an annual 5% inflation rate, your money's worth decreases by $50 the next year, and your money worth $950.

The money invested in a project decreases its worth over time if a country has a positive inflation rate. For this reason, the payback of that investment must be high so as to cover the inflation rate and earn a profit for the investor. This profit differs in different sectors and with different investors; in financial literature, it is called the *expected return*.

The other rate that can be important in the NPV is a *discount* rate. It is a rate that the central bank of each country defines for lending money in the commercial banks. A risk-free investment is lending funds to the banks.

The NPV formula is:

4 The "historic costs" is a name in the accounting literature for showing the value of an asset. This value is the amount of money paid in the past when the asset was purchased. In the financial reports, it shows as an asset of the units.

$$NPV = \sum_{t=1}^{n} \frac{R_t}{(1+i)^t} - I_0 \qquad (4.10)$$

R_t is the return of the investment in the time t; i is the discount rate or an expected rate; t is the time; I_0 is the initial investment.

The return of the investment of R_t is about the anticipated revenues or cash flows generating in the years 1, 2, 3, ... I_0 is about the first money spent to begin the investment; subtracted from NPV because the investor needs to know about the future worth of this investment in the present. i can be either the discount rate *or* the expected rate of the investor; sometimes the discount rate is much less than and can be a good criterion for the settlement of an investment, yet every investor usually uses their expected rate for this calculation.

The other view of this calculation is Net Future Value (NFV). It says if we need X amount of money in the future, how much money must be spent in the present to receive that amount of money. This formula is quite simple and similar to NPV; in contrast, it views the NPV from the future to the present.

$$NFV = NPV(1+i)^t \qquad (4.11)$$

These two concepts are used in the LCC methods for estimating the corrosion cost.

4.2.5 Input–Output Model in Economics

This section is to familiarize you with the *input–output function* that is used for estimating the Battelle input/output method of corrosion cost, where the Leontief method is used.

Professor Leontief was born to Russian and German parents in 1905 in Germany. He lived and studied economics in Germany and received his Ph.D. During World War II he migrated to the US and became a professor at Harvard University. He designed the input–output analysis to quantify systematically various sectors of a complex economic system with mutual interrelationships. This analysis can be as vast as a nation or the entire world; it can also be used for a sector or a metropolitan area [12].

Leontief defined two types of matrix for his input–output model; first, a matrix for supply and demand, and second, a matrix for wages and prices. He believed that an economist could observe every individual behavior, understanding their path in the real situation, and introduce how an economy works.

In 1946 he introduced 42 departments for production, distribution, transportation, and consumption, and then he established a matrix of horizontal rows and vertical columns. All of these departments are on both the horizontal rows and vertical columns. The horizontal rows show how the output of each sector of the economy distributes with others (how one sector produces another sector's inputs). Conversely, the vertical columns show how each sector gets its needs from other sectors (how one sector receives the other sector's outputs). It means each sector's output is the other sector's input [12]. Today, the number of industries reported in this matrix has increased to 405 industries and are shown in three main categories; the use table, the supply table, and the total requirements. "The use table" has "the Use of Commodities by Industries" information; the supply table shows "the Domestic Supply of Commodities by Industries" information; and the total requirements consist of three tables. The total requirements have "Industry-by-Commodity Total Requirements, after Redefinitions" information, "Commodity-by-Commodity Total Requirement, after Redefinitions" information, and "Industry-by-Industry Total Requirement, after Redefinitions" information. There is also some level of collection for all of these subcategories; "Sector level (1997–2019)," "Summary level (1997–2019)," and "Detail level (2007, 2012)." With such vast calculations, economists try to explain the massive interactions between parts of an economy. I choose the simplest and most informative table of these categories; "Industry-by-Commodity Total Requirements, after Redefinitions" in the collection of "sector level (1997–2019)."[5]

As you can see in Table 4.6, the percentages of the sector outputs give us the ratio. This ratio shows the rate using the output of one sector with other sectors. For example, "the agriculture, forestry, fishing, and hunting sectors" using 0.014504 of "the mining industry" output in their production processes.

The numbers of this table are between zero and one, or between one and two. In this section, we first explain what these numbers mean, and then explain how many models have this matrix.

4.2.5.1 Technical Coefficients

In the nation-wide subdivided sectors to $n + 1$ sectors they produce $n + 1$st final demand. Imagine we have two sectors, with the names i and j; the output of sector i is x_i, and sector j absorbs this output as its input with x_{ij}. Also, there are many other sectors to absorb x_i, thus the final demand of sector i is x_{in+1}, and its quantity is identified by y [12].

The output quantity of sector i absorbed by sector j per unit of its total output j es described by the symbol a_{ij} and called *the input coefficient* of the product of the sector i into the sector j [12].

$$a_{ij} = \frac{x_{ij}}{x_i} \tag{4.12}$$

5 If you need to see these collections click the link: https://www.bea.gov/data/industries/input-output-accounts-data

Table 4.6 Industry-by-industry total requirements, after redefinitions (in producers' prices).

Name	Agriculture, forestry, fishing, and hunting	Mining	Utilities	Construction	Manufacturing	Wholesale trade	Retail trade	Transportation and warehousing	Information	Finance, insurance, real estate, rental, and leasing	Professional and business services	Educational services, health care, and social assistance	Arts, entertainment, recreation, accommodation, and food services	Other services, except government	Government
Agriculture, forestry, fishing, and hunting	1.308343	0.014504	0.006629	0.0233435	0.089756	0.005423	0.0074	0.010609	0.007606	0.003462	0.0063303	0.008431	0.014725	0.008773	0.014143
Mining	0.029744	1.106993	0.069481	0.0366479	0.098012	0.006586	0.007145	0.012726	0.009148	0.005009	0.0064083	0.00988	0.01137	0.010489	0.021741
Utilities	0.01689	0.019423	1.039667	0.0087643	0.016841	0.010009	0.019478	0.013306	0.006716	0.016434	0.0064966	0.009436	0.019942	0.007636	0.009477
Construction	0.010933	0.012108	0.017967	1.0043815	0.008275	0.005835	0.008658	0.011155	0.005161	0.029059	0.0042869	0.005618	0.007341	0.008308	0.028253
Manufacturing	0.362952	0.247929	0.112698	0.3848273	1.603123	0.089434	0.088702	0.178209	0.12593	0.054612	0.0855832	0.141419	0.143237	0.144401	0.211474
Wholesale trade	0.148236	0.05427	0.028439	0.0830633	0.123149	1.040735	0.030848	0.041324	0.029293	0.013718	0.0194902	0.035663	0.034636	0.030326	0.041381
Retail trade	0.007517	0.00472	0.00621	0.0527151	0.008639	0.002814	1.007375	0.012913	0.002742	0.004332	0.003444	0.002685	0.010467	0.016389	0.004184
Transportation and warehousing	0.06677	0.060643	0.076907	0.0469812	0.078289	0.057397	0.062132	1.141404	0.030579	0.017566	0.0282698	0.024822	0.026819	0.022671	0.038654
Information	0.018664	0.028123	0.021878	0.023881	0.027753	0.039151	0.043343	0.033862	1.170978	0.030402	0.0538464	0.036918	0.038686	0.039473	0.046268
Finance, insurance, real estate, rental, and leasing	0.169548	0.147883	0.073805	0.0918533	0.093325	0.154394	0.204048	0.180835	0.104931	1.27121	0.1289613	0.17422	0.166838	0.152452	0.08525
Professional and business services	0.093134	0.183594	0.111761	0.1238129	0.156891	0.210588	0.197624	0.150462	0.211976	0.141723	1.2191624	0.165303	0.191192	0.122492	0.127952
Educational services, health care, and social assistance	0.000558	0.000395	0.000871	0.0005784	0.000584	0.001668	0.004765	0.00096	0.000585	0.000358	0.0006611	1.013837	0.001801	0.005028	0.008646

(Continued)

Table 4.6 (Continued)

Name	Agriculture, forestry, fishing, and hunting	Mining	Utilities	Construction	Manufacturing	Wholesale trade	Retail trade	Transportation and warehousing	Information	Finance, insurance, real estate, rental, and leasing	Professional and business services	Educational services, health care, and social assistance	Arts, entertainment, recreation, accommodation, and food services	Other services, except government	Government
Arts, entertainment, recreation, accommodation, and food services	0.00886	0.009383	0.010558	0.007192	0.010584	0.013541	0.014534	0.025424	0.036963	0.019559	0.025723	0.027588	1.034283	0.015908	0.012957
Other services, except government	0.008549	0.007116	0.004953	0.010367	0.011284	0.019002	0.015802	0.021107	0.009739	0.009329	0.0116407	0.014007	0.015972	1.015067	0.01261
Government	0.02209	0.020474	0.033863	0.0143762	0.025504	0.022637	0.024013	0.048217	0.018364	0.022138	0.0133602	0.016811	0.021682	0.018961	1.017299

Source: Bureau of Economic Analysis, 2019.

The matrix arranged by all of the a_{ij} items is called *structural matrix* of that economy.

There is a balancing between the total output and the combined inputs of the product of each sector. We can show these in a set of n equations:

$$\begin{array}{llll} (x_1 - x_{11}) & -x_{12} & - \cdots & -x_{1n} = y_1 \\ -x_{21} & +(x_2 - x_{22}) - & \cdots & -x_{2n} = y_2 \\ -x_{n1} - \cdots & -x_{n2} & - \cdots & +(x_n - x_{nn}) = y_n \end{array} \tag{4.13}$$

In this set of equations, $x_1, x_2, \cdots x_n$ are a set of outputs, and $y_1, y_2, \cdots y_n$ are the set of the final bill of goods absorbed by the economy such as households, government, industries, and so on. Combinations of the (4.12) and (4.13) make the following equations:

$$\begin{array}{llllll} (1 - a_{11})x_1 & -a_{12}x_2 & - & \cdots & -a_{1n}x_n = y_1 & \\ -a_{21}x_1 & +(1 - a_{22})x_2 & - & \cdots & -a_{2n}x_n = y_2 & \\ \cdots & \cdots & & \cdots & \cdots & \cdots \\ -a_{n1}x_1 & +a_{n2}x_2 & - & \cdots & +(1 - a_{nn})x_n = y_n & \end{array} \tag{4.14}$$

If we solve these equations for *unknown* x's, we can find the number of goods absorbed by parts of an economy are $y_1, y_2, \cdots y_n$. Thus, Eq. (4.14) becomes:

$$\begin{array}{l} x_1 = A_{11}y_1 + A_{12}y_2 + \cdots + A_{1n}y_n \\ x_2 = A_{21}y_1 + A_{21}y_2 + \cdots + A_{2n}y_n \\ \cdots \quad \cdots \quad \cdots \quad \cdots \quad \cdots \\ x_n = A_{n1}y_1 + A_{n2}y_2 + \cdots + A_{nn}y_n \end{array} \tag{4.15}$$

For our example these equations mean total demands for x_1 are combinations of all demands of these inputs in all parts of the economy, a sector y_1 with its coefficient A_{11}, plus sector y_2 with its coefficient A_{12}, and so on.

The constant A_{ij} shows, one unit increase in x_i (of the ith sector) if the y_i increase in their demands. The output x_i for $i = j$ is directly affected (and also indirectly) and for $i \neq j$ affected indirectly. Sector j increases in their outputs, while sector i provides additional inputs for other sectors including sector j as y_j. The computational view tells us each magnitude of coefficient A in the solution equations, (4.15) in general, depends on all the input coefficient a appearing on the left-hand side of the system of equilibrium Eq. (4.14) [12].

The matrix A in the mathematical language appears as

$$\begin{bmatrix} A_{11} & A_{12} & \dots & A_{1n} \\ A_{21} & A_{22} & \cdots & A_{2n} \\ \vdots & \vdots & \ddots & \vdots \\ A_{n1} & A_{n2} & \dots & A_{nn} \end{bmatrix} \tag{4.16}$$

This matrix shows the constants appearing on the right-hand of the solution (4.15), and these constants are the *inverse* of this matrix:

$$\begin{bmatrix} (1-a_{11}) & -a_{12} & \cdots & -a_{1n} \\ -a_{21} & (1-a_{22}) & \cdots & -a_{2n} \\ \vdots & \vdots & \ddots & \vdots \\ -a_{n1} & -a_{n2} & \dots & (1-a_{nn}) \end{bmatrix} \tag{4.17}$$

Because there are many outputs, $x_1, x_2, \cdots x_n$, which are inputs for other outputs, $y_1, y_2, \cdots y_n$, the inverted matrix, A_{ij}, are *nonnegative*. There is a sufficient condition for this,

$$\begin{bmatrix} a_{11} & a_{12} & \dots & a_{1n} \\ a_{21} & a_{22} & \cdots & a_{2n} \\ \vdots & \vdots & \ddots & \vdots \\ a_{n1} & a_{n2} & \dots & a_{nn} \end{bmatrix} \tag{4.18}$$

and of all its principal submatrices,

$$1-a_{11} > 0 \begin{bmatrix} (1-a_{11}) & -a_{12} \\ -a_{21} & (1-a_{22}) \end{bmatrix} > 0 \cdots \begin{bmatrix} a_{11} & a_{12} & \dots & a_{1n} \\ a_{21} & a_{22} & \cdots & a_{2n} \\ \vdots & \vdots & \ddots & \vdots \\ a_{n1} & a_{n2} & \dots & a_{nn} \end{bmatrix} > 0 \tag{4.19}$$

It means this matrix should be positive; this is the so-called Hawkins–Simon condition. If this condition satisfies for one sector, then it satisfies for all other sectors, too. The material interpretation of this condition is if each sector function of an economic system absorbs the output of other sectors, directly and indirectly, it can not only sustain itself, but can also make positive delivery to final demand for smaller and smaller subsystems and so on, and this can make a sustainable production system . However, if one of them cannot pass the test, it makes abound and cause a leak that will destroy the sustainability of all systems [12].

In other words, the condition of the sustainability of an economy is the sum of the coefficients of *each column* and its structural matrix should be *less than or equal to 1*, if at least one of the column sums is less than 1 [12].

In an open economy, exports can be positive and imports can be negative for components of final demand.

4.2.5.2 Price and the Input–output Table

Another item of an economy determining an open input-output system is price, and there are a set of equations to show the price of each productive sector of the economy. These equations consist of prices received per unit of its output and the total outlay incurred in the course of the production; these two items must be equal. The equations below show the total outlay of an economy output, these outlays include both purchases of inputs and also value-added for other exogenous sectors [12].

$$\begin{array}{ccccc} (1-a_{11})p_1 & -a_{21}p_2 & - & \cdots & -a_{n1}p_n = v_1 \\ -a_{12}p_1 & +(1-a_{22})p_2 & - & \cdots & -a_{n2}p_n = v_2 \\ \cdots & \cdots & \cdots & \cdots & \cdots \\ -a_{1n}p_1 & +a_{2n}p_2 & - & \cdots & +(1-a_{nn})p_n = v_n \end{array} \tag{4.20}$$

In these equations, p_i is the price of each product of endogenous sectors, and each of these equations describes the balance of a receiving price and payments for output. This price usually contains wages, interest on capital, and entrepreneurial revenues of households, taxes, and other final demand sectors [12].

Similar to solving Eqs. (4.14) for x_i's, in these equation we solve for p_i's to determine the prices of all products from the value added of each sector [12].

$$\begin{array}{l} p_1 = A_{11}v_1 + A_{21}v_2 + \cdots + A_{n1}v_n \\ p_2 = A_{12}v_1 + A_{22}v_2 + \cdots + A_{n2}v_n \\ \cdots \quad \cdots \quad \cdots \quad \cdots \quad \cdots \\ p_n = A_{1n}v_1 + A_{2n}v_2 + \cdots + A_{nn}v_n \end{array} \tag{4.21}$$

In Eq. (4.21) the price p_i is a dependent variable of sector j with value added, v_i, earned per unit of its output in sector i. In these equations too, the constant A_{ij} measure these dependent variables, p_i's [12].

There is a fantastic relationship between these two set of equations; the a_{ij}'s in the rows of q. (4.14) and corresponding column of the q. (4.20) are related to each other. This means the A_{ij} coefficients in the row of output matrix build the corresponding coefficients in the columns of the price solutions matrix, q. (4.21).

The following identity is derived from qs. (4.15) and (4.21), because of the internal consistency of the price and the quantity of relationships within an open input–output system.

$$x_1v_1 + x_2v_2 + \cdots + x_nv_n = y_1p_1 + y_2p_2 + \cdots y_np_n \tag{4.22}$$

This equation's sides show two main principles of an economy. First, there are some endogenous parts of this system and exogenous parts. The left-hand side shows paying out of endogenous parts of the system to the exogenous sectors of this system; on the right-hand side it combines values (quantities × prices) of their respective products delivered by all endogenous sectors to the final demand. It shows the accounting identity between the receiving and spending national income.

The *price analysis* has another viewpoint relating to the *stocks* used in the production processes. Stocks are a particular kind of goods such as machine tools, industrial buildings, and "working inventories" of primary or intermediate materials. These stocks are produced by industry i and employed by industry j, and measure its value based on per-unit of the sector j output. To show this, capital requirements represent a matrix B. Each column of the matrix B describes the *physical capital* requirements of a particular industry. In the same way, the corresponding column of matrix A has its "current inputs" requirements [12]. The capital coefficient b_{ij} is:

$$\mathbf{B} = \begin{bmatrix} b_{11} & b_{12} & \dots & b_{1n} \\ b_{21} & b_{22} & \cdots & b_{2n} \\ \vdots & \vdots & \ddots & \vdots \\ b_{n1} & b_{n2} & \dots & b_{nn} \end{bmatrix} \tag{4.23}$$

The price analysis can advance by splitting the right-hand side of the equation into two parts. The value-added is divided into two parts: capital, invested in building, machinery, and other stocks of goods required for the production of output; and wages. In addition, there is a return on capital representing the value of all productive stocks multiplied by the given rate of return [12].

Finally, the price of goods and services has a relationship with wage rates, the rate of return on capital:

$$P = (1 - A' - rB')^{-1}W \tag{4.24}$$

where W is a column vector of wage costs paid by different industries per unit of their respective outputs [12].

4.2.5.3 Dynamic Input–output Analysis

To describe and analyze the process of economic growth, we employ a set of linear differential equations representing dynamic input–output relationships.

$$X(t) - AX(t) - B[X(t+1) - X(t)] = Y(t) \tag{4.25}$$

In this equation, the column vector $Y(t)$ represents the amounts of various goods and services in year t produced by sectors to households and other final users. $X(t)$ and $X(t+1)$ are the column vectors representing the output levels of different industries in times t and $(t+1)$. A is the matrix of input coefficients in Eq. (4.16), and B is the matrix of capital coefficients in Eq. (4.23). The capital stock produced in year t is used in year $(t+1)$.

These sections are to familiarize you with the main principles of the input–output model introduced by Professor Leontief. In the following section we discuss Corrosion Economics and the input–output model by Battelle Institute, which extended these primary principles for estimating corrosion cost. To describe and analyze the process of economic growth, we employ a set of linear differential equations representing dynamic input–output relationships.

4.2.6 Depreciation, Consumption of Fixed Capital, or Corrosion

Depreciation, fixed capital consumption, or corrosion; which one is correct? The answer is 'it depends.' This short section shows the exact differences between these three terms.

Depreciation and CFC – In accounting terms, depreciation is used for a historical cost. The CFC is used for current value and determined by the future [11].

In the accounting system, an asset purchase finance manager calculates the asset depreciation based on a standard depreciation method. At the end of every year, the finance manager decreases that asset value without any observation in real degradation or loss. This method is used for business accounts. For example, when machinery is purchased with \$2000 US and is estimated to work for 10 years with a fixed depreciation rate, its stock value will be \$500 US; this calculation is:

$$\text{Current value/stock value: } 2000 - 500 = 1500$$

$$\text{The fixed depreciation rate for 10 years: } 1500/10 = 150$$

Every year decreases the value of this machinery \$150 US.

Unlike depreciation, the CFC must calculate based on the *remaining benefits* of a fixed asset in current prices. Computing the CFC is strict in practice because the capital that is purchased in the past must make the value in current prices as its benefits for the future.

For these reasons, depreciation as recorded in business accounts may not determine the right amount for calculation of CFC.

Depreciation and corrosion – In financial literature, these two different words are used for showing degradation in the materials. Sometimes these two words are used interchangeably, but there have some specific differences between them.

The word 'corrosion' derives from the verb 'corrode,' from Old French 'corroder' (14c.), and directly from Latin 'corrodere.' Corrodere means "to gnaw to bits, wear away," from assimilated form of com- + rodere "to gnaw." Figurative use from the 1630s [13].

The word' depreciation' derives from the verb 'depreciate,' and from the Latin word 'depretiare.' Depretiare means "to undervalue, under-rate" from de "down" + pretium "price." From the 1640s intransitive sense of "lessen the value of, to lower in value." The intransitive sense of "to fall in value, become of less worth" is from 1790 [13].

As you can see, the word depreciation relates to the price of a substance, and the word corrosion relates to its chemical degradation. Thus, when we need to speak about the value of material we must use the term depreciation; for this reason, the financial reports use the word depreciation. Consequently; when, we speak about gnawing and chemical literature we must use the term corrosion.

4.3 Corrosion Economics

The global corrosion cost in 2013 was about 3.4% of its GDP, this percent is about $2.5 trillion per year [1]. This amount was about 77% of US gross savings and 14% of US GDP in 2013.

In economics, there are some investment models to estimate the cost of capital. These investment models design on time, the real interest rate, the capital depreciation, capital price change during the time, changes in output, tax policies, uncertainty about the future probabilities, and discount factors. These models cope with a firm investment in capital and its cost that affects the exogenous variables in the real. It also requires an infinite rate of investment [14]. These factors create complicated models that help economists to estimate economic investment and use them in their policymaking.

However, it is crucial to measure a project cost even if it is about a petrochemical refinery, but managers need the simplest and most applicable methods; these types of complicated investment models are not suitable. Thus, I bring the two most useful methods for corrosion cost measurements in this chapter.

4.3.1 Input–output Model in Corrosion

One of the most important methods for estimating corrosion cost is the input–output model (I/O). This model was introduced with Battelle Columbus

Laboratories. Mr. Gordon Battelle was the founder of Battelle Memorial Institute, who lived from 1883 to 1923, and the Battelle Columbus Laboratories is one of the headquarters of the Battelle Memorial Institute. During his life, Mr. Battelle invested in lead mining and smelting operations; he later became acquainted with Professor George Waring and made a successful commercial laboratory for an economic appraisal for him [15]. The I/O method was generated with the Battelle institute in 1976 and sent to the National Bureau of Standards [16].

The I/O method applies the input–output model in economics to corrosion cost estimation. There are two broad dimensions to illustrate the I/O model. First, using and developing the input–output Leontief model, and then using three conceptual worlds for this input–output matrix. Finally, we discuss this report suggestion method for estimation of corrosion cost.

The three conceptual worlds make three scenarios for an economy; World I is an economy as it exists, World II is a hypothetical economy without corrosion, and World III is the economy with effective corrosion-control practices. The I/O model is constructed to describe each of these economies by using its matrix.

The most significant principle in corrosion cost control is about the *avoidable* cost. Some of the cost is unavoidable because the gnawing is the nature of the materials. Although this cost is natural, there is some avoidable cost which can reduce it with more professional corrosion control, and the initial level for this control is familiarizing yourself with them.

As we see in the input–output of economics, we have a matrix for technical coefficients called matrix A, a matrix for capital coefficients called matrix B, and a matrix for output, X. In this section, we have three basic matrices but will divide them into their components. In the Battelle I/O method, we use the same matrices as Leontief used for his model, but the Battelle institute used it for estimation of corrosion costs, while. Leontief used it for all parts of the economy.

4.3.1.1 Matrix of Technical Coefficients

Matrix A is the matrix of direct technical coefficients. These matrix cells show the dollar worth of inputs that is required to produce one dollar of output. The worth of inputs are shown in the row, and the worth of outputs are shown in the column of this matrix. The sum of column coefficients in this matrix is inputs *per* dollar of its output.

To design matrix A we add three items together to obtain the matrix cells; these items are value-added, social savings, and foreign trade. These three rows are outside the matrix A.

Value-added is the intermediate matrix speaking about this item's nature in economics.

Social savings is a new row added to the transaction table and matrix A, and it is a hypothetical item. In World I, which is the real world, this row is empty. This item measures the labor and resources not used in the absence of corrosion; in other words, the reduction in the GNP that would take place if our world were corrosion-free. It has an indirect influence on the sizes of all the intermediate direct coefficients [16].

Foreign trade consists of the imports of intermediate and final demand for goods and services. These imports subtend *competitive* and *uncompetitive*. The items included in the Battelle model are "noncompetitive" imports and show in the matrix as a row across the table. Because the competitive imports constitute a large proportion of the intermediate demand, these imports are not included. Demands such as crude oil contained the competitive items in this matter, for they became heavily negative total outputs [16]. To modify this problem, the Battelle carries all imports as a row outside the intermediate and inverted matrices. Thus, the total input becomes:

$$TI = TDI + DVA + IM \tag{4.26}$$

where *TI* is total input; *TDI* is total domestic intermediate input; *DVA* is domestic value added; *IM* is imports.

Regardless of domestic or imported, every direct coefficient in the *A*-matrix is *technologically* correct. An import column calculates by multiplying each sector's output by the sector's import coefficient. These calculated values can add as a column to the table so that:

$$TDO = TIO + TFS - IM \tag{4.27}$$

and

$$TDO = TDI = TI - IM$$

where *TDO* is total domestic output; *TIO* is total intermediate output; *TFD* is total final demand; *TI* is total inputs [16].

Finally, the column sum of the *A*-matrix plus the sum of the three above items (value-added, imports, and social savings) is equal to one since the coefficients are in terms of proportions. The sum of the column coefficients of matrix *A* is equal to per dollar of the output of the column sector's use of its domestic intermediate inputs.

In World I, the industry's *total output* is equal to the column industry's *total inputs*, also equal to the sum of *cells of matrix* and *value added* and *import rows* filling with its dollars.

4.3.1.2 Matrix of Capital Coefficients

Matrix *B* is the matrix of stocks used in the production processes and called *the capital* matrix, as you learned in the Leontief input–output model. The previous matrix, *A*, measures dollars worth of input per dollar of output, while *B* matrix measures dollars worth of capital per dollar of output. However, because there are many other definitions for the capital stock matrix, the capital matrix used in this study is for estimating this purpose.

In this matrix, the sectors listed across the top of the matrix show capital users, sectors listed across the side of the matrix indicate capital producers. There are a great many empty rows in this matrix because the capital producers are low.

Although the capital stocks excess is for more than one year and include durable goods, they need to *grow* and *replace*. The growth in the demand for an industry's output is the reference of the growing need for capital stocks, regardless of whether these growths are demand for final or intermediate goods, or foreign or domestic. In addition, capital replacement may relate to technological change or the age of existing capital. For these scenarios, matrix B changes in this manner:

$$\overline{B} = B(G + R) = BG + BR \tag{4.28}$$

where B is the standard capital coefficient matrix (expressed in the form of ratios of capital stock to output); G is matrix of annual growth rates of total capacity (capital); and R is matrix of annual replacement rates of capital.

The growth matrix – Growth in demands calls for growth in the capital matrix to determine this capacity. We need the ratio g from the first approximation growth rates of total output (X). If using K as the capacity, we use g in this manner:

$$g = \frac{K_1 - K_0}{K_0} \tag{4.29}$$

We can substitute a comparable ratio based on X (which approximates capital formation change based on this total output of the sector of the economy) so that it changes approximately equal to the direct or indirect growth of general demand.

$$g = \frac{\overline{X}_1 - \overline{X}_0}{\overline{X}_0} \tag{4.30}$$

The replacement matrix – All the stock capital has a life period; this age-structure relates to the steady growth at the rate g. In other words, the replacement rate is a function of the growth rate and the replacement life of that stock. The replacement rate (r) of each sector is a combination of both replacement life expectancy and sector growth.

Matrix of replacement lives (U) is a full matrix, like B-matrix. Every U_{ij} corresponding with a nonzero cell in B show a replacement life expectancy value. These values may take from a given number of years or a range of years. Each column has an average annual growth rate, g_j, derived with Eq. (4.30) and used to describe the entire cycle of replacement lives.

To show the current deterioration of stock capital which has formed during a period u years, and assuming smooth growth, we must consider the reasons for growth and replacement during this period. In this sequence of years, K_0 is the first year and K_{u-1} is the end of this period. The total value of the current capital stock can show as $\sum K$:

$$\sum K = K_0 + K_1 + K_2 + \cdots + K_{u-1} \tag{4.31}$$

where K_0, K_1, ... is the growth and replacement capital purchase in year 0,1,2, ... respectively, and U is the life of first replacement of the capital stock.

Now, to obtain the annual replacement rate of capital for these sectors we have:

$$r = \frac{K_0}{\sum K} \tag{4.32}$$

In a given annual growth rate of the sector g, each K becomes:

$$\begin{aligned} K_1 &= K_0 * (1+g)^1 \\ K_2 &= K_0 * (1+g)^2 \\ &\vdots \\ K_{u-1} &= K_0 * (1+g)^{u-1} \end{aligned} \tag{4.33}$$

As you can see, K_0 is common to all terms, thus:

$$r = \frac{K_0}{\sum K} = \frac{1}{1 + (1+g) + (1+g)^2 + \cdots + (1+g)^{u-1}} \tag{4.34}$$

Then this can transfer into:

$$r = \frac{g}{(1+g)^u - 1} \tag{4.35}$$

To build the replacement matrix, R, we establish a value of r_{ij} for each corresponding value of u_{ij}.

Now is the time to set up the basic I/O model equation with recent principles together.

4.3.1.3 Input–output Model

The Leontief input–output equation can be written with this standard matrix notation:

$$X - XA = FD \tag{4.36}$$

where X is total output; A is the direct coefficient matrix; and FD is total final demand.

The rephrase of this equation says to produce the total output, so we must cover both intermediate input requirements and final demand.

Taking account of the investment demand rewrite Eq. (4.36) in this manner:

$$X - XA - X\overline{B} = \overline{FD} \tag{4.37}$$

where $\overline{B}$ is total capital coefficient matrix [which we saw in Eq. (4.28)]; and $\overline{FD}$ is total stipulate (noncapital) final demand.

Thus, Eq. (4.37) can be written:

$$X - XA - XBG - XBR = \overline{FD} \tag{4.38}$$

where XBG is growth capital and XBR is replacement capital, then the sum of these two terms is the total capital.

Then we extract the total output from Eq. (4.38):

$$X = [I - (A + B(G + R))]^{-1} * \overline{FD} \tag{4.39}$$

4.3.1.4 Final Demand

The third major component of the I/O model is final demand. This final demand is the *consumption* we learned of in the first section (in estimating GDP), personal consumption, government purchases, exports, and net inventory change. In theory, the final demand summation is equal to GNP; but in reality, it needs to be adjusted. The adjustment of GNP done by adding gross private domestic investment to and subtracting imports from the consumptions.

In the final demand, the assumption is that we have full employment in the economy [16]. One of the four main concepts of economics is the employment rate. In economic principles, full employment occurs when 5% of working-age people, aged18 to 65, are unemployed, and 95% of people are employed.

The consumption of households, government expenditure, exports, and inventory change are the four types we have discussed. In the following paragraphs, I bring these types of consumption components in the Battelle final demand.

In the Battelle approach, we have a three-dimensional matrix for household consumption. These dimensions are *income behavior class*, *age of head of the family*, and the *family size*. This concept can estimate the level of consumption by a family and the entire economy as a whole [16].

For this estimation, the Battelle report designated a separate analysis to measure how families with different income and different size characteristics spend their income in certain categories.[6] A regression equation expresses how many average families in a given income class spend their consumptions in one year. The total

6 When the Battelle model was designed there were 42 major categories.

amount gathered with this method uses a PCE column in the I/O table's final demand sector. The governmental expenditure is this expenditure in the normal situation, not about the emergency programs, for example when a country is engaged in a war. Foreign trade in the final demand matrix is about the *gross exports*. In the *A*-matrix we have a row for imports.

The last item contained in the final demand is the *inventory change*. It is about the change in the stock goods over many years. This issue determinate is crucial in practice; thus, economists use econometrics methods to calculate this part. In other words, the inventory change means how much goods every industry has at the end of the year in their warehouse, and this amount changes year by year. You can see this account in the table of GDP expenditure approach.

4.3.1.5 World I, World II, World III

As mentioned in the beginning of this section, there are three worlds for an economy; World I is the existing economy, World II introduce a hypothetical world without any corrosion, and World III is a hypothetical world with efficient corrosion control.

To introduce these worlds we reduce the information of World I to coefficient and range change. This means the World II and World III flow is the adjusted coefficient of World I. Adjustment of capital/output coefficient is expressed as a percentage, excess capacity for an industry, or for an entire economy; World I. Adjustment of replacement life is expressed as a change in the range of years; for example, if for World I we use 10–30 years, for World II use 20–30 years. Changes in final demands are expressed as a percentage change of World I values [16].

The differences between World I and World II is the *national* cost of corrosion and shows resources wasted cause of corrosion. The differences between World III and World I shows the *avoidable* costs of corrosion. These avoidable costs could not occur if economically preventive practices were used throughout an economy. Moreover, the differences between World III and World II measures presently *unavoidable* costs [17]. In all, there are differences between industries in this concept, and it relates to the users of goods and producers of goods.

4.3.1.6 Estimating Corrosion Cost by Battelle

We introduced the Battelle input/output model and described its components, in the following, we bring the Battelle models to estimate the corrosion cost in the industries.

Disaggregation – You saw in the direct technical coefficient matrix that certain cells within the final demand vectors disaggregate. This disaggregation is about the sectors producing durable goods and the corrosion prevention materials, and maintenance and repair of construction sectors. These sectors are disaggregated into corrosion-related sectors and non-corrosion-related sectors. Then, we introduce some annual purchasing methods for these kinds of goods and their replacement and growth.

$$\begin{aligned} Annual\ Purchase\ of\ social\ capital_t = {} & Replacement\ Purchases_t \\ & + Growth\ Purchases_{t-1,t} \end{aligned} \tag{4.40}$$

Social capital/infrastructure is defined as consisting of final demand of durable goods that last for more than one year.

The growth in the purchase of capital is:

$$Growth\ Purchases_{t-1,t} = g(SK_{t-1}) \tag{4.41}$$

where SK is social capital (in abbreviations we usually use K for capital).

Thus:

$$SK_t = (1+g)SK_{t-1} \tag{4.42}$$

and

$$SK_{t-1} = {}^{SK_t}\!/_{(1+g)} \tag{4.43}$$

This equation can be rewritten in this way:

$$Growth\ Purchases_{t-1,t} = \frac{g\,SK_t}{1+g} \tag{4.44}$$

g is the growth purchases in one year, for u years this growth becomes a rate of change:

$$r = \frac{g}{(1+g)^u - 1} \tag{4.45}$$

The annual replacement of social capital is:

$$Replacement\ Purchases_t = SK_t \frac{g}{(1+g)^u - 1} = (SK_t)(r) \tag{4.46}$$

Substituting qs. (4.44) and (4.46) into (4.40) makes:

$$Annual\ Purchase\ of\ Social\ Capital = \frac{g\,SK_t}{1+g} + (SK_t)(r) \tag{4.47}$$

Now solve Eq. (4.47) for (SK_t) because we need to know about the social capital in time t:

$$(SK_t) = \frac{(Annual\ Purchase\ of\ Social\ Capital)(1+g)}{r(1+g)+g} \tag{4.48}$$

Cost parameters – There are four different parameters for estimating the cost (i) the direct cost of corrosion per unit of output, (ii) the total direct costs of corrosion per sector, (iii) the direct and indirect cost of corrosion per unit of output, and (iv) the total direct and indirect cost of corrosion for each sector.

4.3.1.6.1 The Direct Cost of Corrosion per Unit of Output

Comparing production requirements in Worlds I and II and III show the direct per-unit cost of corrosion for each sector [16]. This production requirement is the intermediate requirement (direct technical coefficients including value-added) per unit of the sector's output, capital requirements per unit of output of the producing sector multiplied by the sector's growth rate, and the annual replacement of capital per unit of production. These component summations show a total requirement necessary to produce a unit of a given sector's output. Below are the steps to represent the differences between these three world requirements.

Step 1. Calculate the direct requirements per unit of production:

$$\tau_{wj} = 1 + g_j C_{wj} + \rho_{wj} - SS_{w,j}; (w = 1, 2, 3) \tag{4.49}$$

where g_j is the annual rate of growth of sector j's output; $SS_{w,\,j}$ is the social savings entry for sector j in World w (SS_j is zero for World I); C_{wj} is the total capital/output ratio for sector j in World w; and ρ_{wj} is the total annual replacement of capital required to produce one unit of section j's output in World w.

In Eq. (4.49), $1 - SS_{w,j}$ is equal to the sum of direct technical coefficients plus value-added, and is a more direct approach to obtaining that sum.

Step 2. Calculate total direct per unit costs:

$$\tau_{1j} - \tau_{2j} \tag{4.50}$$

Step 3. Calculate reducible direct per unit costs:

$$\tau_{1j} - \tau_{3j} \tag{4.51}$$

A devised account for keeping the cost of various corrosion-related expenses is the value $SS_{w,\,j}$. This account contains any reductions in operating costs that those reductions relate to; both intermediate inputs and value-added. The input changes are determined through industry surveys, and the value-added adjustments estimated in the following ways.

Value-added adjustments – There are several adjustments in each industry's value-added coefficients, and it shows the differences between World I, World II, and World III. These adjustments group in the common term; Z_{wj}. This value factors the sector j in World I (w shows the number of the world). This value Z_{wj} is the value-added that subtracts from sector j in World I and adds this value to row sector, *SS*, world *w*, and sector *j*. *SS* is the cumulation of net corrosion savings in the column of direct technical coefficients. Thus Z_{wj} is defined as:

$$Z_{wj} = A_{wj} + \Delta_{wj}; (w = 2, 3) \quad (4.52)$$

A_{wj} shows the reduction in the cost of holding capital that the total capital requirement reduced. This occurs when the prime interest rate reflects the fact that some firms cannot obtain capital at the prime interest rate.

$$A_{wj} = \left(C_{1j} - C_{wj}\right)n; (w = 2, 3) \quad (4.53)$$

Δ_{wj} shows the reduction in a net accumulation of the depreciation account. *Gross* using of the capital and the gross accumulations in the depreciation account are measured in *D*. The value of replacement accounted for final demand is measured in ρ. The difference between them represents the net accumulation of in the depreciation[7] account. Δ shows the reduction in the net accumulation in World I to Worlds II and III.

$$\Delta_{wj} = \left(D_{1j} - D_{wj}\right) - \left(\rho_{1j} - \rho_{wj}\right); (w = 2, 3) \quad (4.54)$$

ρ_{wj} represents the annual replacement of capital that is used by sector *j* in World *w*:

$$\rho_{wj} = \sum_i r_{wij} c_{wij} \quad (4.55)$$

where r_{wij} is the World *w* annual replacement rate of capital produced by sector I and used by sector *j*; c_{wij} is the World *w* capital produced by sector I and required to produce a unit of sector *j*'s output.

Then the average replacement rate, RR_{wj}, can be applied to sector *j*'s total capital:

$$RR_{wj} = \rho_{wj}/c_{wj} \quad (4.56)$$

where C_{wj} is the total capital/output ratio of sector *j*.

Referring to the replacement matrix we have:

$$RR_{wj} = \frac{g_j}{\left(1 + g_j\right)^U - 1} \quad (4.57)$$

where *g* is the growth rate of sector *j*; and *U* is the World *w* average replacement life of capital used by sector *j*.

Now combine qs. (4.56) and (4.57) and solve them for the average replacement life of all capital, U_{wj}, used by sector *j*:

---

7 In my view, the account that can estimate the corrosion on the fixed capital by observation is "consumption of fixed capital account," because the Battelle report uses this account, I write the depreciation account.

$$U_{wj} = Ln\left[\frac{gc_{wj} + \rho_{wj}}{\rho_{wj}}\right] \div [\ln(1+g)] \tag{4.58}$$

It is not possible to calculate the straight life depreciation of capital, D_{wj}, owned by sector j:

$$D_{wj} = \frac{C_{wj}}{U_{wj}} \tag{4.59}$$

4.3.1.6.2 Total Direct Costs of Corrosion

In the previous section, we estimated the cost of corrosion per output of every industry. Now, by multiplying all the outputs of every industry in that amount, we can obtain all industries corrosion costs. To obtain the differences between Worlds I and II, and Worlds I and III, you can calculate in this way too. These differences represent the total direct costs of corrosion.

4.3.1.6.3 Direct and Indirect per Unit Costs of Corrosion

We can obtain the total direct and indirect costs by calculating the total direct and indirect requirements needed to produce the vectors of direct requirements, as shown in the previous sections.

This calculation involves multiplying the vectors of direct requirements associated with each sector by the appropriate World I, II, or III inverse (the inverse in this position refers to $(I-A)^{-1}$). The resulting vectors sum, and for each sector World II summation subtracts from the World I sum., then the World III summations subtract from the World I sum. Mathematically the procedure may be represented by the following steps:

Step 1. Form a V_{wj} of which the ith row is defined as:

$$V_{wij} = a_{wij} + g_j c_{wij} + r_{wij} c_{wij};\ (w = 1, 2, 3) \tag{4.60}$$

where V_{wij} is the ith row of vector V_{wj}; a_{wij} is the direct technical coefficient, World w, row i, column j; g_j is the growth rate for sector j; c_{wij} is the capital supplied by sector I to sector j per unit of j's output in World w; and r_{wij} is the annual replacement by j of capital supplied by i (capital per unit if j's output) in World w.

Step 2. Form the matrix product:

$$[I-A]_w^{-1} V_{wj} = X_{wj} \tag{4.61}$$

where $[I-A]_w^{-1}$ is the inverse matrix for World w; and X_{wj} is the resulting output vector corresponding to sector j.

Step 3. Sum columns X_{wj}:

$$\widetilde{X}_{wj} = \sum_{i} X_{wij} \tag{4.62}$$

where X_{wij} is the ith row of X_{wj}.

Step 4. Calculate total direct and indirect per unit cost:

$$\widetilde{X}_{1j} - \widetilde{X}_{2j} \tag{4.63}$$

Step 5. Calculate reducible direct and indirect per unit cost:

$$\widetilde{X}_{1j} - \widetilde{X}_{3j} \tag{4.64}$$

4.3.1.6.4 Total Direct and Indirect Costs of Corrosion

In these ways, we calculate the total direct and indirect costs analogous with total direct cost calculation. The outputs of steps 4 and 5 multiply by the corresponding World I sector output and show the total direct and indirect costs of corrosion.

4.3.2 Life Cycle Cost (LCC)

The life cycle of a product or service is the estimated cost in the process of their life from birth to death [18]. There have been various models for estimating cost during the years, and there is not a standard LCC model for estimation [19]. The only view in the LCC model is about cost and estimating it from all the viewpoints.

During the life cycle of a product or service, we need to consider the pre-production, production, and post-production costs [18]. Every system has a lifespan, and it differs from system to system. These differences include technical, economical, and usage. Annualizing the costs help to compare the different systems to each other [18]. Although as mentioned earlier, this annualizing may not reflect the actual costs, it is an acceptable method to estimate the cost of corrosion.

To modify these differences between various capitals use realistic views. Lifespan is defined as when a system begins to work, until the end of its usefulness. The cost occurring during a capital life may occur for various factors, such as insufficient maintenance during the life of the product, or those that occur naturally like hydrogeology, precipitation, temperature, and humidity. We can also include some natural disasters in nature, like floods, droughts, or hurricanes that affect the life of assets [18].

4.3.2.1 Life-Cycle Cost Model

In the LCC model, estimation of cost sum up all the costs of a component in the lifespan and makes its life cost. This section brings two kinds of equations to use to estimate the LCC.

4.3.2.1.1 The Dr. Reddy Formulation

This equation was developed by V. Ratna Reddy and his co-author in the "Life-cycle Cost Approach for Management of Environmental Resources" book. This equation is quite comprehensive to show the cost of a system.

The basic functional form of the equation is:

$$LCC_{xt} = f\left\{\sum_{t=1}^{n}(CapExhw_{xt};CapExsw_{xt};CapManEx_{xt};CoCap_{xt};DsCost_{xt};\right.$$
$$\left.IDsCost_{xt};OpEx_{xt}) + CoEExt_{xt}\right\} \tag{4.65}$$

where LCC_{xt} is life-cycle costs of specified products/services; $CapExhw_{xt}$ is Capital expenditure on hardware (initial construction cost); $CapExsw_{xt}$ is Capital expenditure on software; $CapManEx_{xt}$ is Capital management expenditure (rehabilitation cost); $CoCap_{xt}$ is Cost of capital; $DsCost_{xt}$ is Direct support costs; $IDsCost_{xt}$ is Indirect support costs; $OpEx_{xt}$ is Annual operation and maintenance cost; $CoEExt_{xt}$ is Cost of environmental externalities; x represents product or service, and t represents year.

This equation brings the LCC of an asset by either including the expenditure on the capital process of life or the environmental externalities[8] and social sustainability. However, quantifying some of these costs is difficult to determine, as these costs must annualize for standardizing cost. Some of these costs are for the past or future; thus, we must make these costs to the *PV* to compare them with other investments [18]. Accordingly, Eq. (4.65) can be written as:

$$LCC_{xt} = f\left\{\sum_{t=1}^{n} pcf_{xt}(CapExhw_{xt};CapExsw_{xt};CapManEx_{xt};CoCap_{xt};\right.$$
$$\left.DsCost_{xt};IDsCost_{xt};OpEx_{xt})CoEExt_{xt}\right\} \tag{4.66}$$

where pvf is present value factor $(1 + r)^t$; r is rate of interest or inflator and; t is time period.

Using the inflation rate or the prevailing interest rate may be appropriate for estimating the PV or worth [18].

8 https://www.investopedia.com defines externality as: "An externality is a cost or benefit caused by a producer that is not financially incurred or received by that producer. An externality can be either positive or negative and can stem from either the production or consumption of a good or service. The costs and benefits can be both private – to an individual or an organization – or social, meaning it can affect society as a whole." Environmental externalities are about the pollution of the air, water, or environment that occur during factory production. This is not a direct cost of the production, it's the third party of this production.

The other view that Dr. Reddy brings is the risk based in analyzing the cost of a lifecycle. Risk and uncertainty are features of some of the components. Systems may not follow a schedule and fail randomly. Natural factors and unexpected climate events create various risk and uncertainties impressing the estimation of the component's cost.

$$LCC_{xt} = f\left\{\sum_{t=1}^{n} \begin{matrix} pcf_{xt}(CapExhw_{xt}; CapExsw_{xt}; CapManEx_{xt}; CoCap_{xt}; DsCost_{xt}; IDsCost_{xt}); \\ OpEx_{xt}; CoEExt_{xt}[Psf_{xt}] \end{matrix}\right\} \tag{4.67}$$

where Psf_{xt} is probability of risk.

This formulation is appropriate for a wet environment if the groundwater is substantial. This view of the LCC helps to estimate the real-life cost in every climate and location [18].

4.3.2.1.2 The Professor Adeli Formulation

The other view of estimating the life-cycle cost is from Professor Hojjant Adeli and his co-author in the book "Cost Optimization of Structures." They introduce a formulation for measuring the LCC of steel structure, but we can use it for any other components.

$$\begin{aligned} C_{lifecycle} = C_{Initial} &+ \sum \frac{1}{(1+i)^{y_{n1}}} C_{Maintenance} + \sum \frac{1}{(1+i)^{y_{n_2}}} C_{Inspection} \\ &+ \sum \frac{1}{(1+i)^{y_{n_3}}} C_{Repair} + \sum \frac{1}{(1+i)^{y_{n4}}} C_{Operating} \\ &+ \sum \frac{1}{(1+i)^{y_{n_5}}} C_{Failure} + \sum \frac{1}{(1+i)^{y_{n_6}}} C_{Dismantle} \end{aligned} \tag{4.68}$$

where $C_{Lifecycle}$, $C_{Initial}$, $C_{Maintenance}$, $C_{Inspection}$, C_{Repair}, $C_{Operating}$, $C_{Failure}$, and $C_{Dismantle}$ are the total life-cycle, initial, maintenance, inspection, repair, operating, probable failure, and dismantling costs of a steel structure, respectively; i is the discount rate of money; $y_{n1}, y_{n_2}, y_{n_3}, y_{n_4}, y_{n_5}, y_{n_6}$ are the years when the associated costs incur; and $\sum$ denotes the summation of all the costs of the same category during the life of the structure [20].

You read in the PV section that $1/(1+i)^n$ is a factor to convert the cost value to the PV discounted. These two models are two various models for estimating the cost of a life cycle in a system. In the first model, the LCC of an asset, you see system externalities, such as its effect on the climate, being profound for these estimations. In the second model, the main point is bringing all costs to the present, whether they are future or past.

If you need to know about more models or estimation of the cost read the book "Life Cycle Costing for Engineers" by Professor B. S. Dhillon; in chapter 4 he introduces some LCC models and cost estimation methods.

4.4 Corrosion and Sustainability

Corrosion prevention and control is crucial for protecting public safety and the environment [21]. Although there is not the same reason for corrosion in the various areas and materials, it has a strong negative impact on its surroundings. These negative externalities reduce environmental sustainability and have a direct link to pollution in the environment and direct and indirect footprint on society by, for example, waters with chemical corroded particles or chemical the soil.

Using alloys in industries in the past years has had both positive and negative impacts on humans. One of its positive impacts is it helps with types of development and growth, but a negative one is the corrosive materials affecting our life and environment. The main reason for failing in the engineered systems is corrosion [22]. In the following, we discuss some examples of different materials and their impact on the environment and human life.

The largest corrosion threat involves nuclear industry. This industry uses zirconium alloys, and this alloy has its corrosion problems and solutions [22]. In nuclear accidents, most issues are about cooling systems and the role of water and steam. In the Three Mile Island accident in 1979, failure of the steam valve system created the accident. After two days, when the operator needed to compress the gas to the decay tanks, it was instead released into the environment. This accident has not been related to deaths or cancer threats, but increased the risk of cancer approximately 0.4%, compared with the 38–40% of all US citizens [23].

The Chernobyl accident in 1986 was the result of a flawed reactor design and operator error. The interaction of too hot fuel with the cooling water created rapid steam production and substantial damage due to design characteristics, and this accident released 5 tons of the radioactive reactor core into the environment. After this accident, 30 people died, mostly power plant employees and firemen, and about 350 000 people who lived in the surrounding areas relocated their homes. In addition, there were approximately 6500 thyroid cancer patients due to this accident [24].

In the Fukushima Daiichi Accident in 2011, following a massive earthquake, a 15-m tsunami disabled the power supply and cooling capabilities of three Fukushima Daiichi reactors. This accident evacuated 100 000 people from their homes. Additionally, there have been 2259 disaster-related deaths due to the consequences of this accident [25].

After reading about these dramatic accidents, you can review the table 4.7 about fixed[9] and variable[10] costs of nuclear power plants and compare them with other types of power plants.

---

9 Which shows the initial cost of maintaining a nuclear power plant.

10 Which shows the repair and maintenance.

Table 4.7 Power plant operation and maintenance costs in the US in 2019, by technology (2019 US dollars).

Power plant	Fixed costs (in US dollars per kilowatt per year)	Variable costs (in US dollars per megawatt hour)
Biomass	125.9	4.81
Advanced nuclear	121.13	2.36
Geothermal[a]	113.29	1.16
Wind; offshore	109.54	0
USC – 30% carbon sequestration[b]	59.29	10.93
USC – 90% carbon sequestration[b]	54.07	7.05
Hydropower[a]	41.63	1.39
Wind	26.22	0
Solar photovoltaic[c]	15.19	0
Combined cycle – single shaft	14.04	2.54

Release date: 1/1/2020; Region: United States; Survey time period: 2019 Supplementary notes:
[a] Cost and performance characteristics are site-specific, so costs represent least expensive plants that can be built in the Northwest Power Pool region, where most proposed sites are located.
[b] USC: Ultra-supercritical coal.
[c] Cost in terms of net AC power available to grid for installed capacity.
Source: www.statista.com.

This table shows the cost consumption for 1 kW per year, the high level of cost consumption of the nuclear power plant for construction, and its need for repair and maintenance. To compare the nuclear plant with a wind or solar plant shows the justification of using solar or wind for power. Not only these types of energies are inexpensive, they are also environmentally friendly and have zero negative externality for citizens. There is no reason for using nuclear power because it is cost consuming, pollutes the environment, and threatens the citizens.

The marine and offshore industries are another industry coping with corrosion. These environments are harsh for metals and other materials. It threatens human life if one cannot use routine repair and maintenance. For example, during 1990–1991, 300 crewmembers of a ship died because the owner cut its maintenance cost and a 12 mm thick section of the ship was corroded down to 3–5 mm; corrosion was to blame for these lost lives [22]. Another example is a ship that has capsized and sank in the water; its corrosive materials cause much pollution to the sea or ocean floor. Alternately, most oil and gas factories are located offshore, and their corrosion pollutes the water and affects the inhabitants of the environment, and also threats human life when tainted seafood is consumed.

Monitoring the effect of corrosion is an interdisciplinary approach, and it seems essential for receiving sustainable development goals. This approach requires corrosion engineers, environmentalists, sociologists, and also economists. Such a group of scientists can work together toward a main goal, and bring their unique views to the subject. The sociologists search the corrosion effects on the society, the environmentalists search for the corrosion effect on nature, the corrosion managers search for the at least negative externalities of the corrosion on these two areas, and the economist can consult with them regarding what happens when all of their recommendations occur.

4.5 Conclusion

Corrosion is a global problem because it is depleting our resources, and thus needs an international effort to control it. International institutions such as the World Bank or UNDP must be included in rectifying this issue, because by considering this problem, we can help sustainable development goals with the aim of the 21st century international agencies. However, some part of this cost is unavoidable, and recent years' developments can increase the avoidable costs we saw in the 19th century. One weakness of this issue is there has been no sufficient estimation for this matter in the past 20 years. It might be one of the regular global measurements for every country, such as GDP that every country regulates to report. This regulation helps to improve corrosion management and reduce its avoidable cost.

To begin this essential effort, it needs one global agency such as UNDP to create the software with the Battelle method, then representing it to every country and getting their estimations for what can be done. Then, UNDP needs to regulate every country's reports on corrosion annually. This type of monitoring can help to manage our resources and also help to use our materials adequately.

Because the I/O method can contain every industry of an economy, it is a suitable way for governments to estimat the corrosion cost. I suggest the government use this method; first estimate their corrosion cost annually, then try to separate the avoidable and unavoidable costs. Separating the avoidable and unavoidable corrosion costs is the initial way to manage corrosion costs. There is not a strict borderline between these two types of costs, but with some effort, these ambiguities could start to disappear. With the vast amount of corrosion depletion, it is reasonable to spend quite a bit of money on this research as a national or international project. It could reduce the avoidable cost in every country.

For some smaller projects, such as oil or gas refinery factories, it is suitable to use the LCC method. This method can measure the cost occurring while the project exists, as well as when it has finished and its future consequences. The LCC method can contain the negative externalities of a project in the environment and society; using a comprehensive LCC method is essential for receiving sustainable development goals.

Considering sustainability in the last section shows the necessity of the suitable material for several environments and industries. It can also bring some consequences for corrosion effects on the nuclear power plant industry and its negative socio-environmental consequences. Along with its economic view, it can explain about the meaningless of this type of energy use. I think using nuclear energy relates to the politicians' requests for their purposes; they use it for generating force in other countries. Using this energy is the greatest threat to our planets' safety. Energy policymakers must replace their policies with environmentally friendly energy, such as solar or wind. The future of energy, especially for daily energy, is for these types of energy..

4.6 Summary

"Knowledge is power;" to empower the corrosion engineers, this chapter first introduces some of the principles of economics, those an engineer needs to know for understanding the corrosion estimation methods. Then, we try to narrow down these principles to only the corrosion area. After these sections, I bring the two most common corrosion cost estimation methods; input/output, and LCC. In the I/O method, we expand on all phases of this complicated method. For the LCC method, we explain two kinds of this measurement method.

Finally, this chapter brings a short section about the corrosive materials in the industries. Although this phenomenon is one of the most eminent in today's knowledge, in this small section, I take a glance at the importance of this subject and suggest working on this view profoundly.

References

1 NACE International Impact (2013). Online; cited 2021 01 02. http://impact.nace.org/economic-impact.aspx#:~:text=The%20global%20cost%20of%20corrosion%20is%20estimated%20to%20be%20US,the%20global%20GDP%20(2013).

2 Khabar, B.H. (2014). Online. Online; cited 2020 12 20. https://www.khabaronline.ir/news/439706/%D9%85%D9%81%D9%87%D9%88%D9%85-%D8%B4%D9%86%D8%A7%D8%B3%DB%8C-%DA%A9%D9%84%D9%85%D9%87-%D9%87%D9%86%D8%B1-%D8%AF%D8%B1-%D9%85%D8%AA%D9%88%D9%86-%D8%A7%D9%88%D8%B3%D8%AA%D8%A7%DB%8C%DB%8C-%D9%88-%D9%BE%D9%87%D9%84%D9.

3 Colander, D.C. (2008). *Macroeconomics*, 7e. New York: The McGraw-Hill, Inc.

4 Navarro, P. The Power of Macroeconomics: Economic Principles in the Real World. https://www.coursera.org/.

5 Sachs, J. (2020). The age of Sustinable Development; cited 2020 June 15. https://www.coursera.org/learn/sustainable-development.

6 Wikipedia (2020). Online; cited 2020 12 26. https://en.wikipedia.org/wiki/Alfred_Marshall.

7 Farmer, R.E.A. (1997). *Macroeconomics*. Los Angeles, CA: Roger E.A. Farmer.

8 Case, K.E., Fair, R.C., and Oster, S.M. (2012). *Principles of Macroeconomics*, 10e (ed. D. Battista). Boston, MA: Pearson Prentice-Hall.

9 tradingeconomics.com. Online. cited 2020 10 11. https://tradingeconomics.com/united-states/gdp.

10 Bureau of Economic Analysis (2020). Online. cited 2020 10 19. https://apps.bea.gov/scb/2020/08-august/0820-nipa-update.htm#pce.

11 System of National Account (2008). New York: United Nations; 2009.

12 Leontief, W. (1986). *Input-Output Economics*, 2e. New York: Oxford University Press.

13 The online Etymology Dictionary n.d. Online. https://www.etymonline.com/.

14 Romer, D. (1996). *Advanced Macroeconomics* (ed. L. Sutton). USA: McGraw-Hill.

15 Wikipedia. Online. cited 2020 November 29. https://en.wikipedia.org/wiki/Gordon_Battelle.

16 Payer, J.H., Dippold, D.G., Boyd, W.K. et al.1978Economic Effects of Metallic Corrosion in the United States. A report to National Bureau of Standards. Ohio: Battelle Columbus Laboratories, U.S. Department of Commerce. Report No.: 78-600033.

17 Kruger, J.E.D. and Verink, J. (2011). *Uhlig's Corrosion Handbook* (ed. R.W. Revie), 16–30. Pennington: Wiley.

18 Reddy, V.R., Kurian, M., and Ardakanian, R. (2015). *Life-Cycle Cost Approach for Managment of Environmental Resourses Dresden*. Springer.

19 Dhillon, B.S. (2010). *Life Cycle Costing for Engineers*. Boca Raton, Florida: CRC Press/Taylor & Francis.

20 Adeli, H. and Sarma, K.C. (2006). *Cost Optimization of Structures West Sussex*. Wiley.

21 Koch, H. Brongers G.P.H. (2002). MaGTN. NACE international. Online. Dublin; cited 10/01/2021. https://higherlogicdownload.s3.amazonaws.com/NACE/

cedda8a4-c3c0-4583-b1b6-3b248e6eb1f2/UploadedImages/Resources/pdf/ccsupp.pdf.

22 Sastri, V.S. (2015). *Challenges in Corrosion Costs*. New Jersey: Wiley.

23 World Nuclear Association (2020). Online; cited 01/10/2021. https://www.world-nuclear.org/information-library/safety-and-security/safety-of-plants/three-mile-island-accident.aspx.

24 World Nuclear Association (2020). Online; cited 10/01/2021. https://www.world-nuclear.org/information-library/safety-and-security/safety-of-plants/chernobyl-accident.aspx#:~:text=The%20Chernobyl%20accident%20in%201986,in%20many%20parts%20of%20Europe.

25 World Nuclear Association (2020). Online; cited 10/1/2021. https://www.world-nuclear.org/information-library/safety-and-security/safety-of-plants/fukushima-daiichi-accident.aspx.

5

Effective Management of Process Additives (EMPA)

*Mohamedreza Hamedghafarian**

Senior Process Engineer of Gas Plants' Utilities & Off-Sites, South Pars Gas Complex (SPGC), Phases 20 & 21, National Iranian Gas Company (NIGC), Ministry of Petroleum, I.R. Iran

5.1 Introduction

One of the most important parts of corrosion management in the oil and gas industry is to pay close attention to process additives, solvents, and beds. Depending on the design, each of these materials can be used in a single unit operation and perform in such a way that will require the process designer to redesign. Therefore, a process will inevitably lead to one of the following:

- Energy consumption will increase significantly,
- Equipment volumes will be calculated extremely large,
- Material selection will be very expensive,
- The land required to erect the processing site will be very large,
- For achievement of guaranteed quality (and in some cases quantity) of the products, a large amount of money will be spent to make their production non-competitive or impossible.

This chapter aims to provide a model of effective management of process additives (EMPA), chemicals, on the processing site. Industrial cases have been presented with the approach of the effects of potentially ineffective management of chemicals on equipment corrosion. The natural gas processing industry has been chosen as the platform for studying and extracting potential cases and operation histories. Among these, the scope of unit operations and the functionality of each work frame stage of various parameters, such as the interactions between quantity and quality control of products with equipment performance, laboratory analysis, procurement of

*https://www.linkedin.com/in/mohamedrezahamedghafarian-955463a0.

Corrosion Policy Decision Making: Science, Engineering, Management, and Economy, First Edition.
Edited by Reza Javaherdashti.
© 2022 John Wiley & Sons, Inc. Published 2022 by John Wiley & Sons, Inc.

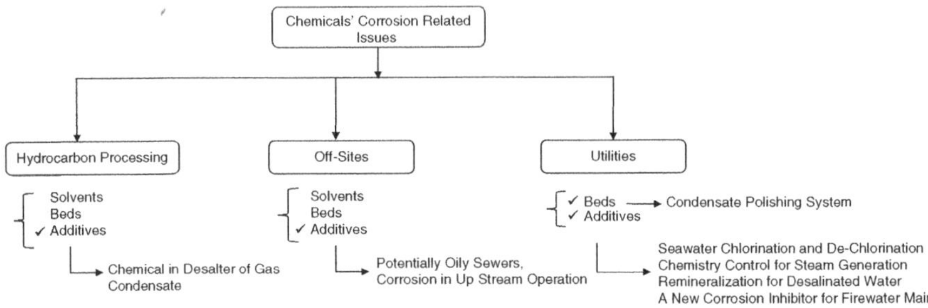

Figure 5.1 Illustration of the limitations for presenting industrial cases.

chemicals, quality control when delivering them, their storage, and consumption in the processing site, and in ultimately reporting their consumption, led me to focus only on the subject of process additives to make it possible to provide real-world cases and further describe their events. A case of a bed, an activated carbon filter (ACF), used for process fluid treatment was also analyzed.

In addition, because the author's experience is limited to utility process engineering and off-sites in hydrocarbon processing industries, these two sections of processing sites have been analyzed in more detail. However, one of the examples in the hydrocarbon processing unit (stabilization of gas condensate), which is a reminder of his presence as a senior field operator of the reception facilities units from upstream, was presented. Figure 5.1 illustrates well the limitations of presenting this chapter.

5.2 A Gas Plant

Here, we describe one of the fields of hydrocarbon processing. Natural gas can be found in parts of the earth's depths; extraction wells may be drilled both on the seabed and on land [1]. Usually the fluid that enters the processing site through the feed pipeline has two major phases; liquid and gaseous hydrocarbons. Liquid hydrocarbons are called gas condensates, which can predominantly contain components larger than pentane ($C5^+$) [2]. Gaseous hydrocarbons include components of methane to butane, gaseous sour agents such as carbon dioxide and hydrogen sulfide, as well as nitrogen and water vapor. It is logical that there are various other compounds within these two phases entering the site. It is also clear that almost all of the components in the feed composition are distributed in both phases. For example, a significant portion of propane and butane in feed are present in condensates and are collected from the top of the distillation column when the Reid Vapor Pressure (RVP) is adjusted, which is called condensate stabilization process.

If the fluid condition of the reservoir or the method of transferring it to the processing site is such that it causes problems such as hydrate formation and corrosion; then it should be expected that the feed entry to the site will also contain hydrate and corrosion inhibitors (CI) [3, 4]. This gas-processing site, in which the reservoir fluid enters to make it usable, is called a gas plant.

First, the reservoir fluid is received through the feed pipeline in a set of facilities which are called reception facilities, and then it is sent to various sections for treatment, extraction of different components, conditioning, and finally export. Each of these sections is called an operation unit, in which one or more processes are designed and operated. This set of units allows the production of hydrocarbon products in accordance with contract or design specifications for sale and export. The block diagram in Figure 5.2 clearly represents the reception, production, and export of hydrocarbon products on a typical gas plant (production equals hydrocarbon processing).

5.3 Utilities

The processing units presented in Figure 5.2 cannot carry out their intended operations alone. This is because their operation depends on two other parts of a hydrocarbon processing plant; utilities and off-sites. Utilities on a gas plant enable production processes to continue working; power for motors, steam required to enhance the reaction, heating and dilution, cooling water, demineralized water, drinking water, utility (service) water, firewater, air required for maintenance activities and, also for instrumentation functions, process air, fuel gas, and nitrogen for different applications are all available for production processes. Off-sites can include processes of a different nature from the reception and production sections. These processes include storage, gas condensate and liquefied petroleum gas, LPG, export, sour water stripping, some types of process additives and solvents storage, and recovery of solvents and treatment of processes' effluents.

There are one or more processes in each operating unit that a gas plant can encounter. A process unit is a series of interconnected activities with a specific sequence that finally leads to a specific product. Table 5.1 has separated the operations related to each designed process in the gas plant, along with the needs of any operating unit for utilities.

Utility units are of particular importance; these distributed utilities exist in almost every area of the gas plant, it is impossible to produce and export products without them. Therefore, it is important to pay attention to the availability of their production equipment and their distribution system. Utilities are the product of processes designed and installed in the utility section of a gas plant. Products such

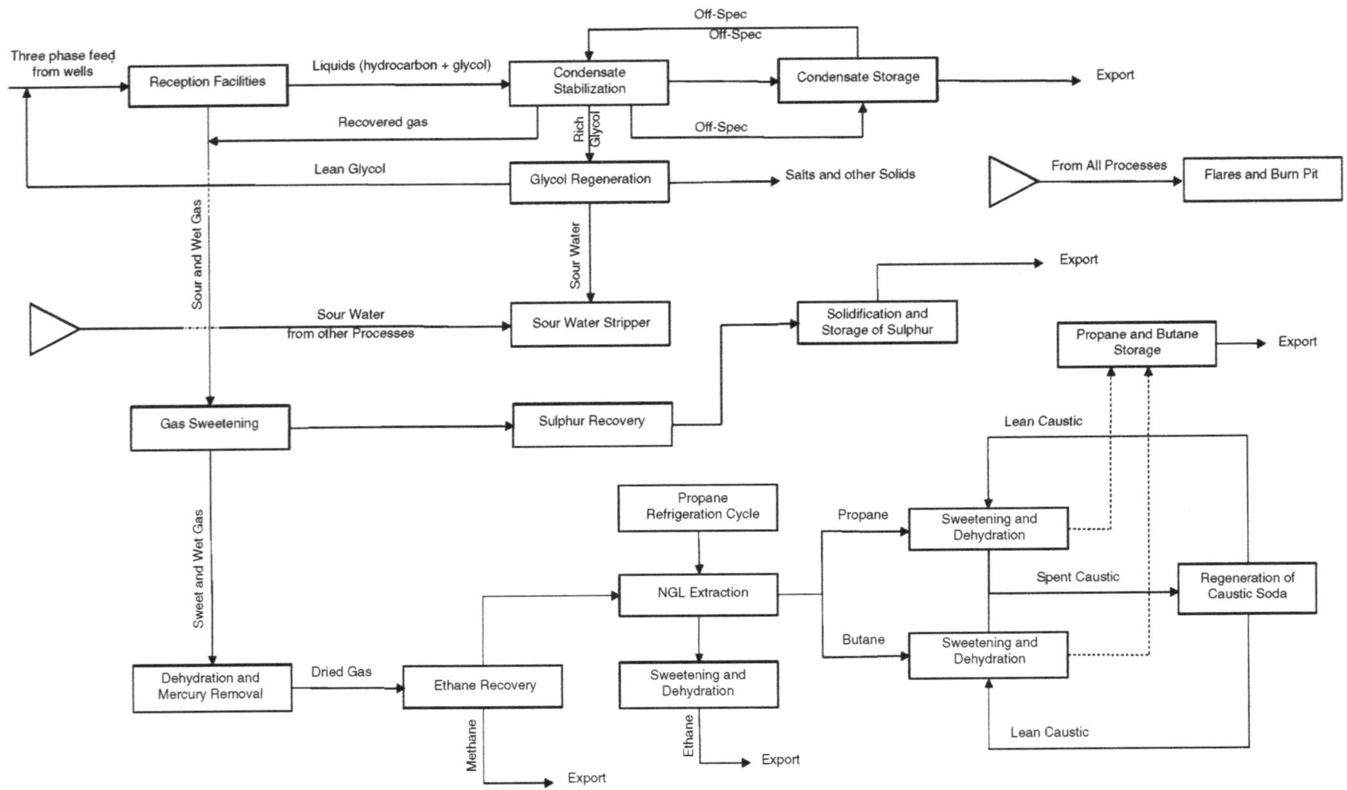

Figure 5.2 The schematic of different operational processing units in a typical gas plant (reception, production, and export).

Table 5.1 Various processes and their needs for different utilities in a typical gas plant.

Units[a]	Processes in operating units	Needs of processes for different utilities[b]	Utility[c,d,e,f]
Reception facilities	Phase separation, filtration, corrosion inhibition, hydrate inhibition, heat exchange	Heating process fluid	Steam
		Dilution of process additives	Desalinated water
		Preventing oxygen contact to process additives, and process fluids	Nitrogen
Gas sweetening	Phase separation, filtration, gas/liquid absorption with reaction, solvent regeneration, solvent purification, Foaming inhibition, heat exchange	Heating process fluid	Steam
		Dilution of process additives	Desalinated water
		Preventing oxygen contact for process additives, and process fluids	Nitrogen
		Cooling process fluid	Seawater
		Recovery of entrained solvent in sweet gas stream	Steam condensate
Dehydration and mercury removal	Phase separation, heat exchange, adsorption, compression, hydrate inhibition, filtration	Heating process fluid	Steam
		Preventing oxygen contact for process additives, and process fluids	Nitrogen
Ethane recovery	Phase separation, distillation, heat exchange, hydrate inhibition, Isenthalpic and isentropic expansion	Heating process fluid	Steam
Natural gas liquids (NGL) extraction	Phase separation, Distillation, Heat exchange	Heating process fluid	Steam
		Cooling process fluid	Power
Propane sweetening and dehydration	Heat exchange, reaction, phase separation, phase dispersion, filtration, adsorption	Heating process fluid	Steam
		Dilution of process additives	Desalinated water
		Preventing oxygen contact for process additives	Nitrogen
		Cooling process fluid	Power

(Continued)

Table 5.1 (Continued)

Units[a]	Processes in operating units	Needs of processes for different utilities[b]	Utility[c,d,e,f]
Caustic soda regeneration	Heat exchange, reaction, phase separation, phase dispersion, filtration	Heating process fluid	Steam
		Dilution of process additives	Desalinated water
		Preventing oxygen contact for process additives	Nitrogen
		Oxidation	Air
Sulphur recovery	Heat exchange, reaction, phase separation, filtration, gas/liquid absorption with reaction, solvent regeneration, solvent purification, foaming inhibition	Heating process fluid	Steam
		Dilution of process additives	Desalinated water
		Preventing oxygen contact for process additives, and process fluids	Nitrogen
		Cooling process fluid	Seawater
		Recovery of entrained solvent in sweet gas stream	Steam condensate
		Enhancing the reaction rate	Steam
			Fuel gas
Solidification and storage of sulphur	Heat exchange, phase separation, solidification	Heating process fluid	Steam
		Dilution of process additives	Desalinated water
		Preventing oxygen contact for process additives	Nitrogen
Condensate stabilization	Phase separation, filtration, distillation, compression, heat exchange, corrosion inhibition, desalting, degassing	De-salting	Desalinated water
		Maintaining pressure	Fuel gas
		Cooling of compressor's jacket	Service water
Glycol regeneration and reclaiming	Phase separation, filtration, solvent regeneration, solvent purification, foaming inhibition, corrosion inhibition, heat exchange	Heating process fluid	Steam
		Dilution of process additives	Desalinated water
		Diluent of effluent	Seawater
		Heating process fluid	Steam

Table 5.1 (Continued)

Units[a]	Processes in operating units	Needs of processes for different utilities[b]	Utility[c,d,e,f]
Sour water stripping	Phase separation, distillation, heat exchange	Dilution of process additives	Desalinated water
Flares and burn pit	Phase separation, combustion, reaction	Auxiliary fuel	Fuel gas
		Reaction	Steam
		Sweep gas	Fuel gas

[a] Units "Butane sweetening and dehydration" and "Ethane sweetening and dehydration" were removed because of similarity.
[b] These are all for continuous use of utilities in normal process operations. For example, water for hydrotest of any kind, nitrogen for pre-commissioning and other purges, and air for line blowing are not considered.
[c] The power required for all electric motors, including pumps, air fans, compressors, blowers, mixers and lighting, is not listed.
[d] The instrument air for the control valve diaphragm is not listed.
[e] The nitrogen used in steam collection systems is not listed.
[f] The firewater distributed throughout the site is not listed.

as methane, ethane, propane, butane, condensate, and sulphur are obtained from hydrocarbon processes that require utilities to continue. Table 5.2 has represented the different types of utilities along with their production process, and main and side applications.

5.4 Process Additives (Chemicals)

As mentioned earlier, this chapter's focus is almost entirely on process additives. In only one case a bed, activated carbon, has been studied. A process additive is a chemical that is continuously or intermittently fed into a process stream. The dose of this substance is a function of the physical and chemical characteristics of the process stream. At the beginning of selecting and/or consuming a process additive, the most basic question is to ask ourselves: what do we want to achieve? The answer to this question is, in fact, to clear the essential for quality control plan. This is further explained in the rest of this chapter, implicitly (Figures 5.10 and 5.20). Process additives in a gas plant have a variety of uses. Some are CIs and others are process stream conditioners. There are almost as many process additives as there are variations in consumption on the utility processing site. Table 5.3

Table 5.2 Utilities in a typical gas plant, their production processes, and applications (main and side).[a]

Utility	Production process	Main application(s)	Side application(s)
Seawater	Seawater intake and coarse filtration	Cooling media, feed of desalination unit	Quench/cooling media
Desalinated water (DSW)	Multi effect distillation – thermal vapor compression	Main feed to other processes to produce DW, UW, and DMW	Feed water to de-salter in condensate stabilization unit
Drinking water (DW)	Remineralization of desalinated water	Potable water	Emergency showers/eye washes, laboratory
Utility water (UW)	Remineralization of desalinated water	Maintenance application	Intermittently used for sampling coolers
Demineralized water (DMW)	Ion exchange of desalinated water	Make up (MU) water to steam generation unit	Solution preparation, turbine blade wash
Steam	Water tube boilers and demineralized water, MU, as feed	Heating media	Temperature control, flare smokeless, reaction enhancing, maintenance
Cold steam condensate (CSC)	Recovered and cooled steam condensate by different types of coolers	Water washing of some gas streams	Solution preparation
Steam condensate (SC)	Recovered and cooled steam condensate to slightly more than its saturation point	Reusing in steam generation cycle	No application
Fuel gas (FG)	A side stream from exported methane (see Figure 5.2)	Fuel of boilers, furnaces, and gas generators	Sweep gas in flare system, assist gas, vessel pressurizing
Air	Centrifugal and air-cooled compressor	Instrumentation, oxidation processes	Maintenance, laboratory
Nitrogen	Air distillation (cryogenic process)	Inert gas for N_2 blanketing and vessel inbreathing/out breathing	Maintenance, laboratory

[a] Figure 5.3 can help to better understand the arrangement of water production units to some extent.
Source: Based on American Water Work Association (AWWA), 2011. Desalination of Seawater, AWWA Manual M61, Denver.

represents the distribution of applying process additives based on the sections; hydrocarbon processing and off-sites, and the number of feeding points in a typical gas plant.

Process additives in the utility section are so diverse and extensive that they require a dedicated table (Table 5.4). Some process additives, such as phosphate in steam systems (waste heat boilers on sulphur recovery trains), CI and biocides (on locally closed cooling systems), are fed into the systems located in hydrocarbon processing and off-site sections. However, due to the non-alteration of the nature of the work, these process additives are considered as chemicals in the utility section.

As the number of chemicals and the variety of additives and their applications increases, a processing site needs to establish an effective model for managing them around related activities. These activities, which are defined as components of an effective work frame, are discussed in the next sections of this chapter.

Tables 5.3 and 5.4 provide the reader with important information about the range and variety of process additives (chemicals) in all three sections of a typical gas plant. The information is as follows:

1) Type of chemicals: traditional chemicals have been written in as they appear in real plants, but the name of chemicals that are licensed or produced by a particular company have been avoided. There are many process additives on the market that any process designer (or field process engineer) can correspond with their manufacturers or suppliers when performing plant design steps; based on their previous design experiences (or change in process fluid conditions).
2) The number of feeding points for each chemical plant-wide: usually utilities and hydrocarbon processing sections are designed and installed in a train-based manner. This is to make the operation of each of these sections more flexible and to increase the reliability of the production. The number of points given indicates the number of feeding points from each process at different sections of a typical gas plant.
3) Chemical feed points: the point at which each chemical is fed into a process stream in an existing unit is given in both tables. These points have a significant number and represent how an individual chemical can be used in different points and applications.
4) Application of each chemical: this column represents exactly what purpose each chemical is feeding into a point. Some of these chemicals have one main application and one or more side applications. In these two tables, the author has tried to focus only on the main application, because the characteristics of process additives from one manufacturer to another can be very different, and consequently, this can confuse the reader. For example, a reverse demulsifier

Table 5.3 Details about process additives in hydrocarbon processing and off-sites sections in a typical gas plant.

Section	Process additives	No. of feed points	Point(s) of application	Application	Feeding regime
Hydrocarbon processing	Anti-foam	8	Gas sweetening	Prevent foaming	Continuous/ shock dosing
			Acid gas enrichment system prior to Claus Unit (sulphur recovery)		
			Ethane sweetening		
	Filter aid	6	Gas sweetening	Solvent purification	Intermittent
			Ethane sweetening		Intermittent
	Methanol	6	Reception	Breaking formed hydrate	Intermittent
			Dehydration/NGL extraction		Intermittent
	Glycol solution (glycol, water, and amine)	16	Reception facilities	Corrosion inhibition and hydrate inhibition	Continuous/ intermittent
	CI	10	Condensate stabilization	Corrosion inhibition	Continuous
	Demulsifier	2	Condensate stabilization	Improve de-salter performance	Continuous
	Catalyst	2	Caustic soda regeneration	Enhance oxidation reactions	Continuous
	Catalyst	2	Sulphur recovery	Degassing liquid sulphur product	Continuous

Off-sites	Anti-foam	6	Glycol regeneration	Prevent foaming	Intermittent
	Inorganic oxygen scavenger	1	Glycol reclaiming package	Reclaiming	Intermittent
	Sodium carbonate	1			
	Sodium chloride	1			
	CI	6	Glycol regeneration	Corrosion inhibition	Continuous
	Reverse demulsifier	2	Sour water stripper	Phase separation enhancement	Intermittent
	Urea	1	Biological Treatment	Micro-nutrient	Continuous
	Diammonium Phosphate (DAP)	1	Biological Treatment	Micro-nutrient	Continuous
	Caustic soda	4	Sour water stripper Neutralization package	Phase separation enhancement and pH adjustment	Intermittent
	Reverse demulsifier	3	API Separator	Demulsifying oils	Intermittent
	Flocculant	3	IGF (Induced Gas Flotator)	Making flocs	Intermittent
	Sulphuric acid	2	Neutralization package	Adjust pH	Intermittent
	Sodium hypochlorite	1	Treated Effluent	Disinfection	Continuous

Table 5.4 Details about process additives in utility section of a typical gas plant.

Process additives	No. of Feed points	Point(s) of application	Application	Feeding regime
Neutralizing amine	8	Deaerators output	Neutralizing carbon dioxide and pH adjustment	Continuous
		High pressure steam letdown stations		
		Low pressure steam letdown stations		
Oxygen scavenger	8	Deaerators output	Removing oxygen and metal passivation	Continuous
		High pressure steam letdown stations		
		Low pressure steam letdown stations		
Phosphate	16	Water tube boilers	Boiler internal treatment	Continuous
		Waste heat boilers		
Sodium hypochlorite (seawater electro-dialysis)	9	Intake chambers	Disinfection, filter aid	Continuous
		Pump basins		
De-chlorination agent	5	MED–TVC packages (multiple-effect distillation/thermos-vapor compression)	De-chlorination	Continuous
		Ion exchange (IOX)		
Sodium hypochlorite (provided chemical)	2	DSW storage	Disinfection	Continuous
		DW distribution		
CI	10	Firewater MU	Corrosion inhibition	Continuous
		Firewater filling		Intermittent
		Closed cooling water system		Intermittent
		Compressor jacket		Intermittent
Biocide	10	Firewater MU	Corrosion inhibition	Continuous
		Firewater filling		Intermittent
		Closed cooling water system		Intermittent
		Compressor jacket		Intermittent

Table 5.4 (Continued)

Process additives	No. of Feed points	Point(s) of application	Application	Feeding regime
Anti-scale	3	MED–TVC packages	Prevent scaling and foaming	Continuous
Anti-foam	3			
Sodium bicarbonate	1	DW and UW unit	Remineralization of DW	Continuous
			Remineralization of UW	
Calcium chloride	1	DW and UW unit	Remineralization of DW	Continuous
			Remineralization of UW	
Caustic soda	2	DMW unit	IOX regeneration	Intermittent
			Neutralization	
Sulphuric acid	2	DMW unit	IOX regeneration	Intermittent
			Neutralization	

can be formulated to regulate the pH, in addition to its performance for separation of oil layer from water, and it has a slight oxidizing property to affect the readily degradable chemical oxygen demand (COD) of oily contaminated aqueous effluents [5–7]. Therefore, adding these properties of chemicals that are formulated for specific purposes may not have a good result in expressing the objectives of this chapter.

5) Feeding regime: one of the important factors that distinguishes the attention of a particular chemical and justifies the establishment of an effective management model is the feeding regime for which the design documents of a gas plant are intended. If a chemical is fed into the system continuously, and has a direct effect on product quality and/or corrosion; then the Process Engineering Department has to make a serious distinction between this type of process additive and the chemical that is used rarely, and only at a certain point in specific period of time (e.g. to maintain the quantity or quality of products). The continuation of this chapter represents that some of these types of chemicals (which are fed only in certain periods of time and have no purpose other than maintaining quality and quantity of some products), have a serious impact on the main operating parameters of a gas plant, of which corrosion is one of the most important (see Figure 5.11).

Figure 5.3 represents a schematic of the feeding of various chemicals into the utility units on a typical gas plant. The number of feeding points is given next to the name of each chemical. Traditionally, the gas plant's aqueous waste collection and treatment unit is a part of the off-sites. However, due to considerations regarding (i) recycling of treated wastewater in the plant and returning it to the hydrocarbon or utility processing section, and (ii) plant layout conditions for conveying gas plant effluents using gravity, which forcibly places the unit along with other utility units (such as seawater desalination and steam generation) in the design of some new gas plants, the gas plant's wastewater treatment unit or plant (WWTP) is considered as part of the utility section. Hence, the arrangement in Figure 5.3 as well as the data in Figure 5.4 are presented based on this fact.

Figure 5.3 also has implicitly represented a comprehensive plan of the arrangement of production units of different types of water, and also steam, their distribution, and then the treatment of aqueous waste streams (effluents).

Note that the type of chemical, the manufacturer/supplier, and the feeding regime into a specific section of the gas plant does not necessarily remain unchanged. The passage of time and operational experiences changes in the products of companies that produce chemicals, the composition of gas plant feedstock, quality control conditions, and plant status in terms of equipment availability are all important and influential factors that tell the process engineering department whether the consumption of a process additive should continue or not, whether the feeding regime should be changed, and many more questions that we need to be prepared to face.

An examination of the contents of Tables 5.3 and 5.4 shows that there are significant differences between process additives in the three sections of the gas plant. These differences are below:

- The number of chemicals used in each section;
- The number of feeding points to which chemicals are fed;
- The effect of feeding these chemicals on the onset and acceleration of corrosion mechanisms;
- Feeding regime of chemicals;
- Having the right to own technical knowledge.

The contents of Tables 5.3 and 5.4 represent an important distinction between the three sections of a gas plant. These distinctions prove how to make the correct decision when it comes to implement a work frame for EMPA and what the priority of implementing this developed model is? These differences are well illustrated in Figure 5.4.

A quick look at this chart reveals that addressing the utility section to implement the EMPA model will be so general that it can cover two other sections of the gas plant.

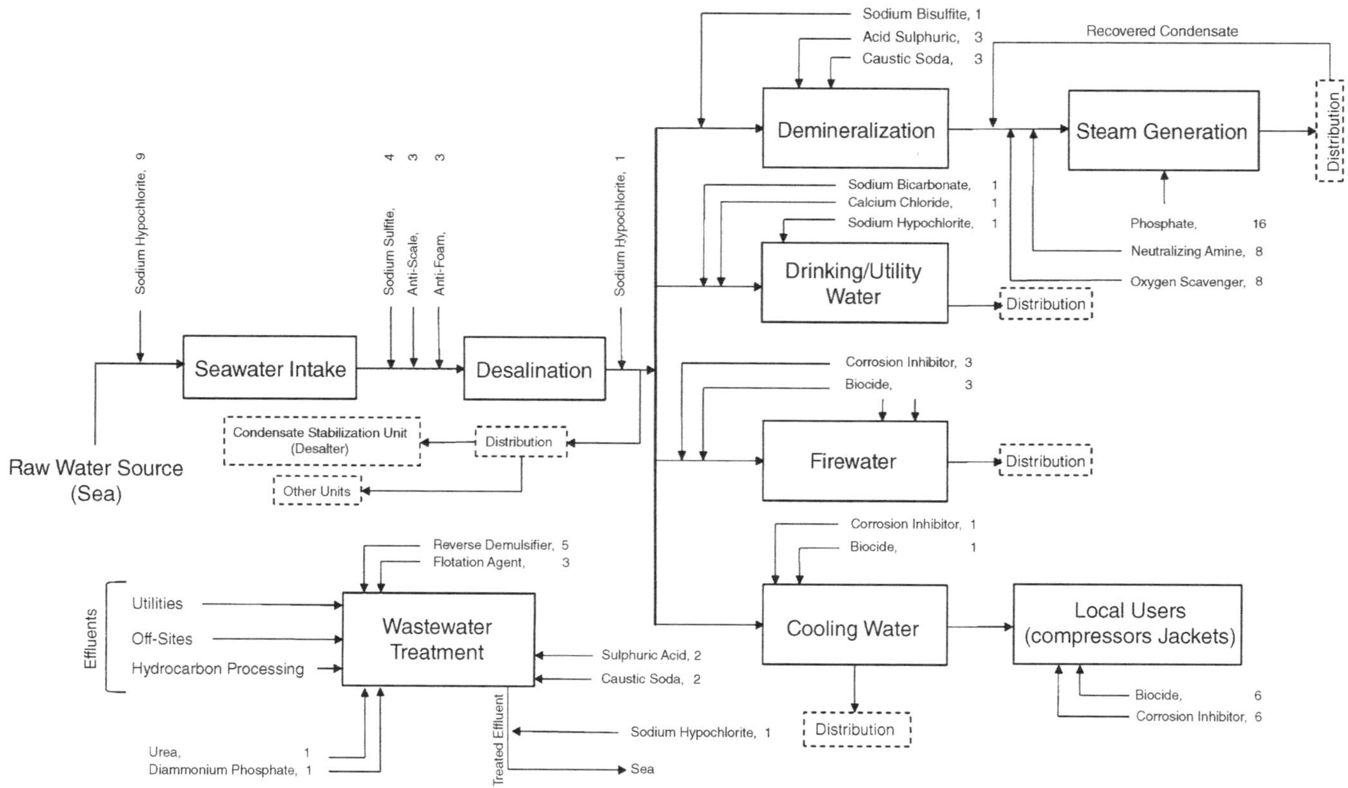

Figure 5.3 A schematic for the feeding of various chemicals into the utility units on a typical gas plant.

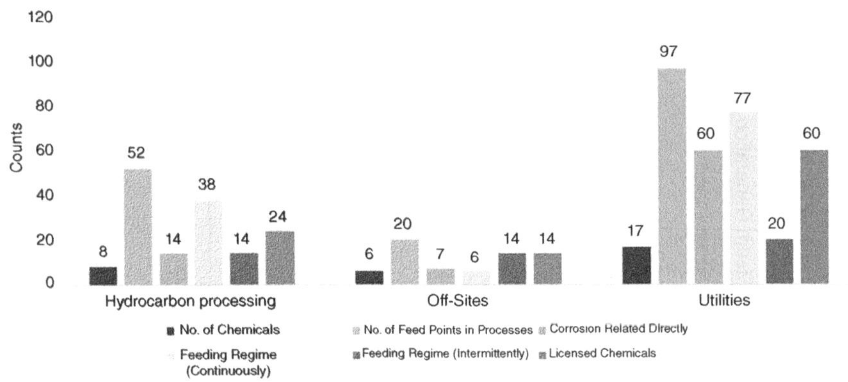

Figure 5.4 Illustrating all process additive related details at different sections in a typical gas plant.

Any reader may analyze this chart in such a way that the priority of dealing with the chemicals in the utility section to establish an EMPA model should be based on an economic analysis and the risk considered for all sections. This is referring to the limitation that it is virtually impossible to care for a gas train if even one gas train is out of service due to corrosion, and if the steam distribution system is damaged due to corrosion. By reducing the production capacity, other gas trains can be applied for compensating the total production, but in case of failure of the steam distribution system, the whole production is practically shut downed.

In the selection of chemicals whose non-feeding has a direct effect on corrosion, only chemicals which directly influence the prevention of corrosion mechanisms have been considered. It is interesting to note that, for example, the chemical demulsifier for desalter may have a final consequence on the corrosion of the distillation column and other parts of the gas condensate stabilization system; but it doesn't directly influence the corrosion of the equipment like a CI, and ironically, this author selected it. This can tell the process engineering department and also the operation team that we are often surprised when the name of the chemical does not actually CI, and there are other chemicals whose mismanagement, from selection to consumption (and, after that, reporting) on the processing site, will have the same impact on plant corrosion, and perhaps with a greater consequence. Figure 5.1 represents that three of the seven industrial cases have this feature.

In counting the number of chemicals that are licensed to produce, it was possible to estimate the number of these chemicals, in which case the hydrocarbon processing, off-site, and utility sections were 5, 5, and 11, respectively. The reason for paying attention to the production license for these chemicals was to show the reader how much the use of such process additives can add to the complexity of their effective management on a gas plant. Therefore, in Figure 5.4, the number of points to which these licensed chemicals are fed is given.

Some process additives have dual applications; for example, glycol-water solution and alkanolamine solution, which is fed at 16 points in reception facilities, is used in some places as hydrate inhibitor and in others as CI. Therefore, in the evaluation of the desired service for each chemical, such cases were also considered.

5.5 Effective Management of Process Additives (EMPA)

Effective management of process additives (chemicals) (EMPA) means the elimination of adverse impacts on the entire operation of an operating (processing) plant in terms of production cost, quality control, corrosion, energy, and environment. If we want to express all the impacts of these chemicals on the performance of a gas plant, then we will encounter the following:

5.5.1 Production Costs

It is clear how much the feeding of chemicals in a gas plant will cost the entire production operation. These costs have a significant impact on the finished cost of different products. Purchasing a chemical can take a lot of time from a process engineering department to a Procurement Department. Carrying out technical selection steps such as advanced and complementary laboratory analysis, pilot preparation, field tests, and environmental permits for new chemicals, as well as procedures for technical–commercial committees to rank different manufacturers/suppliers and then holding tenders are all time-consuming and costly, and affects the purchase of these materials in a gas plant, as well as influencing the production costs.

A closer look at the cost of chemicals can lead us to warehousing, warehouse-to-site delivery, quality control at the time of delivery, after-sales services, and taxes on the use of certain types of chemicals. On the other hand, due to the role of proper feeding of these chemicals on corrosion and depreciation of equipment, the economy of production is once again affected by these chemicals, due to the increasing depreciation rate and failures that lead to undesired trips of equipment and other facilities installed or partially/totally shutdowns on a gas plant.

5.5.2 Quality Control

Laboratory analysis of process fluids to determine the composition of the percentage of components, different inhibitors, free residual chlorine, phosphate, and oxygen scavenger, and also to determine the adsorption status of an activated carbon bed based on the characteristics of its inlet and outlet fluid or direct measurement of some physical and chemical parameters, are among the things that, while affecting production costs, if they are insufficient, they can lead to corrosion. Chemical quality control at the beginning of entering a warehouse or processing site can tell us whether a chemical is of the right quality to use or not.

Selection and consumption of a specific chemical can be based on the cost of quality control, such as sampling according to a predetermined sampling and analyzing regime on the processing site, quality control of chemicals when delivered from truck or train to warehouse, and from warehouse to processing site, use of complcx analytical methods for determining parameters, using expensive online analyzers as well as with their difficult after-sales services for calibration to control the quality of process streams momentary, spending a lot of time by plant specialists, and forgetting more important tasks, as well as many other influential parameters.

Accurate determination of chemical and physical parameters of solvents such as alkanolamines, glycols, caustic soda (bases), and sulphuric acid (acids) are examples of this type of quality control. Some of these analyses seem insignificant at first; but it is enough to know that, for example, the analysis of insoluble matter (IM) from salts like calcium chloride, sodium bisulfite, and sodium bicarbonate that are purchased solidly and stored in the warehouse can prevent the precipitation of IM in the bottom of the dosing tank and unwanted stoppage due to pump failure or sometimes suction loss (because IM, which forms in the bottom of dosing tank, accumulates in the plunger or diaphragm section of the dosing pump, after passing suction strainer). We know how important it is to stop feeding pumps (here, hardness ions) that have salts added to the desalinated seawater or other process streams. Avoid conditions in which feeding pump stops are eliminated; this makes it easier and more stable to regulate the important parameters such as the Langelier stability index (LSI), calcium carbonate precipitation potential (CCPP), and aggressiveness index (AI), based on what are stated in scientific references [8–11]. This is one of the simplest applications of quality control in the application of chemicals; no matter at what stage it takes place and what degree of effects it has on corrosion and production costs.

5.5.3 Corrosion

Sometimes the impact of EMPA at a processing site is mistakenly limited to CIs and biocides in cooling water, firewater, etc. Something else to think about is the deaeration of the boiler feed water (BFW) fed by an oxygen scavenger chemical (external or pretreatment), or the phosphate inside the boiler drums (internal treatment), as these are important parameters that every utility process engineer or any other person responsible for quality should consider for preventing non-uniform or localized corrosion, such as pitting and gouging [12]. However, there are small things that can be just as destructive if not addressed as corrosion-related issues. For example, if the demulsifier of a desalter in a gas condensate stabilization unit has improper control in terms of feeding rate and its performance; it should then be expected that the free water in the distillation column feed will cause corrosion due to water condensation in the upper parts of the distillation column as well as its condenser. Therefore, in dealing with corrosion issues, they should not be categorized into two parts, significant and insignificant; the author of this chapter has seen this mistake repeated on many processing sites.

There are many cases where the proper chemicals are purchased for a processing site and laboratory analysis has misled the unit operator or process engineer to prepare for the corrosion of the equipment itself. How to use chemicals and periodically calibrate equipment related to feeding measurement can also greatly prevent corrosion. Existence of a significant cycle between the results of analyses reported from the quality control parameters, as well as the assurance of the feed that is ongoing, EMPA, can solve a large part of the difficulties in this field. An incorrect evaluation in calculating the retention time of seawater in the transferring pipe to the operating fence of a gas plant or other treatment facilities (utilities) may cause the shock dose coordination time for dechlorination to be incorrectly evaluated, and allow too much free residual chlorine to enter the downstream equipment (stainless steel and copper alloys), which in the presence of seawater can cause severe damage to them [13].

5.5.4 Energy

Energy consumption can be affected by the management of the consumption of a particular chemical at a processing site. On a gas plant, chemicals generally have little effect on energy, but perhaps their impact on the overall consumption in terms of added power for feeding pumps may be considered. For example, the choice of low dilution factor may influence the energy consumption of a processing site, as the feeding pumps in the feeding packages require the use of high speeds and strokes. In this case, if there are more than one hundred feeding pumps in a gas plant, then the impact of these feeding chemicals on increasing numerical power consumption will be significant.

The main effects of EMPA in a gas plant are focused on creating proper management conditions and quality control of various products at the processing site. There are various examples in this regard; (i) forcing to return the deficiently treated oily water effluent to the beginning of the treating cycle due to improper performance of demulsifying and flotation chemicals, (ii) forcing to dump the returned contaminant steam condensate due to improper performance of the deoiling facilities or condensate polishing system installed in the return route of condensed steam (suspect branch), (iii) forcing the return of stripped sour water with high oil content and suspended particles due to feeding and improper selection of demulsifier and pH regulator chemicals can all be clear examples of this. It is obvious that each of these items can increase energy consumption and quality control actions in a processing site.

In hydrocarbon processing units, additional energy consumption may also have a different analysis. For example, when due to poor quality control in the desulphurization section of propane and butane, a gas plant has to send the off-spec products to flare, energy consumption is affected in the following ways: (i) burning of the off-spec products, which in addition to imposing a reduction in production to the gas plant, causes some excess fuel gas to be used for better burning in the flare; or (ii) forcing to use moderate to high pressure steam in flare to ensure smokeless flaring that carries considerable energy (operation of flare without smoke).

Obviously, all energy consumption is inextricably linked to environmental issues, because all of these energy components in a gas plant will eventually lead to greenhouse gas (GHG) emissions. Moreover, with a more detailed approach, we can add the environmental impacts of energy generating units to this increase in energy consumption. For example, if a utility boiler needs to generate more than 40 ton/h of steam to perform smokeless flare operation, in practice, the load applied to the environment from the increase in the continuous blowdown with a temperature of 300 °C, the accompanying chemicals, more fuel gas burning in boiler, and more forced fan duty to handle the need of combusted air into a boiler furnace will be added to increased energy consumption in another unit (i.e. steam generation unit).

5.5.5 Environment

The environment is another issue that can be influenced precisely by working with chemicals in a gas plant. Adherence to an EMPA model is certainly guaranteed to prevent environmental impacts due to consumption of chemicals. These impacts are both directly caused by the entry of chemicals into the environment, and indirectly due to the quality control that is done by the chemicals. The importance of the interactions between the use of chemicals and the environment forces us to address them in industrial detail, but the breadth of the chemicals and the details of each process will complicate matters beyond the scope of this chapter. There are several examples of the impact of chemical consumption on the environment [14–16].

If environmental regulations are not followed properly, a gas plant will be fined by local environmental agencies and must be held accountable. Sometimes these fines lead to a reduction of products and will impose production costs in both directions. It is important for a gas plant to always work in the designed capacity, because in design mode the cost per tonnage of production will be at an acceptable level, and planning for sales and determining the unit price of each product will be done correctly. Some costs in a gas plant are independent of the amount of production, and when the processing site reduces its production the cost per products will increase and cause a monetary loss.

The above issues provide some examples of impacts of chemical consumption on the environment. Almost every action taken in connection with the use of chemicals has an environmental aspect [9, 13, 16–18], however, in order to search for environmental impacts resulting from their use in a gas plant, one must look for process deviations due to inefficient management of chemicals. A good example is given in the energy section, and there are a few more below.

- If a flotation agent chemical at the input of an IGF is not properly managed the effluent from this equipment may enter the environment without sufficient treatment, in the interval between sampling and analysis of dissolved oils (or emulsified oils) and total suspended solids (TSS).
- If the consumption of the neutralizing amine in the feed water entering a boiler and the steam distribution and steam condensate system connected to it is more than necessary, then the composition of continuous and intermittent boiler water (BW) (blowdown), which contains non-volatile ingredients of neutralizing amine (that is a formulated chemical), has a higher concentration and more organic load to the WWTP or will enter the environment directly (for some WWTP arrangements that boiler blowdown is routed to the final point, outfall, and the designer may not have considered that).
- If a film-forming inhibitor has a high concentration of emulsifier materials such as some organic acids in its formulation [18–21], and its consumption is managed in such a way that the residual concentration downstream is not properly monitored, then it can be expected that aqueous and hydrocarbon phase separation do not work properly on downstream, and many organic loads due to the presence of hydrocarbons enter the WWTP, and after their elimination (which may be not enough) the treated wastewater routes the environment.

It is important to keep in mind that the WWTP cannot receive and treat any amount of organic load. On the other hand, there are arrangements in the design to be able to receive and treat pollution loads beyond the design. However, it should be noted that the operation of the WWTP is very sensitive and any conditional changes in it are made with a low factor of certainty in terms of achieving the correct result. Another point in dealing with a WWTP when the incoming effluents deviate from the quality predicted in the design, is that any action to meet the new conditions increases energy consumption and subsequent GHG emission (for example recycling and enhanced biological operations). Therefore, any insufficient quality control in process streams in different units of the gas plant (process effluents), and also for streams inside the WWTP, will lead the WWTP to change its arrangement (temporarily) to compensate for deviations in the quality of process effluents, and impose more pollution on the environment.

Whereas this chapter must deal in detail with the impacts of ineffective chemical management on corrosion, some industrial cases are devoted to the environmental impacts due to corrosion following an improper application of chemicals. The industrial case of contaminated steam condensate in Industrial Case Two (below) can be considered as an example of the impact of improper management of chemicals on corrosion and then environmental impacts [17]. In Section 5.6, the impacts of chemical consumption on the environment are specifically configured.

5.5.6 Process Issues

5.5.6.1 Production Reduction

There are a number of procedures that can reduce production, including the ineffective management of chemicals. Reduction of steam production due to stopping a boiler that had its tubes damaged and improper or incorrect feeding of chemicals used in the pretreatment of BFW to it eventually leads to insufficient supply of steam to processing site, and will then cause an unwanted shutdown of a gas condensate stabilization train, a gas purification and sweetening train, and an NGL extraction train, followed by an LPG purification train. Figure 5.5 represents a better illustration of this path event.

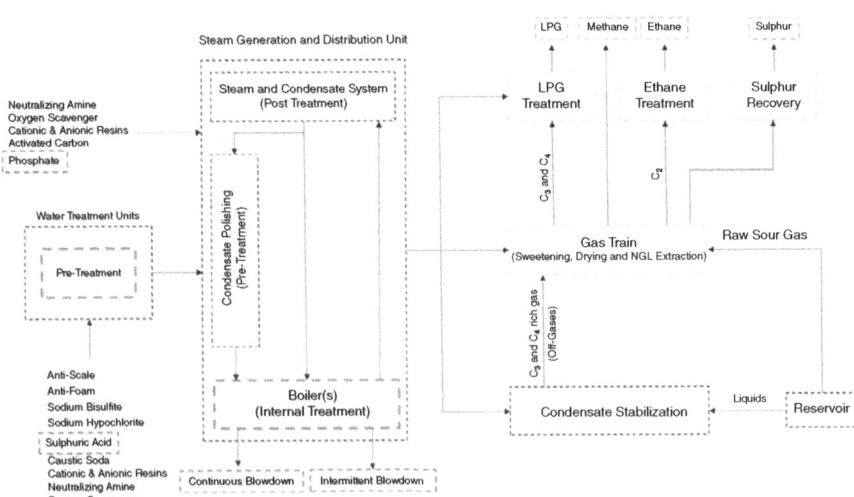

Figure 5.5 Illustration of path event due to lack of generated steam influenced by a chemical, and other consequences on the entire operation of a typical gas plant.

In this case, improper quality control of sulphuric acid (i.e. high concentration of iron), which is applied in the regeneration of strong cation resin, has caused the BFW to be too hard and over time, creates a hole due to sediment deposition and waterside overheating in the tube bank of one of the boilers. Readers can study these references for more details; [12, 18, 22]. From a total of six installed boilers, four were operating at their maximum capacity, and another boiler was undergoing maintenance with the minimum preparation time for its service being estimated at three days. Therefore, steam production was reduced by the capacity of one boiler (250 ton/h) and consequently, one train of gas sweetening and other downstream processing units were inadvertently taken out of service. In Figure 5.5, the chemical that led to this event, the unit that sent its off-spec product (demineralization unit) to the deaerators prior to boilers, and the chemical whose improper control in new conditions or lack of understanding of the need for new adjustment of continuous blowdown rate are shown, they are pre-treatment, Sulphuric acid, phosphate, boilers and two blowdowns. The processing units that were affected by these conditions are shown, they are LPG treatment, Ethane recovery, Sulphur recovery, and gas train. The gas plant whose production had been reduced are shown, they are LPG, Methane, Ethane, and Sulphur. This event eventually led to a reduction in feed to the gas plant from the gas extraction platform, which is shown in gray.

5.5.6.2 Off-spec Products

The method of working with off-spec products in a gas plant is given in Figure 5.6. While every company has a different approach to its off-spec products, the content of this figure is based on the basic design of a typical gas plant and may not be generalizable to the entire natural gas industry worldwide. However, some of these methods of dealing with off-spec products are inevitable, and the presence of normal or special design arrangements will not make a significant difference.

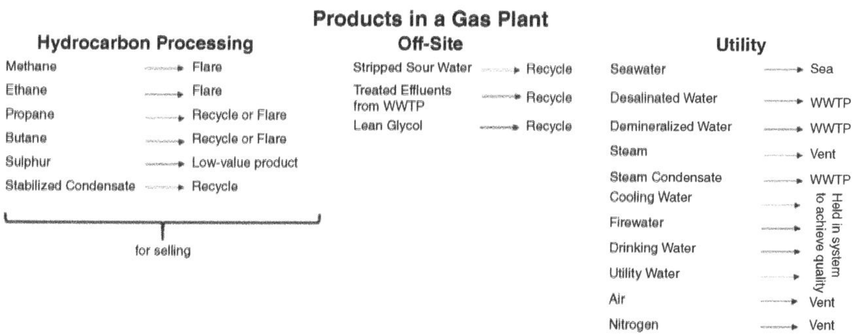

Figure 5.6 How to deal with off-spec products based on design in a typical gas plant.

5.5.6.2.1 Industrial Case One: Is Steam Condensate Deoiling and Polishing Package an Important Unit Operation?

Steam condensate is an intermediate product within the processing site and while not marketable, is valuable. Ideally, it should return to the deaerator with maximum recycling efficiency, and at the right temperature, pressure, and quality. There is a preparation for suspect condensate that may be contaminated by first routing it into a condensate purification or polisher package (CPP), and then reducing the concentration of hardness ions (by ion exchange, IOX) and non-volatile organic matter (by ACF), such as leaking oils from the heat exchanger tubes to the deaerator or steam condensate storage tanks [18, 23–25].

In this case, if ACF in the CPP is not properly selected, quality controlled, and its output stream is not monitored rigorously by reliable online analyzers or a comprehensive laboratory testing regime, according to Figure 5.7, this product should be routed to the WWTP, and after removing (i.e. reducing to regulatory level) the dissolved oils content, it should be dumped into the sea or other water bays. Here, the onset of leakage from the heat exchanger into the steam system or its condensate will occur with non-uniform corrosion in the outer surface of reboiler tubes, which is caused by process streams such as rich glycol (underdeposit corrosion) and raw sour water (acidic corrosion).

Based on the amount of steam pressure used at each processing site, which is a function of the heat load required for the boil-off rate in every fractionating column, as well as the composition of the process streams, return condensate is routed in both suspect (potentially contaminated) and clean condensate. Suspect condensate must pass through the CPP and its potential contaminants (especially non-volatile organic matter) must be removed by ACF, therefore, quality control arrangement in CPP is of special importance.

Quality control of suspect condensate, which is one of the internal and non-marketable products of the processing site, is based on two arrangements of dumping condensate to WWTP. The typical basic design has two online hydrocarbon analyzers at the CPP inlet and outlet. In practice, some industrial arrangements consider only one online analyzer sufficient. Another part of quality control is the laboratory analysis of treated condensate from CPP to measure non-volatile organic matter, while also cross-checking with an online hydrocarbon analyzer at the output stream. In addition, direct analysis of the activated carbon in the bed of each ACF (if parallel trains are installed for supporting continuous operation) and determination of the iodine number [19, 26], and other quality parameters of the activated carbon can be considered as part of strict control to prevent hydrocarbons from entering the BFW system [6, 27].

In order to maintain CPP performance and increase the flexibility of the process to achieve zero hydrocarbon entry into the BFW, CPP is usually designed as train-based. In this case, when there is a need to replace the IOX beds (mixed bed resins)

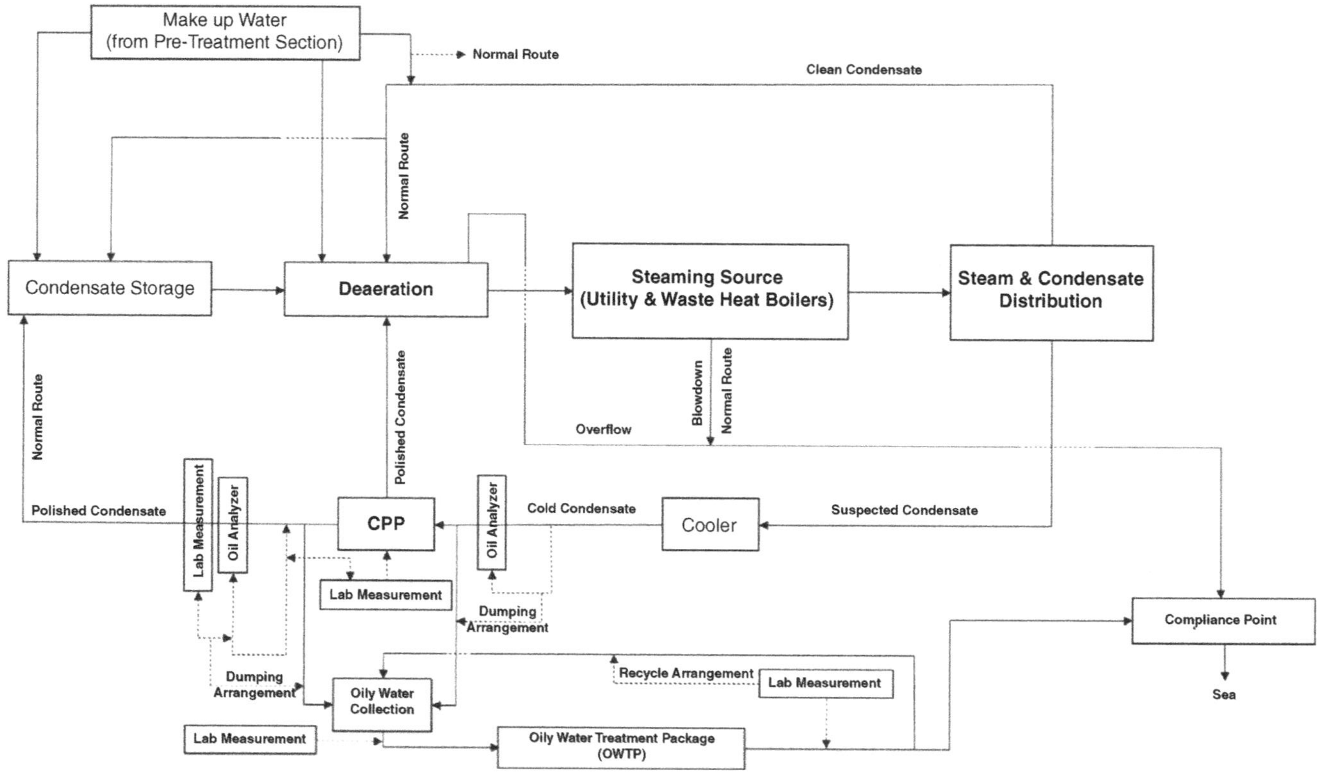

Figure 5.7 Schematic for off-spec condensate (oily polluted) and its dumping arrangement to WWTP.

or their corresponding ACF, another train (in parallel) can be put in service and prevent the CPP from unplanned stoppage, and then unintentionally dumping the suspect condensate to the WWTP or by-passing the CPP and sending it to BFW system.

If the suspect condensate goes to WWTP due to the CPP not being operational, the make up (MU) water must be used to keep the water level inside the condensate storage tanks constant. This imposes the unwanted production of water on the gas plant utilities, and in turn introduces oxygen-saturated demineralized water into the steam generation cycle. MU water enters the deaerator and becomes oxygen-free (equilibrium) by additional consuming steam (because pf lower temperature, therefore more energy consumed, and more GHG emitted), and simultaneously feeding oxygen scavenger. The deaerator is a device that has a high need for stability in operation, and the sudden receipt of MU water or a change in its influence can have a negative impact on its performance. Therefore, not controlling the quality of the suspected condensate to prevent the entry of hydrocarbons (which is done with an ACF) has an indirect impact on the potential oxygen corrosion of the BFW transfer line, boiler, and its condensate distribution system.

On the other hand, its direct impact (i.e. when CPP is by-passed) on corrosion and difficulty in boiler operation should be seriously considered. Depending on the pressure at which the boiler operates (and, of course, at what temperature), nonvolatile organic matter, mainly hydrocarbons, degrade molecularly in the steam generation cycle to produce acidity [18]. Acidic substances consume some of the alkalinity in the boiler and can reduce the pH of the BW and initiate corrosion inside the boiler (in the event of poor internal treatment) [11, 18, 19]. Decreased alkalinity and increased phosphate concentration, which are the most important signs of acid entering the steam cycle [11] in these cases, are observed. Figure 5.8 shows a set of actual trends from such a situation. In addition, the entry of hydrocarbons into the BFW may cause foaming, and in more severe cases, water droplets inside the steam drum to escape into the steam distribution system (carry over), which has marked corrosion impacts [12, 25].

5.5.6.2.2 Industrial Case Two: A Gas Heater Tube Failure that Influenced A Steam Generation and Distribution System

A gas plant was receiving its feed from the extraction platform. The feed pipeline crosses the seabed for more than 75 km and entered the fence of the processing site. Whereas the desired temperature in the basic design to prevent hydrate formation as well as control the parameters of the sweetening unit (mainly foaming) and drying may be deviated; a gas heater was designed and installed on the feed gas inlet pipeline to the processing site.

The film-forming CI was fed into the sour gas in a feed gas pipeline at three points of slug catcher, and its formula also contained an alkanolamine to regulate

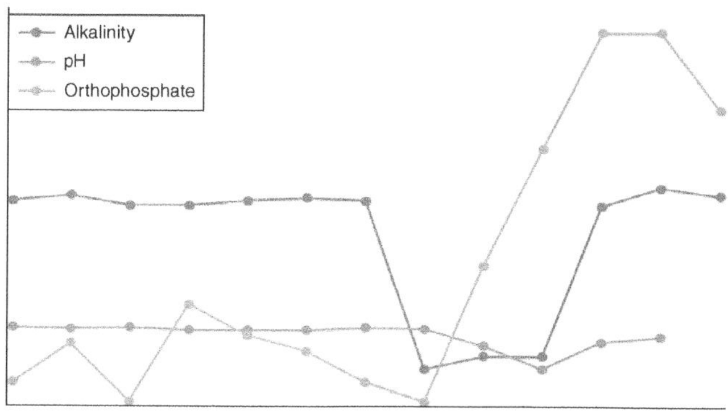

Figure 5.8 Trends of laboratory results due to entry of non-volatile organic matter into a boiler.

the pH. Contrary to the instructions, the alkanolamine was not fed at the correct percentage for three years of operation.

The heater was powered by low-pressure steam. The condensate obtained from this heat transfer entered a surge drum-condenser arrangement to ensure that it was completely condensed (slightly subcooled), and sent to the steam condensate collection and return system via a small pump station. In the basic design, the condensate output from this heat exchanger was considered as suspect condensate; therefore, it joined other suspect condensates, then entered the CPP (see Figure 5.7), and after eliminating the contaminants, entered the BFW system (storage tanks and/or deaerators).

Due to temperature fluctuations during the year, this gas heater was serviced occasionally. The heat exchanger was designed so that the sour gas of the gas plant feed passed the tubes and steam entered the shell. The sour nature of the gas and keeping it out of service (both tube side and shell side) for about six months without observing the principles of out-of-service preservation [28], as well as difficulties in the inlet steam control system to the heat exchanger, eventually led to localized corrosion, especially pitting, and also microbiologically influenced corrosion (MIC), in the upper rows of tubes, both inside and outside. Figure 5.9 represents the internal surface of a tube in a gas heater which was failed (before and after cleaning). After eliminating of corrosion products and other depositions many pits (localized attack) were observed.

This heater was put into service at the beginning of the cold season without anyone being aware of the possible situation of its leakage, and after 48 hours it presented with a complication in the chemistry control of the boilers and the condensate system. The condensate exiting the heater, which was contaminated

Figure 5.9 Internal surface of a tube in a gas heater; corrosion products and black deposition before cleaning (top), and localized attack and high density of pits after cleaning (bottom).

due to a leak, was sent to the suspect condensate collection system after cooling in the condenser and collecting in the flash drum. The designer had provided a sample point on the outlet line of the pumps' discharge header, so that if the condensate became contaminated, it could be discharged through a three-inch-diameter drain from the collecting flash drum.

Insufficient attention to quality control of condensate leaving this pump station caused contaminated condensate (hydrocarbon and sour gas polluted) to enter the collection system. The online analyzer in the route of the suspect condensate entering the CPP was not effectively serviced, and the instrumentation department was in charge of calibrating and repairing after-sales service periods and a number of spare parts. Therefore, the contaminated steam condensate entered the ACF, and after its saturation, contaminated steam condensate entered to the steam condensate storage tanks. The design had only one hydrocarbon analyzer at the CPP inlet.

The regime of laboratory analyses of CPP input and output was not updated with respect to the new condition of the analyzer, which was not in service (because of the ineffective relationship of the maintenance department with process engineering and operation departments). Therefore, for crosschecking the performance of the analyzer, an analysis is performed every three days to determine non-volatile organic matter. Given the non-calibrated status of the hydrocarbon analyzer, it can be clearly understood that it was not sufficient to know the entry point of pollutants into the system.

Simultaneously, the gas plant was preparing a steam condensate storage tank for periodical inspections. Therefore, the suction header of the deaerator feed pumps was placed on this tank to empty its contents as much as possible, and then the rest of its steam condensate (up to 35% height) was to be sent to WWTP. However, the operator, contrary to the existing instructions, had sent the contents of the tank up to 15% to the deaerators. This was while the tank was not boxed up either.

The steam generation unit had an extensive regime of laboratory analyses with different frequencies. Figure 5.10 shows the sampling points and their related analysis of the steam generation site. The table in Figure 5.10 contains only some routine analyses. According to the schedule, the laboratory was asked to perform routine sampling at 05:00; the result of the first set of daily analyses was placed on the software of laboratory information and management system, (LIMS) [29] at 11:30. Boiler phosphate was increasing and the alkalinity inside it was being consumed. The process engineer quickly referred to the sheet related to the results of BFW analysis to check its pH and the results showed that the pH of the BFW did not change; however, the specific conductivity of BFW had tripled.

Because BFW was obtained from the sum of clean and suspect steam condensates, the return condensate analysis sheets were examined, and it was found that the conductivity of the suspect steam condensate was more than five times normal and the pH of the two units had dropped. It was then promptly ordered that the suspect return steam condensate be cut off from the main return header and the MU water, which is highly purified water (demineralized to level of $<0.1\ \mu S/cm$), increased from $15\ m^3/h$ to $75\ m^3/h$.

However, the process engineer (as mentioned in the previous paragraph) did not see a drop in the pH of BFW, because there was a neutralizing amine feeding point after deaeration and the BFW sample point was installed about 25 m after. Therefore, mixing the suspect steam condensate with the clean steam condensate, followed by the feeding of the neutralizing amine, made it impossible to observe the pH reduction for the night shift utility supervisor. This was because the analysis of suspect and clean steam condensate was performed once a day at 05:00 and there was not enough time to react due to the presentation of the related test result at 11:30.

On the other hand, the operator introduced the evacuated and boxed up steam condensate storage tank as out of service and did not allow the sample man to take a grab for analyzing. Therefore, it was not clear whether the tank had contaminants at that height of steam condensate (33%) at 05:00, or whether it was listed in the operating logs of the box-up clock. Of course, steam condensate storage tanks always have the potential for contamination, but sampling could determine whether the high concentration of hydrocarbons entering the boilers was due to the suction of the oil layer created above the steam condensate surface or only to the leak from the heater.

The list of equipment in service in the last 24 hours was quickly reviewed and the gas heater was listed. The initial conjecture was that hydrocarbons and acid gases

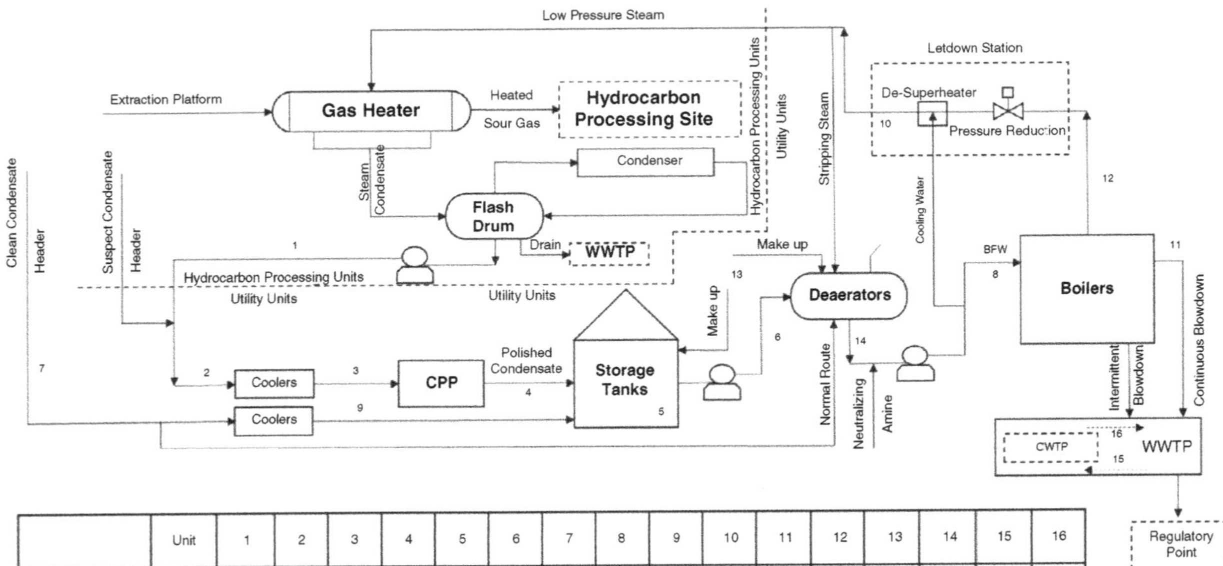

	Unit	1	2	3	4	5	6	7	8	9	10	11	12	13	14	15	16
pH	–	T[1]	T	–	T	T	T	T	T	–	–	T	T	T	T	T	–
Specific Conductivity	µS/cm	T	T	T	T	T	T	T	T	T	T	T	T	T	T	–	–
Non-volatile TOC	mg/L	T	T	T	T	–	–	–	–	–	–	–	–	–	–	–	–
Chloride	mg/L	T	T	–	–	T	–	–	–	–	–	T	–	T	T	–	–
Orthophosphate & Alkalinity	mg/L	–	–	–	–	–	–	–	–	–	–	T	–	–	–	–	–
Oil & Grease	mg/L	–	–	–	–	–	–	–	–	–	–	–	–	–	–	T	T
Dissolved Oxygen	µg/L	–	–	–	–	–	–	–	T	–	–	–	–	T	T	–	–
Cationic Conductivity	µS/cm	–	–	–	–	–	–	–	–	–	T	–	T	T	–	–	–
Total Iron	mg/L	T	T	–	T	–	–	T	T	T	–	–	–	T	T	–	–

Figure 5.10 Event schematic, sample points, and related analyzes.

leaked into the condensate cycle. Based on experience, with an immediate instruction, the continuous blowdown increased steadily, and the rate of neutralizing amine feeding increased slightly. Immediately, the suction of the deaerator pumping station from the box-up steam condensate storage tank (with a height of 28%) was stopped and placed on the other two tanks with a height of 90%. The result was that oil analyzers transmitted in the CPP (on-site) were checked and it was determined that the analyzer has a magenta message in the central control room. This message usually appears when the value being analyzed is outside the range defined for the analyzer, but because of the temporary disconnection of the analyzer from the distributed control system (DCS), no alarm was sent to the control room. In log sheets, where the field operators filled in all the instrumentation readings every two hours, there was no result of this analyzer from the local site display.

The next step was to decommission the gas heater in order to prepare it for inspection and leak detection, but that took more than 12 hours. Therefore, the suspect steam condensate analysis sheet was examined in the reception unit (gas heater location) and it was found that the last laboratory analysis was performed the day before the gas heater was put into service. Immediately, a special conductivity analysis was performed from this point by a portable device, and after observing the high conductivity and the pungent odor of the condensate (due to dissolved sour contents), the outlet of the steam condensate flash drum was opened to close the drain system, and the transfer pumps were taken out of service. After that, sampling of the main header of suspect steam condensate was performed at a frequency of 30 minutes, and after 6 hours it was determined that the contamination had reached zero level. Obviously, total iron experimental results were not similar to this trend.

After 48 hours of troubleshooting, the boiler chemistry and MU water rate returned to normal and the gas heater was put into service by plugging the failed tubes. Subsequent studies showed that the second and third trains of CPP had two problems, respectively; ACF saturation, and defects in the control valve sequence system were not available at the time of this event.

In this example, everything that happened in the form of production costs, energy waste, corrosion in boiler systems and steam condensate, environmental issues, quality control, reduction of production and off-spec products, was in practice, the result of a combination of several important factors, each of which was mentioned separately in the above paragraphs, so it is clear how the use of chemicals such as CI of sour gas feed at the inlet to the gas heater and activated carbon in ACF can have different impacts on the entire operating parameters, and consequently, gas plant performance that Sections 5.5.1 to 5.5.6 referred to. Table 5.5 contains various main occurrences and, their impacts on entire operating parameters.

Table 5.5 Various main occurrences and their impacts on the entire operation.

List of occurrences during the event	Impacts on entire operation
Increase blowdown rate	Production cost, energy, environment, process issue (off-spec product)
Increase neutralizing amine, phosphate, and oxygen scavenger feed rate	Production cost, corrosion, quality control, energy, environment
Draining of contaminated condensate from flash drum	Corrosion, quality control, production cost, energy, environment
Sour gas feed reduction (reduction in gas trains feed and its related products)	Production cost, process issue (production reduction), process issue (off-spec product)
Opening the gas heater and then its repairing	Production cost, corrosion, environment
Closing the main suspect condensate header and its isolation from steam and condensate cycle	Production cost, corrosion, environment, quality control, energy, process issue (off-spec product)
Production of non-compliance steam and related condensate	Production cost, corrosion, quality control, energy, process issue (off-spec product)
Contamination entering in return condensate system	Production cost, corrosion, quality control, energy, environment
pH reduction, alkalinity consuming, and phosphate unbalancing in boilers	Production cost, corrosion, quality control, energy, environment
Introduction of acidity into feed boiler water system and economizer	Corrosion, quality control, environment
Increase MU water flowrate	Production cost, corrosion, quality control, energy, environment, process issue (production reduction)
Replacing two trains of mixed beds (cationic and anionic resin)	Production cost, corrosion, quality control, energy, environment
Replacing two beds of ACF	Production cost, corrosion, quality control, energy, environment
Complementary laboratory analyses to cover troubleshooting	Production cost, quality control, energy, environment

Table 5.5 is a strange table; at first glance, one may think, "The author could have said in one sentence that these events that followed had an impact on all parts of an area's performance on a processing site." By showing the issues, it can show us the results of how to feed a CI and alkanolamine into the sour gas feed, not following of alkaline out of service procedure for gas heater, not checking the gas heater before the start of the cold season and servicing it, while forgetting

to control the quality of the steam condensate output from heat exchanger, and finally neglect of quality control to determine the availability of two rows of activated carbon beds. To clarify the author's view of the contents of the right-hand column of Table 5.5, for example, the interpretation of one of the rows is given below:

5.5.6.2.3 Increase MU Water Flowrate

Production cost: When a suspect steam condensate recovery header is isolated from the main steam condensate return route (to storage tanks, CPP or deaerators), about 60 m^3/h of steam condensate is lost from the BFW system. In order to maintain the water level inside the steam condensate storage tanks, operation teams have to increase the MU water from 15 m^3/h to 75 m^3/h to provide demineralized water for steam generation unit. Decreasing the water level in the tanks can have two impacts on corrosion and reduction in quality of steam, BFW, and return steam condensate.

First, the reduction in liquid level in these tanks will cause atmospheric ingression into them. While storage tanks all have nitrogen blankets, one might ask, so where is the concern? Here are some reasons: The nitrogen pressure control valve (PCV) on a tank has a defined capacity of releasing nitrogen, as well as breaking the vacuum and allowing some air may enter the tank. Another concern is that operation team cannot be sure of the proper operation of these PCVs. Therefore, the probability of carbon dioxide and oxygen ingress into the stored steam condensate increases, and the potential of corrosion also increases [18, 23, 24]. Additionally, despite the removal of non-volatile organic matter by activated carbon beds, in CPP package, it is not possible to ensure proper monitoring of their performance, as well as zero human and instrumentation error, so a thick or thin hydrocarbon layer is always formed on the condensate surface inside the tank. When steam condensate levels in the tanks went to a very low extent, the entry of non-volatile organic matter into the boiler is inevitable. It is also necessary to maintain the height of the steam condensate in the tanks to ensure continuous steam generation. Therefore, more demineralized water, MU water, must be produced to meet the needs of the steam-generating unit, which incurs more costs for plant operations.

Corrosion: When a steam generation unit suddenly receives more than 60 m^3/h of MU water, the functional stability of the deaerators should be considered. The inflow of more MU water to the deaerators is always controlled by a set of level controllers with the summation of returned steam condensate and MU water. However, these controllers need a little (or much) time to reach stability and optimal point, with instability sometimes lasting up to several hours. Often the operations team decides to divert MU water to steam condensate storage tanks for a few hours (and possibly until control returns to normal) to prevent this fluctuation in the deaerators, in which case the tanks will suffer from oxygen corrosion [12, 18, 23, 30]. The difference between temperature of the MU water and the return steam condensate must also be included. On the other hand, MU water is saturated with oxygen. We know this because, by adding bisulfite, all free residual chlorine remaining in the demineralization unit feed is removed to prevent damage to the strong anionic and cationic resin matrix [23, 24, 31]. In case of overdose of bisulfite (for dechlorination), part of the dissolved oxygen (DO) is removed, but this is of little importance and is not economical (and also technical). It is therefore reasonable to assume that MU water is saturated with oxygen. This excess oxygen must be removed from the deaerators, and the oxygen scavenger dose must be increased at this time to moderate the impact of large amounts of oxygen entering. Experience has shown that all of these cases have a certain reliability and with increasing MU water flowrate, the equipment will definitely corrode.

Quality control: This was implicitly explained in the two sections above, and all analyses performed to ensure proper quality in the various sections, from putting a desalination unit (for desalinated water production) in service, to deaerators (for oxygen removal), are costly. During this work, fatigue and high workload of laboratory personnel may reduce the quality of the analysis, and also focus less on the analysis of other processing in live sections of the processing site. As explained earlier, the quality control of almost all sections has been affected by this event.

Energy: Increasing the MU water flowrate clearly increases the energy consumption. The desalination unit, which is of multiple effect distillation-thermo vapor compression (MED–TVC) type, and the desalinated water production unit must be put in service; both of these units consume more energy.

MU water has a temperature of 40 °C, and if it enters the deaerators, steam consumption will increase significantly. This is to restore the function of the deaerator and bring it to 117 °C to ensure the stripping of oxygen and other non-condensable gases such as carbon dioxide [18, 23, 32].

Environment: All of these actions and additional production have direct impacts on the environment. Energy consumption emits more GHG. Even more stringent sampling regimes from different sectors emit more GHG. Because in such conditions, the laboratory has to quickly use four diesel cars at the same time to get to different parts of the processing site and analyze, while under normal conditions, routine samples are collected and analyzed at a much lower frequency.

Process issues (production reduction): A gas plant can produce a certain amount of industrial and drinking water; the size of this production is quite clear when viewing the distribution to various consumers. A seawater desalination unit and a demineralization unit, which are in series, should provide water to different units such as drinking water, service water, gas condensate stabilization, cooling water, and firewater. When the steam-generating unit suddenly needs 60 m^3/h of MU water due to the constant need of other parts for water, it should be expected

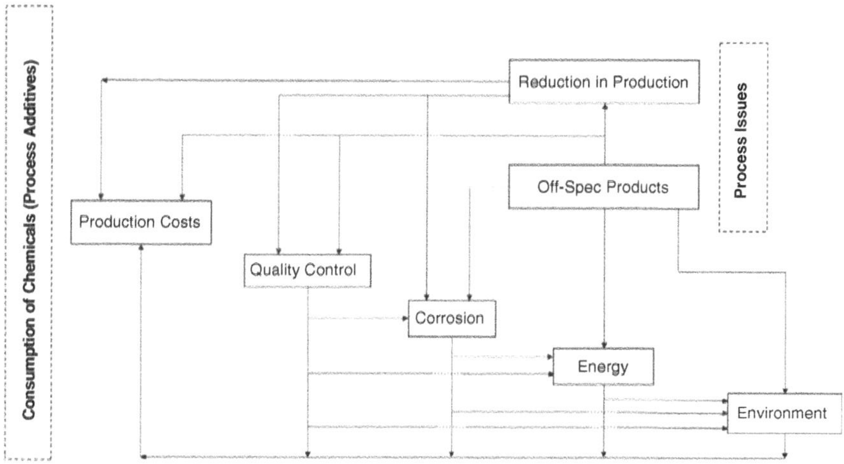

Figure 5.11 Relationship between various impacts for consumption of chemicals on entire operation and gas plant performance.

that the height of the liquid in the desalinated water storage tanks will decrease. Sometimes this production compensation is not possible, and to keep the steam generation site alive, some consumers are temporarily taken out of the consumption circuit, which also has its own adverse impacts on entire operation.

At the end of this section, Figure 5.11 clearly represents the impacts of chemical consumption on a gas plant performance, in this figure, the relationship between the various impacts that are in coordination with EMPA is well represented. In the above sub-sections and two industrial cases, an attempt has been made to implicitly represent the relationship between all these components.

Other cases are given in the last part of this section, in order to clarify the meaning of these interactions on each other with more focus on the potential corrosion of the equipment.

5.5.6.2.4 Industrial Case Three: Corrosion in a Stabilizer Column due to Poor Chemical Performance in WWTP

A gas plant has an independent WWTP to receive and treat its industrial aqueous effluents. One of the parts of this WWTP is the package for receiving and treating oil-contaminated aqueous effluents, and is called the oily water treatment package (OWTP). Various potentially oily sewers (POS) and oily water sewers (OWS) are routed in this package, and two chemicals are fed into the American Petroleum Institute (API) separator and IGF before entering [20, 33]. This package consists of sections that are covered in Figure 5.12 in appropriate detail.

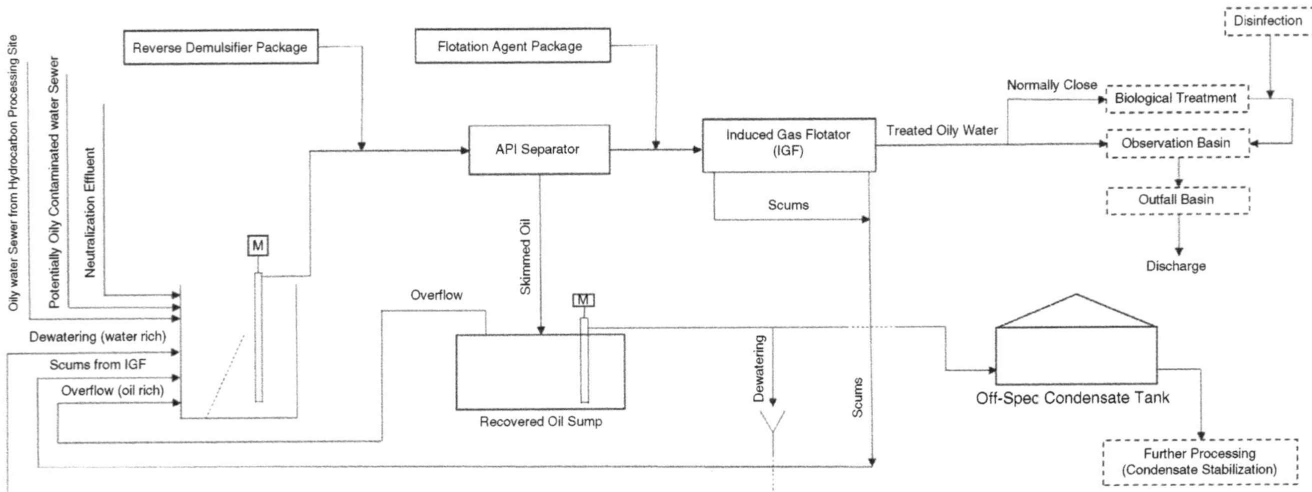

Figure 5.12 The schematic for influent streams and OWTP.

The normal operation of OWTP is that oil droplets and suspended solids are separated from the oil-contaminated effluent by gravity, and by using two polyelectrolyte chemicals. The oil-free effluent then enters a biological treatment [or biological oxidation (BIOX)] package for further treatment and reduction of biodegradable COD to allow for environmental release or advanced treatment and reuse [5, 34]. In the API separator, the oil layer created on the surface is skimmed by two scrappers (one at the beginning of the flow channel and the other at the end), and sent to the recovered oil sump (ROS) [7, 20, 35, 36]. Additionally, scums eliminated from entering stream to IGF are sent to the beginning of the package, oily water inlet sump (OWIS).

OWIS is always a recipient of POS and OWS from different parts of the gas plant. The operation of OWTP is of the batch type. Whenever the POS and OWS level in OWIS exceeds a certain level, the set of effluent transfer pumps, API separator, and IGF, as well as their chemical feeding packages, will be put into service. This operation will continue until the effluent level in OWIS reaches an acceptable value. Each gas plant has its own operation characteristics; therefore, it can be expected that the OWTP has daily, weekly, or even monthly operations. Factors such as rainfall, firefighting drills and/or performance tests, and major maintenance jobs can cause OWTP to remain in service for several weeks. On the other hand, in normal operation, this package can be held out of service for more than a month.

The nature of some POS and OWS streams is such that they can be separated into two phases after a period of time without applying any chemicals, such as demulsifier or mechanical operations. Therefore, it is possible that in some gas plants with very low production of POS and OWS, there is an oily layer on the surface of the effluent in OWIS. This oily layer can act as a magnet for other droplets of oil entering OWIS and become thicker after a while.

On the other hand, the presence of two other factors in the OWTP operation can also contribute to such an event:

1) Improper selection of the reverse demulsifier chemical or its overdose to the feed stream of API separator causes a significant amount of chemical to remain in the skimmed oil by the scrapers. These skimmed oils (water–oil mixture) are routed into the ROS, then with a dewatering mechanism, the water in the mixture is removed based on the density difference (location of the lift pump suction) and is returned to the OWIS. Consecutive analyses determine the time to reach the oil and water concentration ratio of 50%/50% (minimum), by volume, then the dewatering operation is stopped and all the water–oil mixture with a ratio of at least 50%/50% is sent to the off-spec condensate tank.

During dewatering, a significant amount of reverse demulsifier chemical enters the oily water OWIS (beginning of OWTP). This chemical mixes with the raw effluent contained in this sump and over time forms an oily layer on it (like what happens in the API separator).

2) In conventional API separators, the skimming operation is appropriate because while the OWTP is in service, the field operator must check the condition of the API channel and the oil layer approaching the inlet and outlet scrapers of the channel every one hour (at least); if the oil layer formed and was approaching the scrapers, then it should guide the scraper to a lower level (by virtue of a vertical rotation mechanism) and collect the oil layer, and after that immediately return the scraper to its original position. This skimmed layer is gravitationally sent to ROS.

ROS has a constant volume. In this case, if it is full, it will overflow and according to the common designs, the overflow will go to the beginning of the OWTP. What comes out of this sump is a water–oil mixture that contains a few demulsifier chemicals, which go back to the OWIS.

If the chemical is not well selected and/or fed, or the field operator is not careful enough to skim the API separator in a timely manner, a significant volume of oily water effluent (dosed with the chemical) then goes into the ROS. Because the oil content in this skimmed effluent is very low, during the dewatering operation all ROS content must be sent to the OWIS to return to the treatment cycle. On the other hand, what goes to the off-spec condensate tank as a dewatered water–oil mixture (at least 50%/50% ratio) is considered as part of the feed of a gas condensate stabilization unit.

The stabilization unit is designed to receive and adjust the RVP for gas condensate (see Figure 5.2). This unit consists of a set of heat exchangers, flash and gravity separators, desalter, distillation column, and compressors. Whenever the operating conditions of this unit are such that it cannot receive gas condensate from the extraction platform, these gas condensates are temporarily directed to the off-spec tank. If, for operational reasons, the product of the stabilization unit is not of proper quality to be sent to the storage section (e.g. not adjusted RVP and not filtered, correctly), this off-spec product (Figure 5.6) is then sent to the off-spec tank. Depending on the design, the off-spec tank can receive the entire gas condensate feed or off-spec product, which is produced by the stabilization unit for a specified period of time.

After reaching stable operating conditions, the contents of this tank are received by the gas condensate stabilization unit, so that the tank is ready (empty) for further events. Due to the limited capacity to receive and stabilize gas condensate in the feed of a gas plant, it is obvious that by receiving gas condensate (and water–oil mixture) from the off-spec tank, the amount of production from the gas extraction platform must be reduced. Figure 5.13 shows the gas condensate stabilization unit with the off-spec tank; although in this figure the three sections shown in gray do

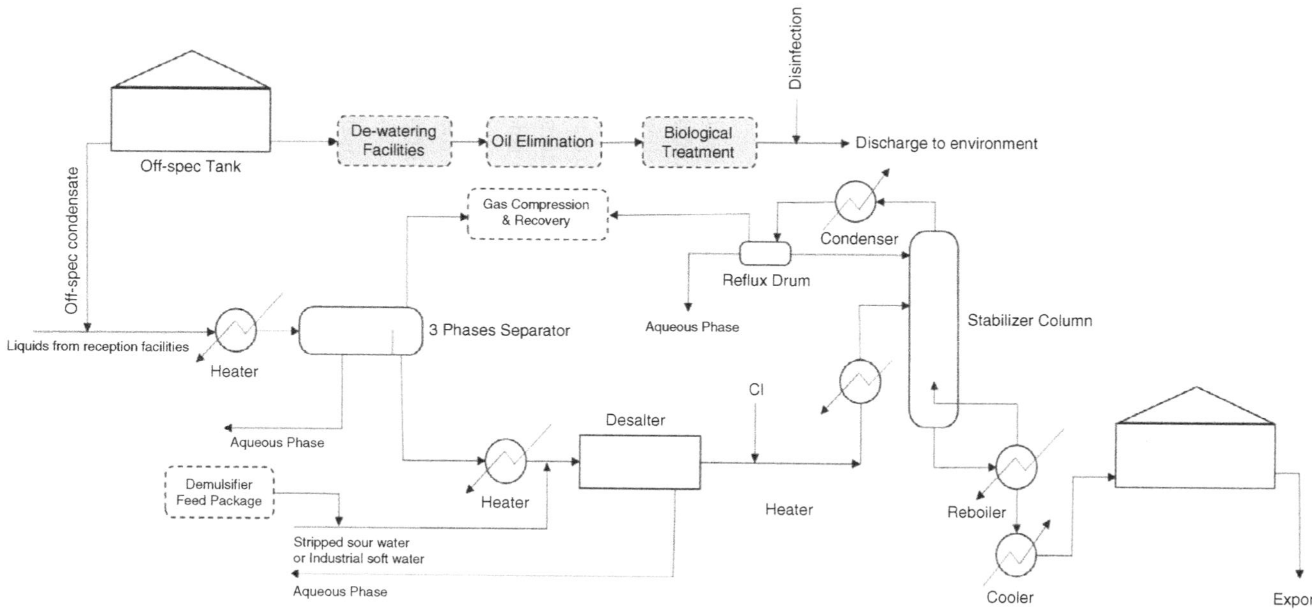

Figure 5.13 Gas condensate stabilization unit with off-spec tank.

not exist in some gas plants. This design is to prevent (or minimize) the water in the bottom of the off-spec tank (which is the product of water–oil mixture input from ROS) into the gas condensate stabilization unit. High concentration of some readily biodegradable aqueous phases like glycols, small flow, presence of heavy metals, and major pollution by oil forces the designer to select sequencing batch reactor (SBR) for BIOX [5, 37]. Gas plants that do not have these three gray sections have more capacity to receive water in the three-phase separator, and the water in the bottom of the off-spec condensate tank enters the gas condensate stabilization unit, directly.

The process is described in the gas condensate stabilization unit as follows:

First, liquid feed enters the unit from the reception facilities (slug catcher and manifold). The fluid flow then enters a heater for better performance of the phase separation in the three-phase separator. In the separator, a gas phase enters a header to collect valuable gases and then goes to the gas compression and recovery section. An aqueous phase present in the liquids entering the unit is formed in the separator, which is sent to the aqueous phase recovery units, such as glycol regeneration or similar facilities. Gas condensate also separates as an organic phase in the separator and enters a heater. This heater is for warming the gas condensate before entering the desalter.

Because the entry of water and mineral salts into the distillation column causes severe corrosion, these two components must be removed from the gas condensate stream before entering the column. Therefore, in order to better separate the aqueous phase from the gas condensate (organic) phase, a water-rich stream (oil-free water) is first used along with the demulsifier chemical to facilitate the separation of the aqueous and organic phases, and implicitly salt (inorganic) in extract aqueous phase (inorganic). A set of high voltage electrodes inside the desalter make it possible to separate the aqueous layer (and also salts) and the organic phase in a high degree [20]. Concentrated water with salt collects in the lower layer of liquids in the desalter and is sent to downstream units to purify or recover materials such as glycol-based solvents.

After the amount of water and salt in the gas condensate stream reaches the allowable limit, the gas condensate outlet stream from the desalter passes a heater to preheat before entering into the stabilization (or distillation) column. In the column, the vapor phase is supplied by a reboiler at the bottom of the column, and the liquid phase by a condenser at the top of the column. The final product of this column, stabilized gas condensate, is taken out from the bottom of the column, and after cooling, stabilized gas condensate is sent to storage tanks to be exported to tankers and oil refineries. The gases released from the reflux drum are also recompressed due to the presence of light hydrocarbons such as propane and butane and sent to the beginning of the gas trains for NGL extraction (see Figure 5.2).

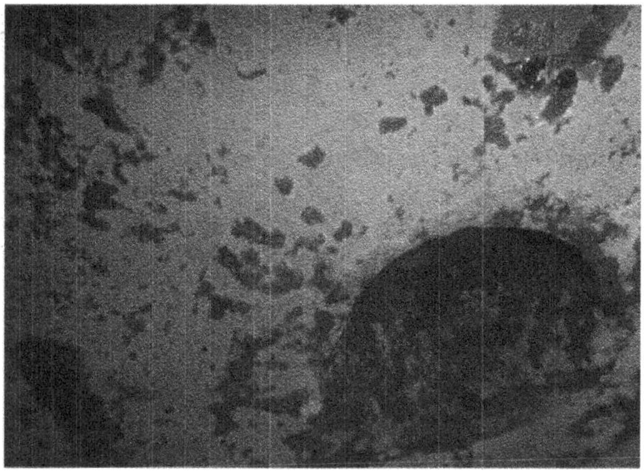

Figure 5.14 Acid corrosion at top of stabilization column (vapor outlet piping to condenser).

The gas plant had an OWTP and two trains of gas condensate stabilization. An acid corrosion was observed in the upper parts of the column and its condenser (Figure 5.14). When these corrosions were first observed in the overhaul, the definitive statement was that the chloride ion in the gas condensate hydrolyzed at column temperature to form hydrochloric acid. Acid droplets had condensed on the inner surface of the column at the top, creating localized corrosion.

According to the design, a film forming amine-type CI was fed into the gas condensate stream entering the column. In addition to its film forming properties, this CI had neutralizer (and volatile) amine ingredients in its composition to neutralize some of the possible acidic agents created during flashing into the column and the boiling of the feed in the column. The technical conclusions have reached a point where they make the neutralizing amine in the CI more concentrated and volatile. After one year of operation, localized corrosion was observed again in the same sections without changing the extent of the attack and the depth of the pits.

Engineering studies have shown that not paying attention to the routine measurement of water content in the gas condensate entering the distillation column has caused the volume of incoming water and the amount of mineral salts in it to increase. Investigations also showed that the operating team did not routinely draw water off from the top of the column, and that this standard operating procedure was largely forgotten.

Further investigation showed that incorrect feeding of the reverse demulsifier into the feed of the API separator and its incorrect operation caused a large amount of untreated effluent dosed with the reverse demulsifier chemical to return to the OWIS and raise the effluent level inside it, the field operator has to send the API separator effluent to ROS and then pump it to the off-spec condensate tank after dewatering (which was practically useless, knowing that the water concentration in this mixture is more than 80%). The high rate of effluent received in ROS and the very low concentration of oil in it had caused the height of the sump to increase rapidly, and due to the volume limitation of the off-spec condensate tank, the gas plant had to receive some part of the off-spec condensate tank content in gas condensate stabilization units, daily.

Calculations and other site evidence showed that in a three-year period of operation, the volume of water received in the gas condensate stabilization unit was more than 27 times the design specification (unit turndown was 130%). Therefore, the equipment that was intended to eliminate the water phase (three-phase separator, desalter, and water draw-off facilities on column) could not remove this excess water from the design and water entered the distillation column. Examination of the laboratory databank showed that the analysis of the concentration of water entering the column, which was an important part of the quality control plan of this unit, could not be measured due to the molar ratio of water outside the defined method test, and the laboratory has not attempted to replace the test method nor to inform the operations team.

An intelligent reader may ask "Why did the operator of the gas condensate stabilization unit not gradually receive feed from the off-spec condensate tank?" (so that she/he could better control the concentration of water entering the unit and not get too far away from the turndown). The answer is; it was not possible to reduce the feed of the gas plant from the extraction platform for this unit. On the other hand, the volume of the off-spec condensate tank was limited to hold the off-spec condensate, and a significant amount of water–oil mixture had to be entered daily.

Investigations continued and it was found that one of the causes of improper separation of oil from water was not paying attention to the proper dose of reverse demulsifier. The dose recommended by the manufacturer was 3–10 ppm and the consumption calculations in the three months leading up to the study showed that the dose was greater than 500 ppm This extremely high dose of the chemical resulted in the formation of a turbid mixture and a stable emulsion of water and oil at the API separator and IGF outputs, which in no way allowed phase separation (Figure 5.15). Volume constraints on ROS, OWIS, and off-spec tank allowed the OWTP field operator to send it to a BIOX package, regardless of the quality of the effluent treated at OWTP.

Regulating the pH at the beginning of the BIOX process, as well as the biological degradation of the demulsifier by the microorganisms in related suspended and attached growth arrangements, caused the oil in the effluent of OWTP (feed of the BIOX) to separate from stream and float on the clarifiers as a free-oil layer. Figure 5.16 represents an example of this oil contamination in the BIOX package introduced by the effluent from the OWTP.

Figure 5.15 Presentation of overdosing of a reverse demulsifier chemical into a stream of oily contaminated sewer water. While gas condensate presents equally in each bottle, the interphase layer does not see it because gas condensate is colorless.

Figure 5.16 Free-oil layer on clarifier next to a BIOX filter (BIOX filter was trickling super high rate [5]).

The WWTP also had a spent caustic soda neutralization package (SCNP) used in the propane, and butane–mercaptan removal or sweetening units [7, 36]. The spent caustic in this package neutralized with sulphuric acid, and after regulating its pH in the range of 7–8.5, the neutralized effluent went to the OWIS. Further investigation showed that the feeding of sulphuric acid was stopped due to the entry of contamination into its storage and dosing system (operation history 12); therefore it was not possible to feed acid into the reactor in the SCNP. The effluent from the SCNP was mixed with the stripped sour water (pH = 2.7–3.5) and its pH barely dropped to 11.5 (from original range of 12.3–13.5). Due to the entry of un-neutralized effluent from the SCNP alone, the field operator had to increase the dose of the demulsifier and thought that with increasing the chemical dose, the separation would be better.

The OWTP mass and hydraulic balance calculations were carefully reviewed and it was concluded that the total oil (and some entrained water) skimmed by the scrapers from the API separator surface should have been 190 kg/h at shock load of 1000 ppm oil in feed (design case). However, a review of DCS data (level transmitters) for the volume received in ROS and OWIS showed that the volume of water–oil mixture skimmed by scrapers entering these two sections was more than 6000 kg/h (the study period of the trends was six months). Studies have shown that poor operation and lowering of scrapers instead of routinely checking the site had caused such a situation.

5.5.6.3 Operation History 1

Another gas plant had a very small volume of POS and OWS. The design and arrangement of the units on this plant was quite similar to the processing site mentioned in Industrial Case Two. At this gas plant, the operation of OWTP was inappropriate and a thick layer of oil was formed on the OWIS. This layer of oil acted as a cover on the stagnant water below, so that the water layer below could not come into contact with air and sunlight (anaerobic condition). Additionally, the management of sludge formed in the lower part of the water layer was very poor and the sludge was not removed from the system for a long time.

Measurements showed that sulfate reducing bacteria (SRB) were present in stagnant water layer with an intensity of 100 000 CFU/ml. This water layer is sent to the off-spec condensate tank according to Figure 5.12 and from there enters the gas condensate stabilization unit, after which it proceeds to the glycol recovery and sour water stripping units. The recovered glycol, which contains at least 30% by weight of the same water, was pumped to the platform to prevent corrosion and hydrate formation of the feed pipeline to the gas plant. Part of the stripped sour water was also used in the desalter [38]; hence, the SRB in OWIS extended to the entire upstream part of the gas plant and included various units.

A standard coupon was tied with a non-metallic string and placed under the oil layer formed in the OWIS, and after six months, several very sticky biofilm accumulation sites were seen on it. After removing the biofilms, black residuals were observed inside the pits which represents iron corrosion products of MIC [18, 39]. Figure 5.17 represents the coupon after cleaning and acid washing.

Figure 5.17 A corrosion coupon placed under oil layer in OWIS with occurred MIC.

Investigations showed that a set of heat exchangers in the reception units, which had general corrosion and minor pits, were completely destroyed after a short period of service, and laboratory and technical studies showed that the predominant corrosion mechanism of this equipment was MIC.

5.5.6.4 Operation History 2

Another gas plant reported that the transfer of water–oil mixture to the off-spec tank was stopped due to both occasional and severe clogging of the transferring line. Investigations have shown that improper feeding of demulsifier and improper operation of scarpers in the API separator caused the transfer of large volumes of effluent to the ROS, and the continuous operation of the lift pumps in this sump caused their frequent breakdowns. Therefore, within three weeks when the main pump and spare pump were out of service, the water–oil mixture in the pipeline was coagulated and the transferring line was severely clogged.

The opening of the transferring line took a long time due to its length and large number of piping details, and the effluent retention in OWIS caused both gradual phase separation and growth of SRB in the aqueous layer. Given that this gas plant also had a design very similar to the previous two processing sites, it can be expected that the equipment upstream of this gas plant also had a risk of MIC.

The water–oil mixture transferring line was about 1500 m long and had a slope of 15° from the ROS to the off-spec condensate tank. Therefore, when pumping from ROS stopped and dewatering was performed, this line was emptied from the dewatering drain valve. Prolonged dewatering operations and forgetfulness of the field operator to close the dewatering valve until the ROS was refilled caused a check-valve to be installed after the dewatering valve. A three-year monitoring of the processing site showed that the problem of transferring the water–oil mixture to the off-spec condensate tank was completely eliminated, and due to the reduced effluent retention time in OWIS, the possibility of SRB formation and microbial corrosion risk was greatly reduced. Figure 5.18 shows the installation location of this check valve, that is installed on a pipeline with a length of 1500 meters.

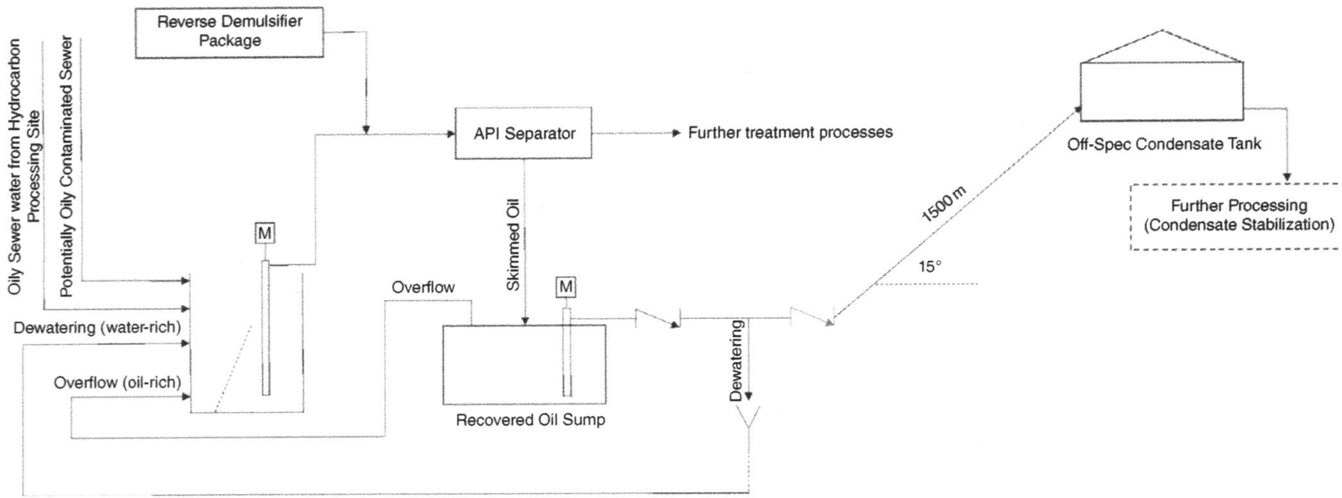

Figure 5.18 Location of OWTP, off-spec condensate tank, and an installed check-valve on transferring line.

5.5.6.4.1 Industrial Case Four: Corrosion in Seawater Thermal Desalination (Chlorination and Dechlorination of Seawater)

There was a set of utility facilities at a coastal gas plant. Water was supplied from the sea and entered a set of MED–TVC packages for desalination. Incoming seawater was also applied to cool large parts of the processing site (process heat exchangers). The water produced from this package was sent to other processing units such as demineralization, cooling water, firewater, and drinking water for various applications throughout the gas plant. Sodium hypochlorite was fed into seawater in two locations on seawater intake facilities (SWI) [40]. The intake basin was of equilibrium type [8], therefore, there was a set of pipes (chambers) connecting the sea and the water intake basin. Sodium hypochlorite was continuously fed into the beginning of these intake chambers (location one). After the chlorinated seawater entered the receiving basins, a set of lift pumps directed this water to the coarse filters and then sent it to the gas plant at a distance of 3 km. Sodium hypochlorite was also continuously fed into the seawater in the lift pump chamber (location two).

To prevent organisms and marine animals from adapting to the fed dose of sodium hypochlorite, four doses of shock (each lasting 15 minutes) were fed at both locations. Sodium hypochlorite was produced by electrodialysis of seawater (at a concentration of 1–1.5% by weight). It inhibited the growth of organisms and biofouling, as well as the growth of marine animals in the transferring pipes and heat exchangers at the gas plant. Figure 5.19 represents a heavy marine animal growth on the condenser feed channel and on a side wall of the evaporator in a

Figure 5.19 Marine animal growth in a MED–TVC.

MED–TVC due to poor feeding of sodium hypochlorite. Additionally, continuous sodium hypochlorite feeding increased the efficiency of seawater filtration in the SWI and minimized the operational difficulties caused by the entry of solid particles into the downstream facilities.

The desalination unit was located downstream of the seawater pipelines. The packages in this unit are made of stainless steel and copper alloy. When highly chlorinated seawater if fed to the evaporator inside, the bromine gas is released from the seawater and causes the following problems; one is a corrosion problem on copper alloy and stainless steel, and another is degradation of desalinated water quality of conductivity and lower pH. Therefore, chlorinated seawater must be dechlorinated before entering evaporators in MED–TVC. Dechlorination was performed with a sodium bisulfite feed package. Based on the logic defined in it, this package operated in full coordination with continuous doses, and four daily shocks designed in the SWI. The maximum prescribed amount of residual chlorine after sodium bisulfite feeding was considered in the plant's documents as a maximum of 0.15 ppm, and another source reported this amount as 0.25 ppm [13].

Seawater feed should be deaerated prior to entering the MED–TVC, and the oxygen level should be lowered to 20 ppb [13, 41, 42]. The required dose of sodium metabisulfite for complete elimination of 1 mg/l oxygen is 6.5 mg/l [18, 19].

In the first inspection, corrosion due to the presence of residual-free residual chlorine was observed in each of the MED–TVC packages. Studies on the amount of sodium bisulfite feeding showed that quality control in this area had several drawbacks, which are as follows:

- The fed dose of sodium hypochlorite in the SWI had not been reported to the gas plant (the concentration of sodium hypochlorite in its storage tank and its feed flow to different locations of the SWI had not been measured and reported).
- The residual-free chlorine concentration calibration table was not prepared and was not regularly updated (using the measurement of residual chlorine at the inlet of seawater to the gas plant, measuring the residence time of chlorinated water in the transferring pipe based on the volume of the pipe and its flowrate, and, finally, measurement of chlorine demand at calculated residence time).
- At the gas plant (utility section), the equivalent stoichiometric dose for dechlorination along with the safe dose was not calculated (sodium bisulfite concentration in the tank was not measured routinely and its feeding pumps were not calibrated, also quality control for sodium bisulfite at the beginning of entering the warehouse was not performed) [19, 43].
- The control logic related to chlorination shock at SWI was not connected to the gas plant main control system for one year after commissioning. After logic connection to the control system, the timing of chlorination shock pumps (in SWI)

and dechlorination shock pumps (in gas plant's utilities) showed a significant difference.

Figure 5.20 shows a view of the SWI and its position relative to the gas plant. Quality control points and other measuring elements such as flow meters and sampling points are shown.

5.5.6.4.2 Industrial Case Five: Corrosion in the Boiler Due to Improper Chemical Control in the Face of an Operational Error

A utility processing site had eight water tube boilers with a capacity of 2400 tons per hour. MU water was supplied through a thermal desalination unit and then an IOX unit with a mixed bed. Chlorinated seawater entered the utility processing site and was converted to industrial water in thermal desalination with a specific conductivity of less than 20 μS/cm. Due to the fact that the operating pressure of the boilers was more than 62 bar, this water was not suitable for entering the steam production system (as MU water), so two mixed beds with strong anionic and cationic resins were applied to bring the water quality to lower than 0.1 μS/cm.

One night in the winter of 2015 at 23:40, there was a sudden demand of 230 m^3/h for desalinated water by the firewater unit. Two thermal desalination packages (out of a total of four packages) were serviced simultaneously to compensate for the need for industrial water. Traditionally, at the beginning of each package, approximately 15 minutes from the production of the package was sent to the outfall basin (where rejected to sea) to achieve the desired quality (conductivity of less than 20 μS/cm), and then all the desalinated water produced was sent to desalinated storage tanks and to general distribution (with a normally close arrangement).

The thermal desalination package had a quality control logic, which sent the produced (desalinated) water with a quality of less than 20 μS/cm to the desalinated storage tanks and to general distribution. If the specific conductivity was more than 20 μS/cm, then this water would be sent to the outfall basin to reject to the sea (Figure 5.21). Due to a malfunction in one of the valves of this control logic in one of the packages, the untreated seawater entered the desalinated storage tanks directly, for 10 minutes. This water was pumped to the demineralization unit where there were two mixed bed columns. One of them was checking the quality of the resin and had an unacceptable pressure drop. The other column was in service and its IOX capacity was quickly exhausted. Due to the unavailability of its parallel train (which was checking resin quality), the demineralization unit was taken out of service and contaminated water flowed from the desalinated water storage tanks to the steam condensate tanks in the steam generation unit.

The entire steam and condensate system was contaminated with seawater and after 6 hours, the chloride ion concentration in the BW (continuous blowdown)

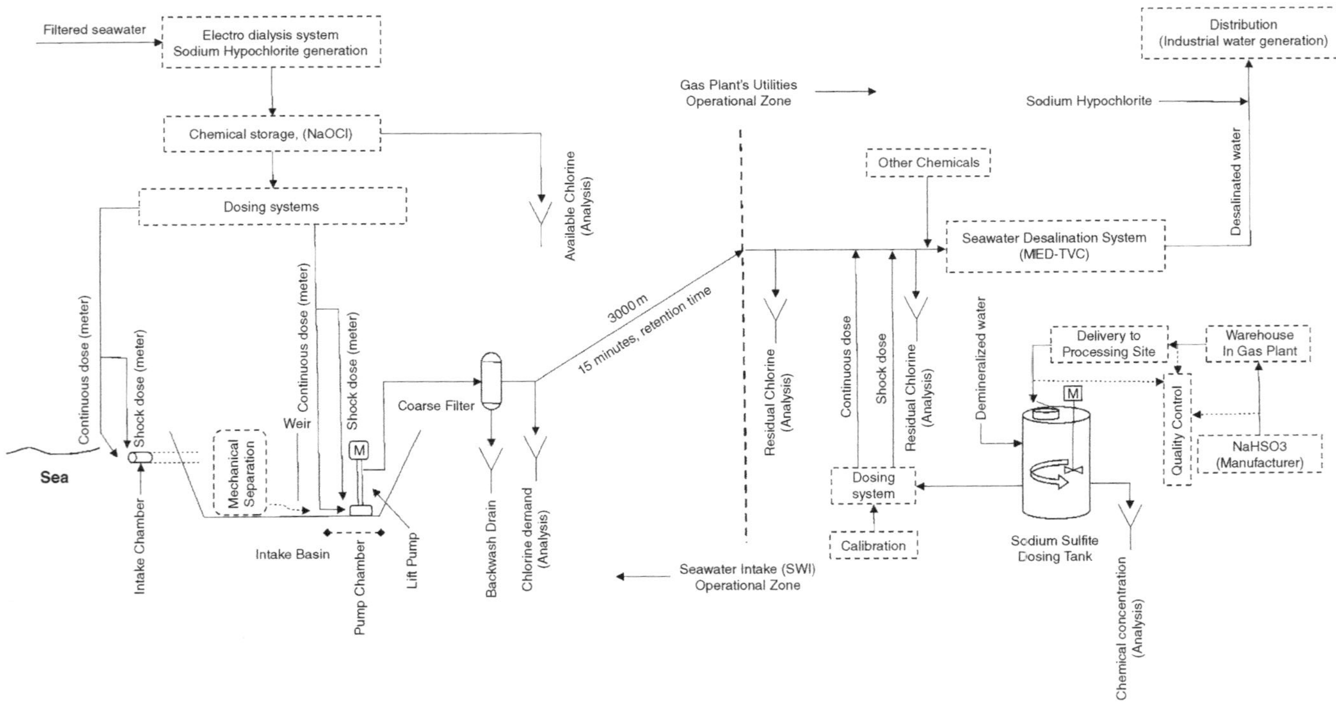

Figure 5.20 Seawater intake facilities and gas plant's utilities, chemical feed and quality control points.

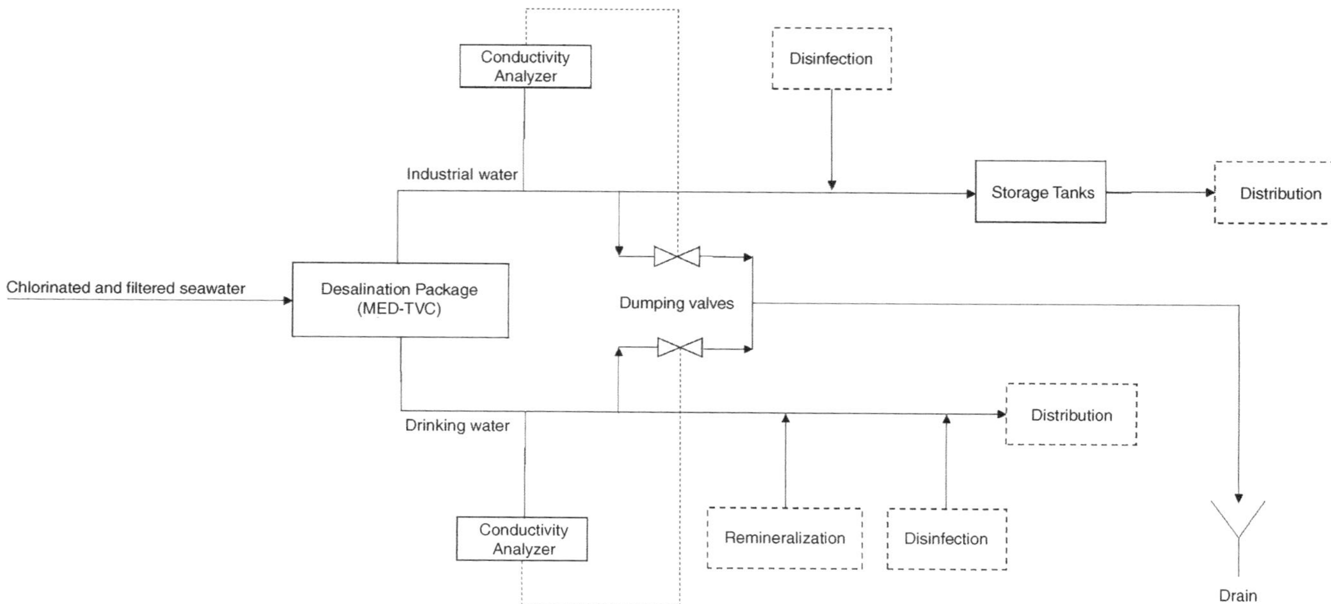

Figure 5.21 Seawater desalination package, its quality control logic and dumping arrangement.

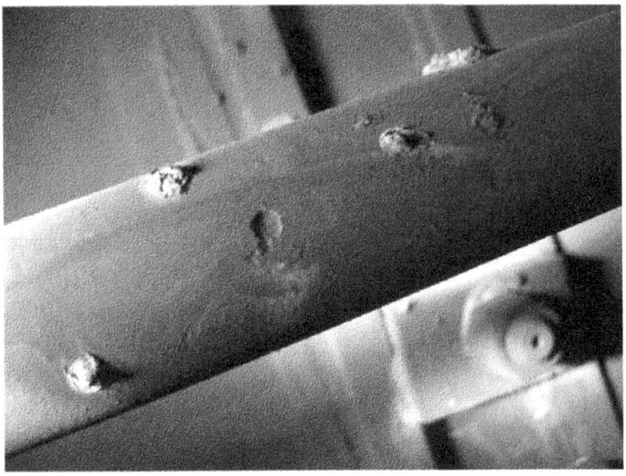

Figure 5.22 Underdeposit corrosion in a water tube utility boiler.

reached 110 ppm (which was previously 1 ppm) and the pH of the BW dropped sharply for more than 3 units. A series of measures were taken to modify and recover the boiler chemistry as soon as possible. The utility engineering department had decided to increase the phosphate in the boiler to eliminate the acidic effect caused by the presence of chloride ions. Increasing phosphate feeding and concomitant applying of NaOH to raise the pH of the system caused the phosphate concentration in the boiler to become unbalanced. This method was incorrect for recovery of the boiler chemistry (as a single and complementary action) and in the first inspection of one of the boilers, it was found that the inside of the boiler was of underdeposit corrosion type due to the heavy deposition of phosphate-salt compounds (Figure 5.22).

Considering that it was necessary to maintain steam production and there was no opportunity to empty and clean the tanks, it would have been better if the utility engineering department had performed the following actions to prevent further damage to the boilers and other equipment (Figure 5.23 presents all details and decision opportunities):

- Immediate truncation of MU water flow from desalinated water tanks and direct receipt of desalinated water from desalination packages (means distribution system).
- Increasing the neutralizing amine feeding into the condensate network as well as BFW (after deaerator).
- Increasing the continuous blowdown flowrate.
- Reducing the production of each boiler (means: heat load) and compensating the total demand of production by putting spare boilers into service.

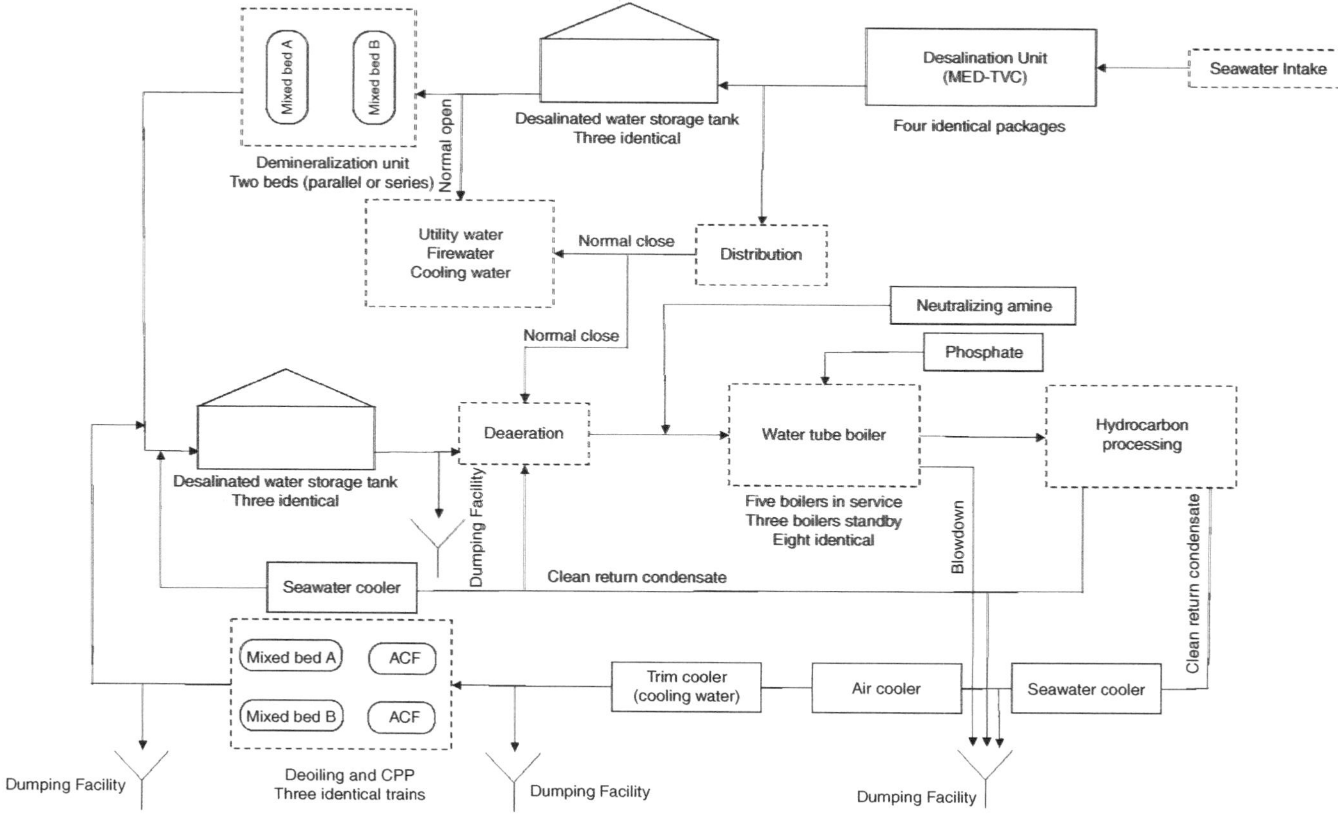

Figure 5.23 Seawater desalination, desalinated water distribution, steam generation, and condensate purification-deoiling system.

The incident lasted four days and the boiler chemistry was inadequately controlled during this time. One of the main reasons for the prolongation of the boiler chemistry recovering was that the utility process engineer did not monitor the return condensate with a flowrate of 1300 tons per hour, and the condensate returned to the steam condensate storage tanks with a pH of 6.5. The delay in detecting the contamination of the desalinated water storage tanks also caused the contaminated water in these tanks to enter the steam condensate storage tanks for two days, and the chloride ion to enter the steam generating system.

5.5.6.4.3 Industrial Case Six: Corrosion in Desalinated Water Distribution Pipelines Due to Remineralization Problems

The service water and drinking water of a gas plant were supplied by two seawater desalination packages. This water was aggressive due to the lack of salts, and imposed significant corrosion on the water main components, which were made of galvanized steel. The desalinated water entered the drinking water/utility water production unit, where the salts needed to adjust the LSI and AI, and CCPP parameters were fed into the desalinated water stream [9, 18, 19, 30].

In order to eliminate the aggressiveness of this water, two chemicals, calcium chloride and sodium bicarbonate, were fed into the desalinated water stream. These two chemicals were purchased as salt and were fed into drinking water and utility water at the processing site and solved in water at the processing site as prepared solutions. There was also a sodium hypochlorite feed point in the drinking water main that disinfected the water entering the drinking water main and prevented the growth of biofilm in the distribution.

Due to the ineffective management of calcium chloride and sodium bicarbonate, a set of interruptions in the feeding of the solution of these two salts into both drinking and utility water was created. The turbidity of the water and the concentration of iron in it both increased. Chlorine demand in the water main also tripled, indicating the presence of free metal ions and biofilm in the system. Many leaks were observed in the drinking water transferring pipes and inspections confirmed the presence of corrosion on the inner surface of the pipes. Figure 5.24 shows a corroded metal pipe of transferring line in desalinated water distribution system.

Further studies showed that (i) supply of poor-quality chemicals and (ii) inadequate monitoring of chemicals along with (iii) delays in the purchase of chemicals were the three major factors in causing these corrosions. Site evidences showed that in one sample the calcium feed package was completely out of service for more than two years. Additionally, the laboratory did not routinely analyze the chemical parameters required to calculate the water stability indices. Figure 5.25 represents the daily monitoring of the LSI and AI parameter in drinking water. The drinking water production plant that was considered in this industrial case was named as plant A in the figure.

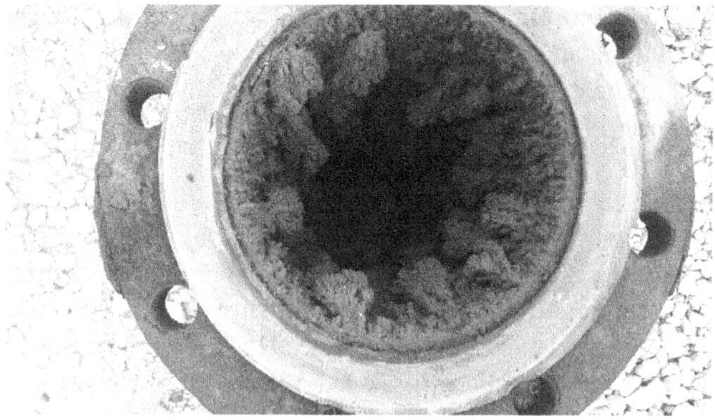

Figure 5.24 A corroded metal pipe in desalinated water distribution system.

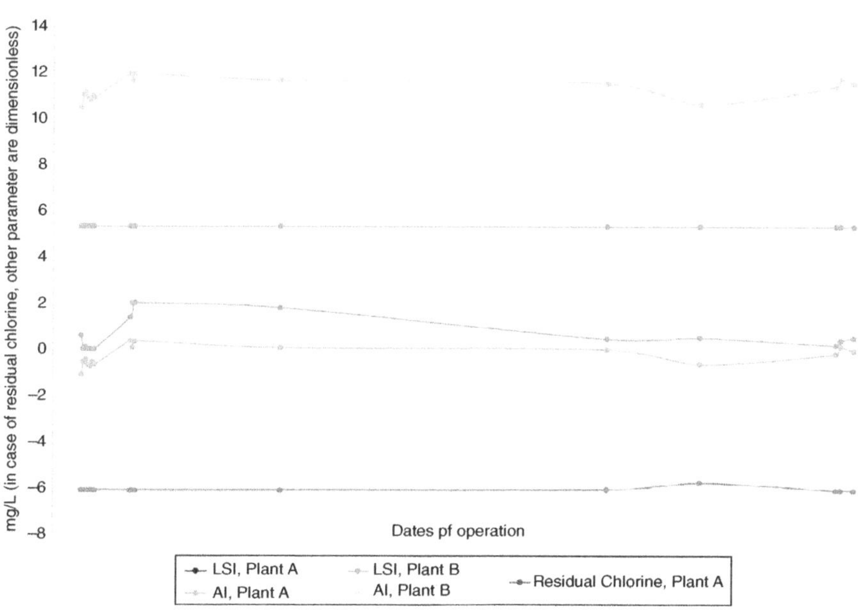

Figure 5.25 Daily monitoring of LSI, AI, and residual chlorine in two identical drinking water plants.

As can be seen in Figure 5.25, very inappropriate LSI and AI values were controlled due to improper feeding of chemicals into drinking water in plant A. The values of residual chlorine are also shown in this figure, indicating that there was no effective management of sodium hypochlorite feeding. Fluctuations in residual chlorine levels and improper control range in the drinking water distribution network lead to biofilm formation, and ultimately MIC. It is important to create suitable conditions for MIC in the event of improper control of stability parameters and electrochemical corrosion [18].

5.5.6.5 Operation History 3

The LSI and AI parameters for another drinking water production plant (B) that had a better situation regarding chemical feeding were given in Figure 5.25. At this plant, the quality control status of the chemicals, upon delivery from the warehouse, as well as the solution preparing were both correct. However, the treatment was not satisfactory because the dose of chemicals fed into the desalinated water stream (feed of drinking water plant) did not match the feed pump calibration diagrams.

The measurement of the parameters required to calculate the stability indices and residual chlorine was performed by the same laboratory. As can be seen, there is an unacceptable time interval between the reported values, as this is while the analysis regime was requested once in each shift. However, the laboratory has changed the analysis regime based on the number of its staff in different work shifts and as a result its ability for covering analyses of other sections of the utility plant.

5.5.6.6 Operation History 4

The drinking water main at a utility plant had a landslide fracture. Water consumption was increased to nine times the design case, and it became impossible to feed salt into desalinated water and reach the appropriate values for the stability indices, because the feeding capacity of calcium chloride and sodium bicarbonate salts was not proportional to the amount of desalinated water flowrate. Therefore, the process engineers who was responsible for drinking water quality control had increased the salts concentration in the dosing tanks to four times to achieve the appropriate LSI, and had simultaneously used the maximum feeding capacity of the pumps.

After some time, the feeding of both salts stopped completely. Increasing the feeding rate of the salts solution to achieve more appropriate stability indices was a roadmap for managing change in the system, but ultimately led to a complete halt of the feeding, and thus worsened conditions. The chemical feed nozzles of both salts were opened and complete clogging was observed. Figure 5.26 shows a calcium chloride feeding nozzle within a drinking water pipeline.

Studies have shown that the passing of the salt solutions concentrations from the saturation point and the presence of IM in the salts (too much) caused the feeding nozzles to clog and eventually stop completely. Leak eliminating lasted for two months and during this period the high flowrate of desalinated water entering

Figure 5.26 A completely clogged nozzle of calcium chloride feed prior to drinking water distribution.

the drinking water distribution system, and frequent clogging of the salt solution feeding nozzles caused the drinking water in the distribution system to have unfavorable conditions in terms of controlling stability indices and it resulted in corrosion.

Therefore, it can be concluded that the intended roadmap failed to manage the change caused by the temporary increase in flowrate. Permanent or temporary changes in the operational parameters of a processing site should be accompanied by the selection of a roadmap for managing changes in chemical feeding. This roadmap must be chosen in such a way that it has the appropriate effectiveness, for example, in this operation history, if a pump with a higher feeding capacity was used temporarily (so that there is no need to increase the concentration of the salt solution), then chemistry control to achieve stability indices was successful.

5.6 Misleading Trends with Corrosion Conclusions

It should be noted that similar trends do not always indicate the occurrence of corrosion or the creation of suitable conditions for its occurrence. The following are five industrial cases of such perceptions.

5.6.1 Phosphate Solution Preparation (Boiler Internal Treatment)

The communication structure of the departments in a gas plant was such that the points related to chemistry control were sent as a set of instructions or guidelines from the utility process engineer to the field operator of the steam generation unit. These instructions included preparation of different chemicals and adjusting the feeding rate (using the variable speed and stroke of the feeding pumps). Laboratory results over a week showed that orthophosphate began to rise and then returned to normal range.

Such results usually indicate the entry of acid into the system (and corrosion), which may have different causes. Examination of other parameters such as alkalinity and pH, in addition to residual orthophosphate, showed that these laboratory results were the result of an error made by the field operator of the steam-generating unit who incorrectly prepared phosphate solution in the dosing tank. Chemical charging was done five additional times and, the shift supervisor (after three 12-hour shifts) had noticed this mistake when reviewing the solution preparation logs. In order to reduce the effect of excess phosphate in the boiler, he had to open the continuous blowdown valve (which resulted in reduced specific conductivity), and when he saw this was useless, he decided to reduce the phosphate-feeding pump stroke.

In addition, preparation of phosphate solution and subsequent actions should have been improved by training of personnel; however, it should be noted that the first sign of a change in chemical balance (here orthophosphate and pH) should not always be considered as the starting point for conclusion to corrosion. Figure 5.27 represents trend behavior of some quality parameters in boiler D (forth boiler).

In this graph, the increase in pH is not clearly known, and is represented more appropriately in Figure 5.28. There were six water tube boilers in the steam-generating unit, and internal treatment of all these boilers was done by a common phosphate-feeding package. This error in solution preparation caused the laboratory results of five operating boilers in the same way (one of boilers was in standby mode).

As it turns out, preparing the incorrect chemical solution caused the pH to first fluctuate, and then to stabilize at an inappropriate range. Initially a misjudgment could lead to the conclusion that the increase in orthophosphate concentration in the continuous blowdown line could be the result of acid ingress and, therefore could have caused corrosion in the return steam condensate system, BFW system, and boilers. As shown in Figure 5.28, the incorrect steps taken to return the chemistry control conditions to normal brought the boiler condition closer to corrosion [12, 23, 25, 30].

The data trend which is presented in Figure 5.29 comes from the transmitted results of an online specific conductivity analyzer on a continuous blowdown line of one of the operating boilers (boiler D). This trend represents three important times in which actions have been taken.

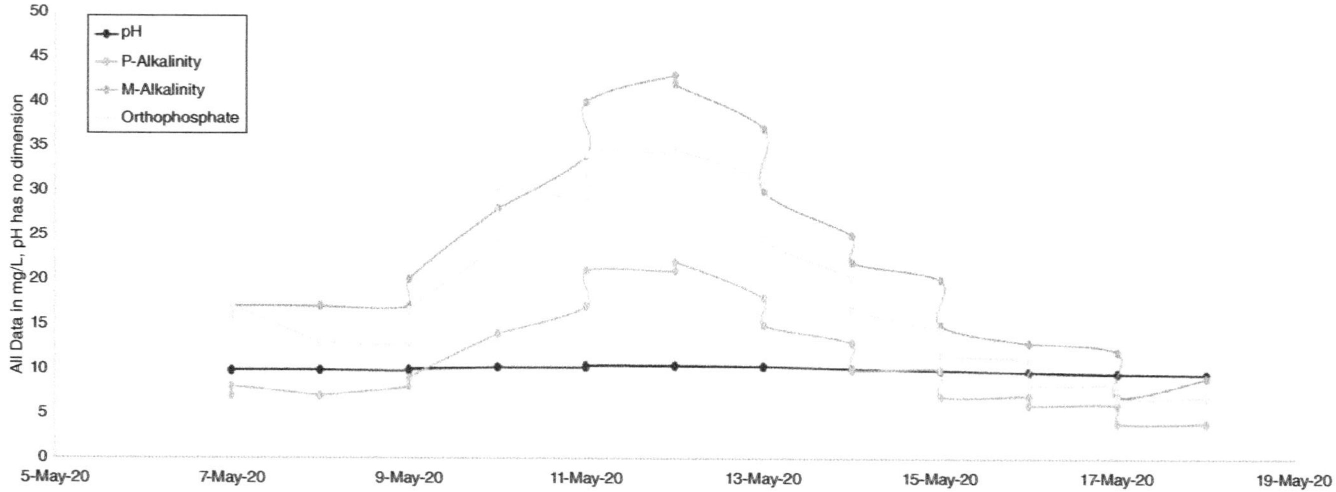

Figure 5.27 Laboratory results for boiler D in steam-generating unit.

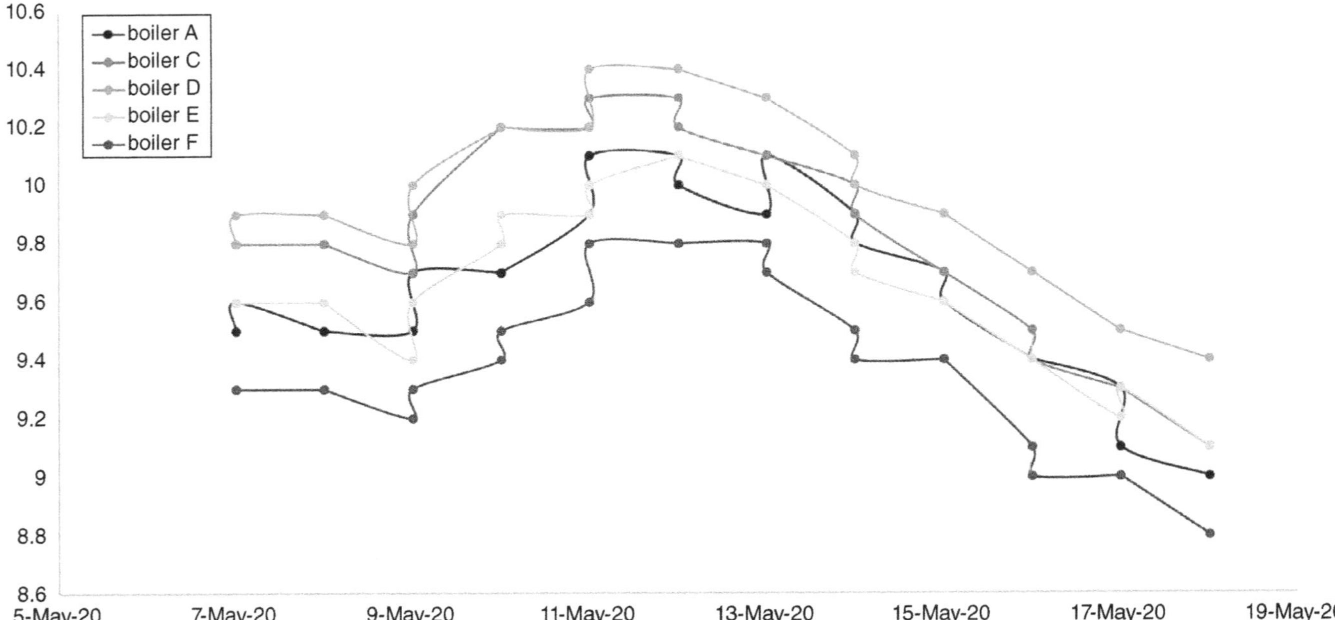

Figure 5.28 Trends of pH in five operating boilers.

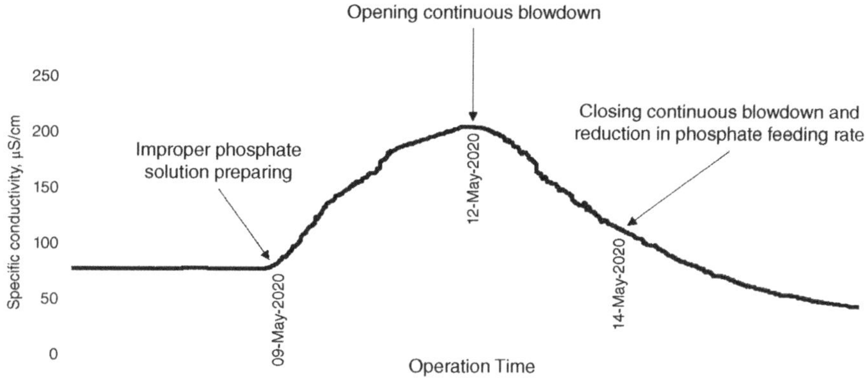

Figure 5.29 Specific conductivity trend for BW in boiler D.

5.6.2 Putting A Kettle-type Reboiler into Service that Has Been Under Maintenance

The return steam condensate in a steam-generating unit had high iron concentrations during three 12-hour shifts. The utility process engineer found the laboratory results to be correct after requesting re-sampling and re-analyzing of iron ions and total iron. Therefore, he examined the following influencing parameters:

- proper operation of deaerators;
- nitrogen blanket of steam condensate storage tanks;
- nitrogen pressure of the return condensate system in its production sections at the processing site;
- feeding of sufficient neutralizing amine and proper pH;
- sufficient feeding of volatile oxygen scavenger to prevent oxygen corrosion in the steam condensate system;
- steam condensate outlet temperature of all condensers inside the processing sections of gas plant.

The results showed that all these parameters were in acceptable condition and the investigation of operation logs did not show any change in their trends. Simultaneously, all laboratory results related to the return steam condensate from 23 points of the processing site were studied. It was found that from five 12-hour shifts before (60 hours), until the iron concentration in the main header of the return steam condensate increased (when the utility process engineer noticed), the steam condensate output from a reboiler in the NGL extraction unit had a high concentration of iron and the operation supervisor of NGL extraction unit did not pay attention to this increase.

Upon further investigation, it was determined that six shifts before (72 hours), one reboiler had been re-installed in the area where the high iron concentration was detected. Several steam tubes in this kettle-type reboiler were damaged, which was serviced to a de-ethanizer column. It took three days to repair and re-service the reboiler; this was while the steam and its condensate paths were properly blocked and pressurized with nitrogen. The return steam condensate system was also kept under nitrogen pressure. After servicing, the iron was introduced to the steam condensate collection system.

According to the operation instructions, the field operator had to open the condensate discharge route to the WWTP to return the steam condensate special conductivity (which is the main indicative parameter of steam condensate contamination) and other parameters to normal BFW specifications. Unfortunately, this action was forgotten and the metal contamination caused by opening and closing the tube bundle, shell, and channel head, as well as the stagnation of steam condensate collection system in that section for three days, had found its way back to the main steam condensate header.

Under such circumstances, simply paying attention to the rising iron concentration in BFW could lead the process engineer to the conclusion that an active corrosion was occurring in the system. While only the entry of a contamination from upstream had led to this event, it should be noted that the entry of contaminants into the boiler cycle can cause precipitation, and in severe cases, can cause some types of corrosion over time [12, 18, 19, 23]. The steam condensate coming out of this reboiler was collected in the route of the suspect steam condensate, because the steam pressure used in reboiler was 26 bar lower than the pressure of the process fluid (ethane, propane, and butane rich gas). The steam-generating unit was designed in such a way that there was no CPP (several trains of ACF-IOX) in the return suspect condensate route, and only one deoiling system (two trains of ACF) was designed and installed.

5.6.3 Problems in Sampling from Deaerator and Oxygen Scavenger Analyzation

The steam-generating unit of a gas plant had two deaerators that were in service in parallel arrangement. Oxygen and other non-condensable gases were removed from the mixture of MU water and returned condensate by stripping steam (deaeration). An oxygen scavenger with the predominant formula, Diethylhydroxylamine, DEHA (chemical elimination mechanism) was used to ensure complete removal of oxygen up to 7 ppb (boilers were operating at pressure of 42 bar) [18, 19, 25] and to increase the efficiency of oxygen separation by steam (physical elimination mechanism). The oxygen scavenger feeding point was deaerated water in the steam condensate-accumulating drum.

Over a period of several months, the inlet steam condensate flow to the two deaerators fluctuated and differed. The two deaerators were installed perfectly symmetrical with respect to the steam, return steam condensate, and MU water lines. However, a problem in controlling the level of one of the deaerators caused the feed to be about 35% less than the other deaerator.

Inadequate monitoring of process parameters resulted in the same feeding rate being selected for oxygen scavenger of both deaerators. Therefore, the DO in the BFW outlet of the deaerators with water flow had more fluctuations and reached 15 ppb in most cases (about 115% increase). Therefore, by studying the condition of the system and the change in the rate of oxygen scavenger feeding into the deaerator that had a higher inlet flow, the problem of DO fluctuation was completely eliminated. This was done by an independent feeding pump.

However, just measuring the DO is sufficient to ensure the absence of oxygen corrosion in the operating conditions of the steam generating unit (water tube boilers with an operating pressure of 42 bar); the oxygen scavenger supplier had recommended the maintaining of a residual of DEHA in range 150–200 ppb in BFW, and monitoring of this parameter was part of the performance guarantee of the oxygen scavenger chemical.

Six months after the control problem for receiving the specified flowrate of MU and returned steam condensate was resolved and the inlet flowrate to both deaerators was equalized, the oxygen scavenger residual fluctuated sharply over a two-week period. While the rate of oxygen scavenger feeding to both deaerators was 4.2 L per hour, the DEHA residual in one deaerator was four times lower than the other. At the same time, the concentration of DO in the BFW output from the same deaerator was reported to be about 10 ppb, and the operation team feared an increase in possible oxygen corrosion.

After several studies, it was shown that the laboratory sent its Orbisphere continuous oxygen meter out of the gas plant for periodical calibration and instead used colorimeters, which can cause errors. It was also found that the lab department had lost the portable residual DEHA measuring device two months before, and the sample was transferred to the laboratory at 30 °C (more than 3-km distance) and analyzed after one hour. This was not enough to cause the result of one of the deaerators to deviate, because this method was same for both; hence, the investigations continued.

Technical studies and subsequent analyses all showed that both the DO in the BFW and the DEHA residue were appropriate from the outset. However, in this sample, all signs indicated the onset of oxygen corrosion conditions in the system.

Since the reactivity of the oxygen scavenger to air was very high, attention was paid to the lack of sealing of the sampling lines. It was found that the deaerator sampling lines, which had less DEHA residual in analyses, had leakage which returned to normal after sealing and tightening the related connections; the piping of these sampling lines was an unusually long design. Further investigation showed that the cooling water valve in the same deaerator sampling cooler also had problems and, due to the impossibility of temperature regulation, the sample man often collected the sample at a temperature of more than 50 °C instead of 25 °C [44, 45].

5.6.4 Problems in Sampling and Analyzing Specific Conductivity from Demineralized Water

A water treatment plant met the needs of the steam-generating unit for MU water. In the pre-treatment unit, the technology used to treat water was IOX. In this unit, two mixed beds, each with a capacity of 100 m^3/h, were designed and installed. These resin beds were of strong cation and anion type and could operate in series or in parallel arrangement. The design specification of product quality, demineralized water, was the specific conductivity and less than 0.2 μS/cm.

To monitor the demineralized water quality produced in this unit, an online analyzer was installed in the output stream route of this unit and sent a specific conductivity to DCS every three seconds. According to the maintenance department policy, an analysis was performed by the laboratory as a cross-check for each analyzer within the utility processing site.

Because the analyzer had failed three times in the past year, quality controllers at the water treatment plant had more confidence in the lab results. Hence, when the result of the analysis of 0.4 μS/cm was reported, (and the many times rinsing and regeneration of the resins did not change this result), they were sure that the resins had reached the end of their life and that a series of initial and then detailed tests or a unit performance test should be run [22, 23, 31].

It was soon discovered that the laboratory had not used a degassing device to measure specific conductivity, which was an important practice [31]. This was while the existing online analyzer sensor had a degassing probe. It was also found that the temperature corrections used in the online analyzer were not the same as the laboratory portable device, and that the change in speed at the inlet of the degassing probe and its inadequate ventilation due to the short length of the siphon tube connected to the probe caused the values to transmit higher than expected. In this case, it was thought that the quality of demineralized water (MU water to the steam generation unit) would definitely cause precipitate in the boilers along with corrosion by deposition of hardness ions in the downcomer tube bank, and damage by overheating [18, 19, 23]. However, studies had shown that water quality is quite good from the start, and the trends reported by the lab, as well as the online analyzer, had been misleading.

5.6.5 An Improper Sample Point and Mistake in Determining Free Residual Chlorine

A utility processing site produced the industrial water it needed by desalinating seawater with MED–TVC technology. The water entering the processing site was chlorinated and filtered seawater. Because residual free residual chlorine could damage desalination equipment (stainless steel and copper alloys), free residual chlorine was first removed with sodium metabisulfite to the lowest possible level (0.15 ppm per design specification). While a safe dose was intended for the dechlorination agent, obsessive thinking that resulted from successive incorrect solution preparation (bisulfite dilution) over a period of time allowed the free residual chlorine remaining after dechlorination to be measured. The utility process engineer named this point "after bisulfite feeding."

It is quite obvious to what extent this non-standard sampling point affects energy consumption and the environment, as well as the overall efficiency of the process. Because it was installed at the end of the seawater supply header, it must be a continuous drain. Because electricity had been used to transfer water from the SWI to the gas plant operation zone, some of this energy had been wasted through this sampling point. The total efficiency of the unit (but not each of the desalination packages) was defined in terms of total seawater pumped and total desalinated water produced, and because seawater was discharged continuously from three similar sampling points (each package = one sampling point), a significant portion the chlorinated seawater returned to the sea (waste of energy and more GHG emission).

A non-standard sampling point was provided for this purpose. There was a drainage-piping element in the desalinated seawater piping at the desalination unit. The utility process engineer, regardless of the piping arrangement and based on trust in the proper functioning of the isolation valve and the existing check valve, selected it as a sampling point and placed sampling and determination of residual free residual chlorine in the list of routine laboratory analyses. Figure 5.30 clearly shows the arrangement within this unit.

The result of laboratory analyses showed that the free residual chlorine in this pipeline was between 0.3 and 0.8 ppm. The vendor of the seawater desalination package, meanwhile, set the maximum working concentration in seawater feed equipment at 0.15 ppm. Therefore, it was concluded that free residual chlorine had entered the packages and corrosion conditions had been provided. The laboratory results for determining the sodium bisulfite concentration in the dosing tank, as well as the rate of its feeding into the inlet seawater stream, were immediately reviewed and fortunately everything was good.

Subsequent investigations showed that the isolation valves (two gate valves) were leaking and chlorinated water had entered the header. Due to the location

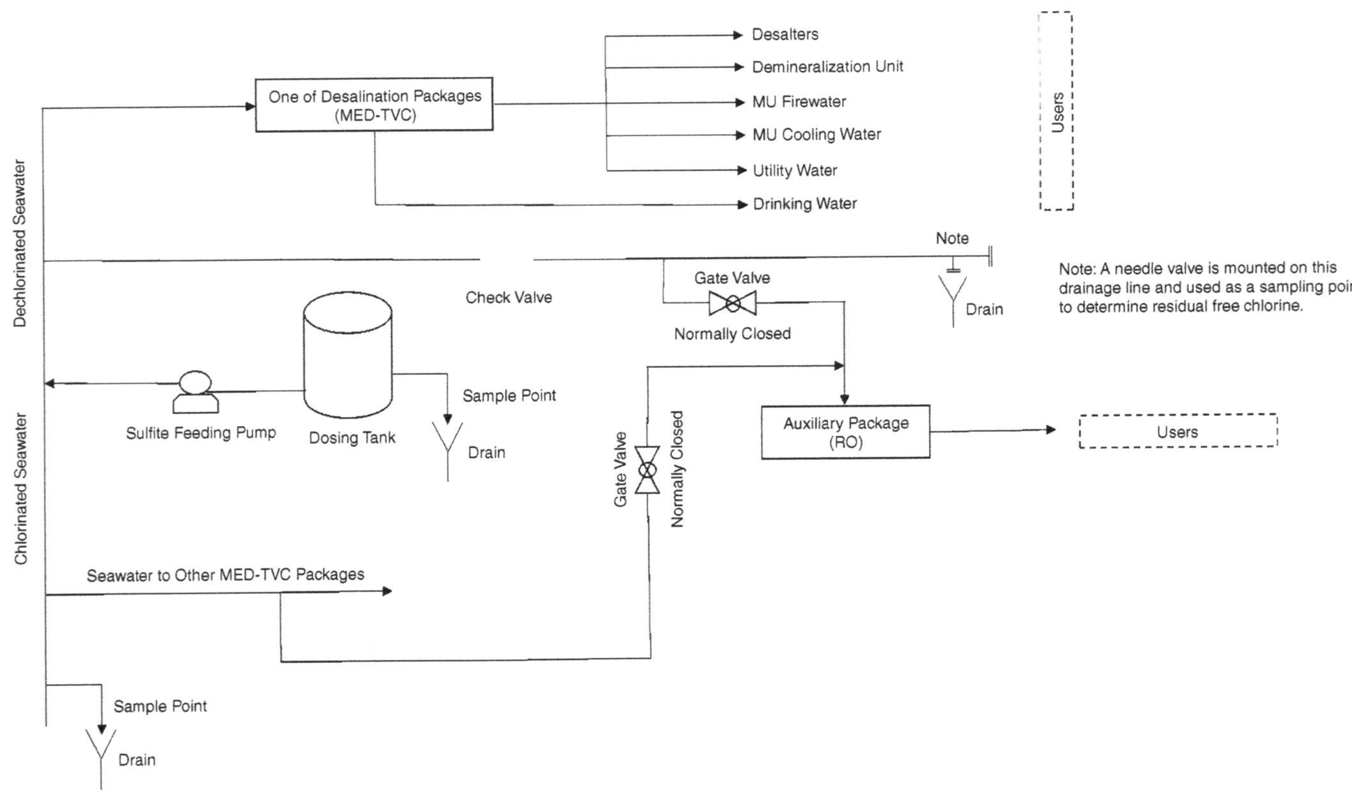

Figure 5.30 Seawater desalination unit, its related sample points, and chemistry control scheme.

of that point at the end of the header, and in addition to the stagnation of water, selecting that point for sampling and determining the dechlorination efficiency could not provide a representative sample to the process engineer. Therefore, an incorrect determination of the sampling location, as well as the application of unnecessary analysis, and due to covering, the sensitivity of proper sodium bisulfite solution preparation had caused the system to be considered subject to corrosion due to more free residual chlorine than design specifications.

5.7 Chemicals, Their Corrosion, and Impacts of Their Corrosions on the Environment

Based on the experiences of presence at different gas plants and their utilities, the impact of chemical consumption on the environment is classified into four separate categories. This classification is given in Figure 5.31, which classifies the environmental impacts of chemical consumption within this chapter of the book.

In this section an operation history is presented where improper use of the chemical caused corrosion, the corrosion then caused one or more of the environmental impacts created.

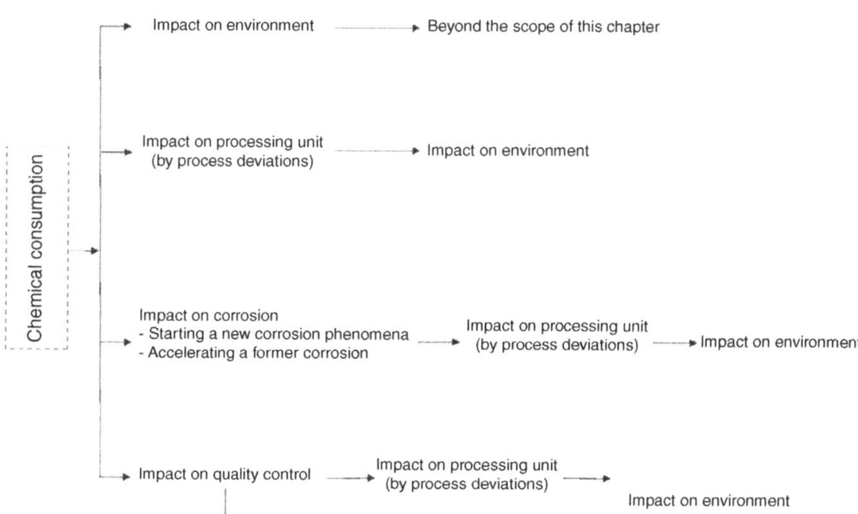

Figure 5.31 Classification the impacts of chemical corrosions and impacts of their corrosions on the environment.

5.7.1 Operation History 5

A sour water header of a gas plant was inhibited from corrosion by a film-type CI. This inhibitor has a specific emulsifying property that, in the event of overdose and/or improper control of the residual at downstream, prevented the hydrocarbon (organic phase) and the aqueous phase from separating in feed-flash drums.

As stated in operation history 23 (Section 5.10.1.5), the increase in the CI dose to the sour water transferring pipeline eventually caused severe localized corrosion in the reboiler of the two sour water stripping columns, and the two trains of stripping were taken out of service. The design had an open steam arrangement that could be continued if the reboilers were not ready. Open steam is a type of generating vapor phase in the fractionating column (here stripping) [46]. Therefore, open steam arrangement was used for six months until two tube bundles were prepared for each of the reboilers.

The use of steam caused the steam to return to the condensate system and the utility plant was forced to supply 23 tons of demineralized water per hour as MU water to the steam generation unit. Open steam also increased the volume of raising vapor phase in the condenser. To condense this high volume of vapor (mainly water) more air fans were put into service in the column condenser (in an air-cooled condenser there are two or more air fans that control the rate of heat exchanging). Obviously, they experienced higher electricity consumption due to the service of more motors for air fans, which caused higher GHG emissions.

Consuming more MU water meant consuming more energy, more chemicals, and more person power, along with more quality control acts, all of which led to more GHG (see Section 5.4). Also see Industrial Case Two in Section 5.5.6.2 (*Increase MU water flowrate*) for study about impacts of MU water on corrosion, energy, and environment.

5.8 Configuring EMPA

In this section, management is defined as the complete knowledge of the work components (different activities) in a system and their proper performance to achieve maximum benefit. Here, the maximum profit is the result of the following conditions:

- minimum costs for chemical procurement;
- minimum costs of delivery and storage;
- minimum consumption in different points of a processing plant;
- minimum direct and indirect impacts on the environment; and
- preparing consumption reports based on facts and operation feedbacks from different points of view, and modifying decision-making processes.

Figure 5.11 represents these different aspects of which corrosion is an integral part in the present chapter. It is important that all of these conditions and related activities are within the framework of one cycle.

A model or framework of EMPA in the utility processing sites of the natural gas processing industries is to include all activities related to working with chemicals in this field (this framework can be generalized for either processing zones of gas processing or the oil and petrochemical industries). Different activities can be identified for this purpose. Each of these activities represents a specific stage of this management or framework. Inside each, a set of consecutive jobs is performed, and joining this chain will provide a suitable framework for managing different activities.

The activities of this model are, in fact, the stages that can be performed sequentially with chemicals, these are selection, procurement, delivery, storage, consumption, and reporting. Like any engineering system or any type of management system to implement a model, documentation will be an integral part of this "effective management." These stages are a separate and general set of steps. Several factors such as the volume of chemical request, the type of chemicals used, number of chemicals, and the geographical location of a certain industry (usually processing site) can cause a number of these both stages and steps to merge or overlap.

In a gas plant, like any other process industry, all the quality control parameters are on the threshold of change. Suffice it to say the temperature of the reflux of a distillation column, the opening percentage of a Joule–Thomson valve, or the flowrate of the inlet steam to the reboiler of a cryogenic column will change slightly. As a result, field process engineers have to wait for the quality of product(s), the quantity of production, or both to be affected. Of course, more comprehensively, other events such as excessive energy consumption, and increased workload of personpower must be expected, which are usually the next priority for a gas plant because of its commitment to production and, after that, export.

If we consider a modern gas plant that is built as an unmanned type, then in three examples given above, the control systems consisting of logics and other components in the control algorithm (sensors, transmitters, actuators, valves, calculators, controllers, etc.) determine the deviation(s) in a certain process. This deviation is due to an error in the control system, a malfunction of its components (such as valves or sensors), or is a response to fluctuations in the system that must be found at the initiating point.

There are obvious differences in chemical consumption with a process control action. Most of the activities in this section are based on human jobs. Clearly, stages such as selection, procurement, delivery, storage, consumption, and reporting are completely human-based, and a set of organizational guidelines and/or engineering procedures are used to perform them. However, the consumption activity of the processing site may have two different approaches; (i) chemical feeding based on manual calculations and issuing guidelines for preparation of solutions and feeding, along with quality control with predicted physical–chemical lab experiments (or online analyzers), or (ii) chemical feeding based on pre-designed control algorithms. In the rest of this chapter (Section 5.10.1.5), more information about consumption details is provided on the processing site.

The above paragraphs clearly show that documentation is a serious need when working with chemicals. Like the digital control systems and histogram features embedded in the DCS, here in most cases, there is no such thing as a histogram (or other type of automatic record-keeping systems). As a result, when you need to refer to DCS records in terms of chemical feedings, only a few limited numbers of all the required information will be available, such as results from analyzers (if they are calibrated and transmit values properly) or numbers recorded from flowmeters. Some processing sites have features such as metering pumps that record the fed flowrate of the prepared solution or neat liquid chemicals [47]. Of course, the numbers recorded by them without regular and effective periodic calibration do not have any technical value for chemicals impacts on processing sites.

Therefore, in order to control all jobs in each activity of this framework, all steps must be effectively documented. Documentation will provide an opportunity to return to the system's histories and evaluate alternative routes, as well as to predict the operating conditions of areas affected by chemical feedings and to be aware of possible corrosion. Documentation is also a powerful tool for analyzing the root causes of process deviations that may have been affected by chemical feeding. Figure 5.32 represents the sequence of these activities in the context of a working model or frame.

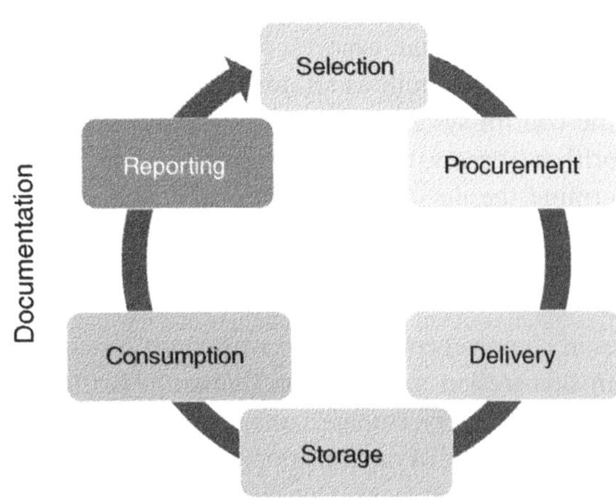

Figure 5.32 A framework for EMPA.

5.9 Setting up an EMPA

Setting up an EMPA can be easily accomplished by following these steps:

- provide a clear definition for each of the activities listed in Figure 5.32 (that is detailed in Section 5.9.1);
- collect field information to match the knowledge associated with each process additive in these defined activities; and
- define a specific job sequence for each activity and then follow it to achieve effective management.

The number of chemicals used in a processing site are limited, hence, the following are expected.

- The knowledge related to the chemicals is quite limited, and it is possible to get a good knowledge of these process additives by spending time, using the experiences of industrial experts, and manufacturers/suppliers.
- The parameters that are affected within a processing site by the use of these chemicals are quite limited and specific, and it is possible to record and classify them in a short period of time. These parameters are unique to a specific processing site and should only be written based on the experiences of the field process engineers and/or field operators on the same site.
- The impacts of using chemicals can be achieved using scientific sources, experiences of industrial experts, and experiences of the same processing industries. If a processing site has a good structure for reporting the performances of its various departments, it can be expected that these impacts are easily identifiable and quantified, and then planned for improvement.

It is important to note that the use of data and experiences even from similar processing sites must be done with great care and rigor. There are many reasons why this should not be a priority decision for chemicals that are beyond the scope of this chapter.

Knowing the above information and the method of obtaining each, it is easy to set up an EMPA. As (i) chemicals, (ii) applications, (iii) affected process parameters, and (iv) the impacts of chemical consumption (subject of Figure 5.11) are well defined, they can be achieved by conducting various sensitivity analyses to the knowledge related to them. Management of change (MOC) complicates this implementation method slightly; the initial suggestion may be that MOC should not be incorporated into these procedures at first, and after an EMPA has demonstrated success and has been properly implemented into the system in a processing industry, we then define MOC and attach it to the current EMPA.

5.9.1 Description of Activities

Here, the various activities of the EMPA are described separately based on Figure 5.32. In order to better explain, this section describes one or more operation histories (limited in description) in each activity. Each of these operation histories includes other activities, but in the process of describing them, an attempt has been made to express the same activity more prominently.

5.9.1.1 Selection

Each unit operation in a processing plant has its own considerations, and to reach an optimal point in selecting a chemical, one must first consider these unique considerations. The most important consideration in choosing a chemical is the quality control that is gained by that chemical. A process additive must first meet the purpose required by the site process designer or field process engineer, then it will be time to evaluate its other features, and it is just as important to address its impacts on the production processes.

It should be noted that the selection of chemicals for a processing site is part of the site developer's commitment to detailed design (detailed design is one of subsections in plant's development contract). Therefore, the initial chemicals selected by the process designers in the project are delivered to the owner as one of the project documents and can be used for the normal operation of the site. This document can be used until the last day of a process industry under the following two conditions; (i) manufacturing/supplying chemicals can be performed with no difficulty, (ii) changes in site processes do not force the field process engineer to change the chemical. Therefore, it can be expected that in the hydrocarbon processing industries, the same basic document, called the "Chemical Selection" will last for the rest of plant life.

5.9.1.2 Operation History 6

An important distinction must be made between "substituting" a chemical with "replacing" it with another. For example, at a gas plant, a film-forming CI was used to prevent corrosion of the sour water pipeline. Some process limitations as well as changes in the chemistry (quality) of generated sour water led the process engineering department to eventually decide to use a neutralizing amine instead of a film-forming CI. In this case, an alternative was made (replacement) because the new chemical was not functionally similar to the previous one.

Figure 5.33 shows corrosion-monitoring coupons after six months in sour water pipeline. As it turns out, when using the film-forming CI, a number of pits are created by the localized corrosion. However, in the case of using a neutralizing amine, a uniform iron sulfide is seen on the surface of the coupon.

The author's experience showed that under normal operation of a processing site, the need to select a new chemical will only arise when it is not possible to

Figure 5.33 Two corrosion coupons in same location of a sour water transferring line. The left-hand coupon represents the performance of film-forming CI, and the right-hand coupon represents the result of applying a neutralizer amine.

procure a preselected chemical. This impossibility can be due to one of the following reasons; discontinuation of production of a chemical, increase in price, change in the formulation of a new product (from the same manufacturer), increase in process costs if that preselected chemical is used, impossibility of supply for some reason (such as the strategy of independency), or new environmental laws.

First, two basic questions need to be answered; Is the chemical mentioned in the chemical selection document already available? Will the current operational needs be met with the same chemical as mentioned in the chemical selection document? If the answer to both of these questions is yes, then there will be no activity called selection in EMPA, but if the answer to one or both of these two questions is no, this stage must be passed.

Here are the key points to consider when selecting a new chemical:

- cost;
- environmental regulations;
- ability to use storage and feeding equipment available on the processing site;
- access of technical knowledge to work with new chemicals;
- production of intermediate and final products of processes with the desired specifications in sale and export commitments, and basic design;
- possibility of chemical quality control (i) in the stage of delivery from the manufacturer/supplier and storage in the warehouse, and from warehouse to processing site, and (ii) possibility of quality control jobs after feeding into process streams;
- compliance with the governance policies of that processing industry in terms of the tendency to depend on non-native manufacturers and related obstacles;

- paying attention to the geography of that industry and examining the possibility of storage with existing facilities, and way of transferring purchased chemicals to processing site.

5.9.1.3 Operation History 7

A gas plant had a firewater system with an approximate storage volume of 100 000 m^3. The total length of the distribution lines in the gas plant's firewater main was more than 36 km, and there were many points where the water remained stagnant for weeks. There was a weekly instruction to return 10% of the firewater main volume to the storage tanks (in 30 minutes), in order to circulate stagnant firewater in the system and mix the chemicals properly throughout the network.

A biocide was selected for this gas plant to prevent biological growth as well as MIC (selection was done in a detailed engineering step of gas plant development). According to the biocide manufacturer's recommendation, the field operator should feed 100 ppm of the mentioned biocide into the tanks once a month and activate the water circulation mechanism for three hours.

According to the basic design, the MU water entering the firewater system was of the desalinated seawater type and did not pass the remineralization stage. Therefore, it did not have enough alkalinity to resist the acidity of biocide (which had a pH of 2.4) and the system suffered severe corrosion. This was while there was no package to regulate the pH on the utility processing site. Therefore, after the low pH (about 5.8), corrosion occurred in the firewater distribution system, so the feeding of biocide (means the source of acidity) was stopped. Figure 5.34 represents a standard corrosion coupon which was corroded by acidic mechanism (about 12% weight loss).

Ignoring biocide feeding caused biological growth in the system. Figure 5.35 represents a non-metal base (made of wood) of corrosion coupon in a rack with a residence time of three months. Coupon base was in cylindrical shape and a biofilm was formed on its surface. However, it was not clear whether this biofilm formed on the base of the coupon could have caused MIC [39]. The present operation history represents how an error in choosing a chemical (here, a biocide) can

Figure 5.34 A standard corrosion coupon in firewater system in low pH condition.

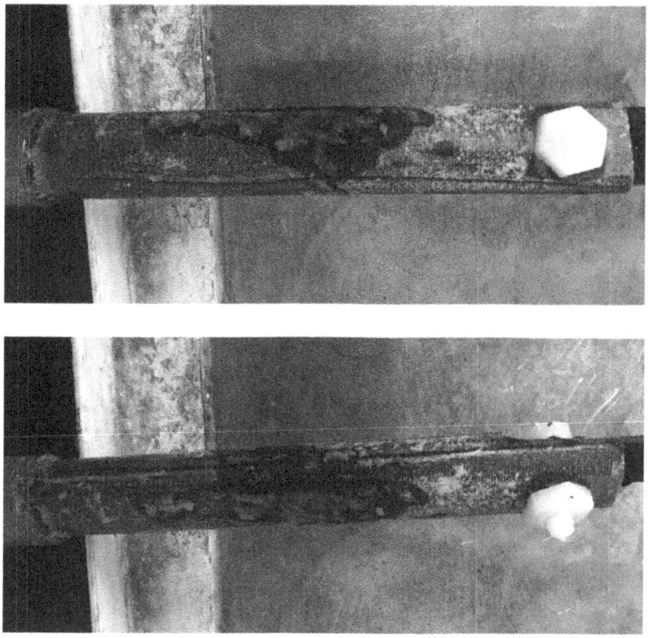

Figure 5.35 Biological growth on both sides of the cylinder (corrosion coupon base).

lower quality control of fluid, advance the system into corrosion by acidic corrosion, and increase the potential of MIC.

5.9.1.4 Operation History 8

A gas plant had a 43 000-cubic-meter firewater system. MU water to the system was desalinated seawater with a conductivity of less than 10 μS/cm and total dissolved solids (TDS) was negligible. The CI in the basic design was formulated based on silicate and phosphate [14, 19]. The corrosion coupons in the system are given in Figure 5.36.

Studies showed that this CI has not been compatible with water chemistry, and the pH required to form a barrier film to inhibit corrosion (~ 9) was not present in the firewater. Due to the unprepared CI feeding package (about three months), the firewater main was filled with aggressive water and there were large amounts of primary corrosion products and suspended iron in the system. Therefore, the modified CI was subjected to regulate pH, and the phosphate to polyphosphate ratio was adjusted to the appropriate value. After applying various changes in the

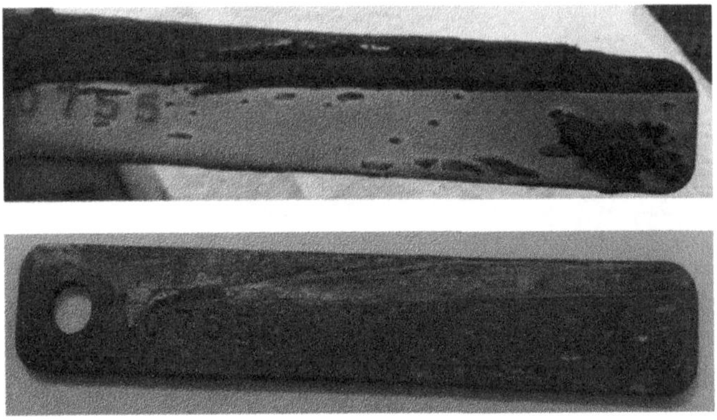

Figure 5.36 Corrosion monitoring coupon in the firewater system before and after cleaning.

Figure 5.37 Corrosion monitoring coupon in the firewater system. There were no precipitations and depositions on the metal coupon surface.

system, the corrosion results reached an excellent level (0.03 mpy). Figure 5.37 shows a standard coupon in the firewater system which was inhibited by modified CI.

5.9.1.5 Operation History 9

An operation history of the selection of a new set of chemicals refers to the replacement of two reverse demulsifier and flotation agents. The design documents for a WWTP suggested two specific chemicals (produced by a well-known company) to improve API separator and IGF performance (see Figure 5.12). Five years after the gas plant was commissioned, the possibility of supplying both chemicals was eliminated. In order to maintain the efficiency of equipment, these two chemicals were substituted. The result of this substitution was a significant benefit in terms of oil recovering capacity, as well as a reduction in application of utilities such as diluent water and electricity and a reduction in person power requirements of 50%. The total consumption of both chemicals decreased from 97 tons to 13 tons, yearly. It is clear that sometimes the impossibility of procuring a chemical will lead to the selection of a better one with fewer impacts on the unit operation (and in some cases on entire processing site).

During the substitution process (selection of a new set of chemicals), three things happened that could help clarify the discussion:

1) A manufacturer was able to successfully pass the laboratory results and pilot test by obtaining a novel combination of polyelectrolytes, then the chemicals required to be used for one year in a WWTP were produced in a chemical plant. The chemical plant was located in the a cold region, and given that the WWTP had placed its order delivery time in February, these two chemicals had frozen due to cold travel conditions (about 1600 km distance) and changed their physical–chemical properties. Conversion to two phases along with some other problems made them difficult to use in WWTP. A large part of the shipment was returned to normal by a corrective action in the chemical plant, and it was decided that after the shipment, the chemical would be made in the same area where the WWTP was located and delivered to the gas plant warehouse.

Therefore, one of the points that should be considered in selecting new chemicals is the geographical location of chemical manufacturing sites. Sodium hypochlorite in particular falls into this category. Perhaps better conditions can be achieved by increasing the cost of chemical formulas and changing the way they are delivered; but the total operating cost of the processing site should also be considered.

2) The reverse demulsifier chemical, when fed into the influent stream of API separator, clogs and blocks the feeding line into the oily water influent pipeline. The reason for this was the high strength, and consequently high kinetic of reaction of the alternative (substituted) chemical compared to the previous one. Therefore, studies have been performed to reduce the strength of the new reverse demulsifier and to use an alternative arrangement for feeding.
3) After using the new arrangement and feeding a lower dose than the previous chemical, it was found that feeding pumps cannot feed values less than 10 ppm. Therefore, it was necessary to produce a low-strength chemical (which was not possible in terms of stability) or to dilute it on-site with fresh water. After two months of working on the formulation of this chemical, the manufacturer of the chemical delivered a sample with the ability to dilute it to the process department in WWTP. The interesting point was that in the imported type of this chemical, it was not possible to dilute with water, and if the contents of the dosing tank were not used up in three days, the entire solution would be converted to two phases and became unusable (because of humidity absorption from atmosphere). Therefore, making of a water-soluble chemical with good stability for the manufacturer and the WWTP was considered a success.

Industrial Case Three represented well how an improper selection and overfeed of a reverse demulsifier chemical can cause excessive water received in the gas condensate stabilization unit and eventually corrosion in the distillation column.

5.9.1.6 Procurement

It should be noted that the chemicals used at the processing site impact on the entire operation (Figure 5.11). Therefore, when procuring, departments should pay enough attention to all the items that have the potential to increase the unwanted impacts of chemical consumption. Developing quality control conditions in the purchasing contract, selecting and introducing the quality control reference laboratory, and requesting the official publishing of the guaranteed quality control parameters are among the measures that must be taken into consideration when purchasing chemicals.

It is expected that some deficiencies will be felt in the initial purchases of chemicals, in which case a return to the sequence cycle of activities (Figure 5.32) is the only way to prevent further recurrence and losses. In other words, the procurement department must reflect the results of the activities cycle (EMPA) in its current and forthcoming purchasing contracts. The most important part of this work is to pay attention to the quality control feedback of intermediate (nonmarketable) and final products (marketable) in processing sites, as well as the impacts of corrosion rate and the environment on plant performance, which are all connected to the procuring of inappropriate chemicals.

When it comes to procuring chemicals, the quality of the chemicals procured is not always a fundamental issue. Sometimes the delivery time and the packaging or delivery method of chemicals are of equal importance, and in some cases even higher priority than the quality of the procured chemicals.

5.9.1.7 Operation History 10

Sodium hypochlorite, NaOCl, is a chemical used in the disinfection and control of the biological growth of seawater, treated effluent, and maintaining of biostatic conditions of raw industrial water produced in the utility units of a gas plant (Tables 5.3 and 5.4, and Figure 5.3) [48, 49]. Storage conditions at the processing site as well as being located more than 1000 km from the nearest sodium hypochlorite generation plant had caused the amount of free residual chlorine remaining in the process streams to be unsuitable, and due to the volatility and instability of this chemical over time, quality control for intermediate and final products of plant could not be done properly. One of the reasons for this was the purchase of sodium hypochlorite in 220-L barrels, which, along with the small amount of chlorine solution preparing/dosing tanks, caused only 15l of a 220-L barrel to be charged in each of the three NaOCl feeding packages and left another 175l in the barrel. This used barrel was not returned to the warehouse due to lack of logistics, and was kept in poor condition on the site. Analyses showed that the commercial percentage of NaOCl was stored in a barrel with the lid open for five days in inappropriate conditions inside the processing site, and the NaOCl decreased from 12% [29, 48, 50] to less than 1.5%.

Subsequent studies showed that despite the purchase of NaOCl from the chemical plant with a commercial percentage of 12% and payment based on this active percentage, the active percentage in the chemical at the time of delivery was a maximum of 7% (and sometimes 5%), indicating the method of transporting the chemical from the factory to the gas plant was unsuitable. Therefore, new clauses were added to the purchasing contract to prevent future events from happening. For example, one of the clauses was that in subsequent purchases, all shipments ordered in 40-L barrels would be delivered to the gas plant, because the site could not dilute the concentrated chemical at delivery time based on guidelines.

5.9.1.8 Operation History 11

A gas plant consumed a reverse demulsifier to treat its oil-contaminated effluents from hydrocarbon processing and off-site units (Figure 5.12). It was a polyelectrolyte sent from a chemical plant at a distance of 1500 km. Because the need for this chemical depended heavily on on-site operation conditions and the production of effluents from overhaul and surface flooding from monsoon rains, purchasing this chemical reduced its efficiency during storage as much as the annual requirement, and in some cases the penetration of moisture into the barrels (or dosing tanks) lead to the absorption of moisture and conversion into two phases of the chemical, which ultimately made it impossible to use.

The first decision to avoid reducing the quality of the chemical and thus increasing the corrosion rate (Industrial Case Three) was to always purchase fresh chemical using a partial shipment contract. Another alternative was to purchase the raw polymer material from the market and dissolve it in demineralized water (solution preparation) at the processing site. Studies have shown that purchasing dry polymer can impose more complexity on the EMPA and possibly not effectively increase the yield of oil removal. Partial shipment costs (like additional transportation jobs) over three years also forced the procurement department to notify the engineering department to reconsider the selection of chemical or oil contaminated effluent treatment process. Therefore, it can be expected that in Figure 5.32, in addition to a large cycle among the six major activities in EMPA, there is also a set of internal feedback among the sequential activities of this cycle.

5.9.1.9 Delivery

Delivery of incoming chemicals to a processing site (i.e. its warehouses) is of great importance. This activity can be implicitly included in the contract of purchasing from the manufacturer/supplier or defined as an independent activity in the

EMPA cycle. It is recommended in industries where there are many different chemicals to define delivery as an independent activity. Sampling of all consignments received according to current standards, analysis of chemicals according to the provisions of the purchase contract, and obtaining analysis sheets from an internal laboratory (in plant) or external reference are the most important parts of the delivery process.

The work in this section lacks any flexibility and decision-making based on the existing conditions. Any deviation in the implementation of written regulations on the steps of delivery of chemicals can lead to severe loss or failure to the processing site, production economy (includes corrosion), and the environment. The working algorithm in this activity is such that if the answer of a step (block) is negative, the flowchart is then stopped and referred to the purchase agreement clauses for how to return the chemicals back to the manufacturer/supplier.

Delivery issues can be due to human error or related to warehousing. It is important to note that achieving an ideal delivery is by no means possible unless the two activities of procurement and storage are done properly. So, (i) delivery of cargo during non-working hours of delivery warehouses (after 16:00) as well as on holidays [47], and (ii) arbitrary charging of materials such as solvents in storage tanks, which are often done incorrectly, are two recurring human factors in the steps of delivering chemicals. In some cases, the driver has been seen filling the glycol tank without informing the chemical delivery operator, and due to non-compliance with the site's current instructions for emptying the road tanker, once the atmosphere contacts the glycol, the glycol is partially degraded due to reaction with oxygen.

5.9.1.10 Operation History 12

A utility processing site had a centralized section for storing chemicals. Sulphuric acid is stored in a horizontal tank and piped to various units of the processing site. Ten road tankers entered the site almost every three months to fill this tank. A strainer and filter arrangement and a road tanker-emptying pump sent the acid into a horizontal tank, and two sulphuric acid transfer pumps sent acid to three points at distances of 100, 500, and 1200 m at the processing site. After some time, the acid transfer became difficult and eventually stopped.

Investigations showed that the acid content of one of the 10 tankers was transferred during non-working hours on holidays, and due to the requirement for a road tanker to enter the site during the workday with a traffic permit, the Utility Operation Department had to accept the acid tanker at 18:00. This clock coincides with the change of operational shift and there is minimal worker concentration at this time. The field operator connected the acid transfer hoses to the horizontal storage tank and put the road tanker-emptying pump into service, but the acid was not transferred to the storage tank. When they checked the condition of the strainer they found that it was completely clogged and should be replaced or cleaned. Strainer replacement was postponed to 20:00 due to the operator leaving the area for handover of the operation logs (shift change), and the driver decided to deliver his cargo quickly; arbitrarily bypassing the strainer.

The acid transfer pipes were completely filled with a paste-type material and the transfer was reduced to zero, disrupting the performance of all three sulphuric acid acceptor units. The main horizontal acid storage tank was opened, and more than 1 m of a paste-type material was found inside. Laboratory analysis of this paste-type material showed that 85% of it was composed of hydrocarbon organic matters. Subsequent correspondence with several other customers showed that the supplier had a history of contaminating several other systems, and did not properly follow the principles of cleaning its road tankers after each shipment was delivered.

Problems in the delivery of sulphuric acid to three consumers caused the acid acceptor systems (to which it was not possible to transfer the acid now) to malfunction, and the quality control to be such that the system itself or the downstream units would corrode, or increase the risk of corrosion in them. For example, one of these consumers (processing units) was a demineralization unit (see Section 5.5.6.1). The acid was used in this unit to regenerate strong cation resins. The demineralized water produced in this unit was used as MU water for the steam generation unit as well as the demineralized water required for washing the turbines blades. The impossibility of transferring acid to this unit made it difficult to regenerate the resins and ultimately caused the product to deviate from the expected quality. It is obvious how water with higher specific conductivity (or TDS) for a boiler with a certain pressure can be effective in creating precipitation, deposition, and corrosion damage [18, 19, 23].

5.9.1.11 Operation History 13

A utility processing site had written the allowable amount of IMs in calcium chloride required in the purchase contract and emailed it to the winner of the tender. The contract clearly stated that the representation of product analysis sheets is insufficient to determine IM and must be sampled at the time of delivery of the entire consignment. It was decided that the delivery of chemicals in the warehouse would be done by following the sampling instructions when receiving the shipment from the truck. The instructions have not been studied properly by the delivery operator and sampling is limited to opening only one bag and analyzing it. The inconsistency of the shipment arrival time with the presence of laboratory personnel had caused resistance to the analysis of more samples (from other bags). Therefore, a shipment of calcium chloride with an unallowable percentage of IM was in warehouse and was gradually sent to a processing site to be used in the remineralization of desalinated seawater (post treatment of desalinated seawater). The feeding system consists of a solution preparing/dosing tank and two feed pumps (one was a spare pump). Calcium chloride salt was poured manually and desalinated seawater was piped into the tank for dissolving. A stirrer was placed into service at tank for at least 15 minutes, dissolving all the salt in the desalinated seawater. There was a level controller on the dosing tank which, at a height of 25% of the solution, sent a trip signal to the feed pumps.

After about one month, the feed pumps repeatedly broke down to the point that the spare pump did not respond to the high number of failures. Therefore, the feeding was stopped and the desalinated seawater was sent to the downstream units without any post treatment.

Investigations have shown that both feed pumps failed due to the suction of sludge-like materials and insoluble solids (or IM). Examination of the calcium chloride in the warehouse showed that the IM in more than 50% of the delivered shipment was about five times the allowable limit written in the text of the purchase agreement. In some bags, the amount of impurities was so high that debris and gravel particles were visible to the naked eye.

The document, which was communicated to the field operators as EMPA on the processing site, emphasized that before sending any chemical from the warehouse to the processing site, all the required analyses are performed and if the quality was appropriate, the chemical would transferred to the processing site. The EMPA document also stated that the contents of the solution preparing/dosing tank should be completely emptied after a maximum of five times of charge or solution preparing, and the bottom of the tank should be washed with water.

In this example, the failure to follow the three activities envisaged in the EMPA, namely (i) delivery instructions for the main warehouse, (ii) instructions for sending the chemical to the processing site, and (iii) instructions for washing the solution preparing/dosing tank, caused complete failure of the feed pumps and then stopped the calcium chloride being fed into the seawater desalination plant. It is quite clear to what extent the non-feed of this salt into desalinated water (in order to achieve water stabilization) caused electrochemical corrosion in metal piping and other equipment [9]. Figure 5.38 represents the results of stoppage in remineralization of desalinated water.

The two cases given in this section clearly point to the important role of proper delivery in achieving proper quality control. Depending on the structure governing the activities of a processing site and the complexity of chemical quality control jobs, delivery can be extended from one step to two or more steps.

5.9.2 Storage

Both selection and procurement activities have a major impact on chemical storage. Selecting chemicals that fit well within the storage facilities in the industry, as well as the type of purchase contract (partial or total shipment), can ensure the

Figure 5.38 A standard coupon in glass holder (top left) and removed from that after six months (top right), corrosion products on a laboratory scale (bottom left), and the cleaned and dried coupon (bottom right).

correctness of storage activities. In storage, some important points should be considered, such as; storage of chemicals based on the conditions in the technical documents related to each, analysis and preparation of periodic reports of the status of chemicals in warehouse (including inventory), and management of input and output dates for each chemical.

5.9.2.1 Operation History 14

A utility processing site purchased the dry calcium chloride, $CaCl_2$, it needed in 25 kg, double-layer bags. This chemical was delivered in 400 kg pallets (including 16 bags stacked on top of each other). Unloading the pallets from the truck and transporting them to the warehouse, moving the chemical into the warehouse, tearing the bags for sampling (without following the instructions for simultaneous sampling and delivery to the processing site), and the moisture adsorption of calcium chloride all caused a loss of quality for 70 tons of this chemical in three years.

Due to the absorption of moisture, the solubility of the chemical in water was severely reduced and created two problems for the utility processing site; (i) it was not possible to make pre-calculated concentration for calcium chloride solution and to feed more of the solution into the process stream, and (ii) decreased calcium quality caused a significant amount of IM to settle in the solution preparing/dosing tank and caused the suction strainer of diaphragm pump to suck in particles and shut it down. In this case, the impossibility of feeding sufficient calcium chloride into the process stream reduced the quality of the product (desalinated seawater) as well as the electrochemical corrosion of all related equipment.

5.9.2.2 Operation History 15

A gas plant had a storage space disproportionate to the volume of chemicals it needed. The delivery of a shipment of 500 tons of activated carbon had left the liquid catalysts needed for the Merox™ process (a typical unit operation for elimination of sulphur from propane and butane [2]) stuck in corners, and the warehouse staff had lost access to the 4.5-L catalyst containers. Along with to the simultaneous need for these catalysts with major overhauls and the lack of forklifts to move at least 30 activated carbon pallets, the warehouse worker had given the new catalyst shipment to responsible operation personnel in another part of the warehouse. Forgetting to reduce the inventory by removing older chemicals, the five gas plants were faced with 28 tons of expired liquid catalysts after one year, all of which had expired due to the lack of simple storage tips (more than two years difference between new and older materials expiration date).

5.9.2.3 Operation History 16

One utility processing site required 17 tons of sodium hypochlorite, NaOCl, per year. This processing site purchased all the sodium hypochlorite it needed from a chemical manufacturing company 1000 km away. Due to the location of the utility processing site in the tropics, as well as the lack of facilities (such as cold rooms and initial dilution to maintain the quality of the chemical), its commercial concentration was reduced from 12.5% to 5% in the first analysis after delivery of this chemical (about two months later).

Decreased effective concentrations of sodium hypochlorite and chemical deficiencies eventually led to poor quality control of products such as desalinated chlorinated seawater as industrial feed water. With a residence time of approximately six days from production, this water provided the feed of a demineralization unit that purified the MU water of the steam generation unit. Water entered the steam generation unit without biological stability and microbial measurements such as TBC and SRB showed that there was microbial growth in the steam-generating unit, and as a result there was a risk of microbial corrosion, especially in dead points like cold condensate piping and bottom of cold steam condensate storage tanks. Biological slimes fouled in these sections may decompose in the boiler to form carbon dioxide and chlorides [24]. Chlorides should be limited in BW cycle [25]. It is obvious that the instability of desalinated seawater fed to the demineralization unit, in addition to increasing the growth potential of microbes in the steam generating unit, will cause biofouling in the mixed resins, which sometimes make their recovery and use very difficult or completely impossible [24, 31]. Sodium hypochlorite also has a slight oxidizing effect on some metal ions and residual desalinated organic matter in water [6, 49], which is an important factor in creating stable and non-reversible fouling on charged resins in the demineralization unit. Oxidation of some organic matters by sodium hypochlorite may prevent the increasing level of biological growth and other anions in a boiler.

Finally, a change in the method of chemical delivery (here, partial shipment), the use of road trucks with cooling facilities, the use of a canteen with cooling facilities to create suitable storage conditions in the central warehouse and the warehouse of the processing site, change in the amount of storage per barrel from 220 L to 40 L, and a change in some consumption instructions, including how to dilute and use chlorine analysis instead of measuring residual chlorine, all improved the conditions and established a suitable EMPA for this chemical. The sum of these changes and the establishment of EMPA caused a 45% reduction in cost and a 70% increase in quality compared to the previous conditions (the study was conducted over a period of six months and the presented percentages were calculated based on the average figures reported in the six months before the study).

5.9.2.4 Operation History 17

A gas plant used a heat exchanger and a desalter to condition the gas condensate and then sent it (dehydrated) to the stabilization column for adjusting the RVP (for storage and then export). The gas condensate was first heated to a certain temperature and then entered into a desalter. The desalter had a demineralized water feed system and a transformer to create an electrical field for better separation of water (the aqueous phase or the phase in which salt was extracted). Therefore, it could be expected that the desalting process at this unit (gas condensate stabilization) was in fact a water removal process. In order to enhance the separation of the organic layer (gas condensate) and the aqueous layer (water inside the condensate, demineralized feed water, salt, and other minerals present in the gas condensate), a demulsifier chemical was fed into the gas condensate stream entering the desalter at a specified feed rate. Eventually, dehydrated gas condensate entered the stabilization column and lost its volatiles constituents such as propane and butane, and the bottom product of the column was sent to the floating storage roof tanks after degassing. If there was water in the stabilization column (i.e. distillation process) at the operating temperature of the column and in the presence of some mineral contaminants such as chloride (and carbon dioxide), it causes acidic properties and localized corrosion in the upper parts of the feed line to the column, especially in the condenser.

The demulsifier chemical was charged in the dosing tank at the condensate stabilization unit and was in service for six months. Measurements showed that salt removal and water content were in very good condition. There was then a problem with the desalter performance at the same time there was a need for chemical cleaning of one of the heat exchangers. Hence, there was a three-month break in chemical consumption. The contents of the dosing tank became two-phase after the start of the processing site and completely lost its efficiency.

Investigations showed that there was a weakness in the readying of EMPA for this chemical; the processing site did not have temporary storage facilities and the moisture absorption of the demulsifier was not considered by the processing site supervisor. Therefore, an instruction was added to the existing set of documents so that in such events the contents of the dosing tank were transferred to closed barrels with suitable facilities or, if possible, use of a nitrogen blanket in the same dosing tank to prevent the atmosphere from entering the dosing tank.

5.9.2.5 Operation History 18

A utility processing site stored approximately 21 tons of calcium chloride for two years, which was needed for remineralization of its desalinated water. The amount purchased had no proportion to the amount of the chemical consumed daily. Many of the double-layer plastic bags had rotted and damaged due to incorrect sampling, forklift handling, and sunlight. The cargo was stored in an open warehouse in a tropical region with an average humidity of 65–85% and a temperature of 15–48 °C.

Calcium chloride is a hygroscopic chemical and should be stored in a dry place [29, 51]. The long-term adsorption of moisture and its subsequent condensation in the resulting crystals released a significant amount of moisture under the calcium chloride pallets in the warehouse (Figure 5.39).

Figure 5.39 Improper storage of calcium chloride in an open warehouse at a tropical region.

5.10 Consumption

This activity plays a very important role in having an EMPA, and almost all activities in the EMPA cycle will be the result of this section. Proper use can accurately highlight the strengths and weaknesses of the process and/or chemicals and positively influences all other activities. In this section, several operation histories will be given to better illustrate the concept of consumption and its place in EMPA. The consumption discussion is expected to address at least the general issues encountered; determine the chemical analysis of each process stream, issue dosing/solution preparing guidelines, periodically calibrate feed systems, routine chemical analysis of each utility or any other process stream, and receive feedback on treatment performance.

Figure 5.40 is a true block diagram of all the steps to be followed in the consumption activity. The most important part of this activity is the documentation, which must be carefully compiled under a proper classification protocol, and the various documents of each protocol must be loaded in the intended category. The documentation, which will be described in an independent section in this chapter, will be an important aid in reviewing system feedback, as well as studying its records. Classifying and preserving documents in all the activities in Figure 5.40 (as well as Figure 5.32) is the only way to distinguish process deviations from process difficulties, each of which requires its own solution method. In the absence of proper and classified documents, all correspondence made and forthcoming with chemical manufacturers/suppliers will be practically useless and can in no way be used to apply the legal clauses in the purchase contract (such as after-sales service).

The following paragraphs provide examples to show how not following any of these steps in the consumption activity has resulted in poor quality control and

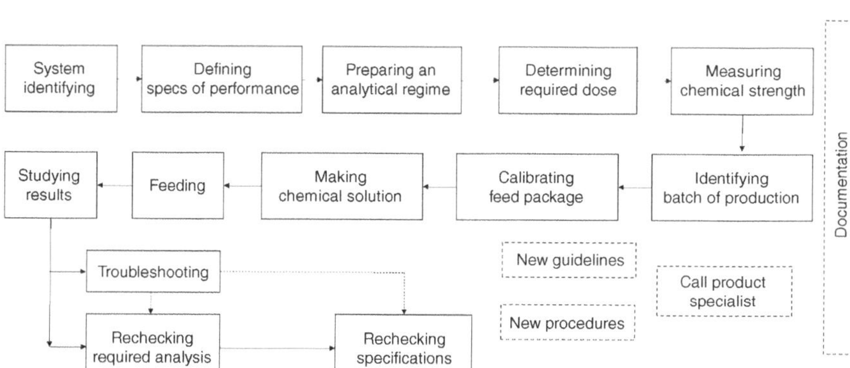

Figure 5.40 All steps in consumption of a chemical in a processing site.

ultimately increased corrosion risk (as well as other processing-site operation parameters represented in Figure 5.11).

5.10.1 Operation History 19

System Identification: a utility processing site (in a petroleum refinery) has a steam-generating unit inside. The returned steam condensate from the hydrocarbon-processing site is divided into two branches, one called "clean steam condensate" routes directly to the deaerators or steam condensate storage tanks; and the other branch called "suspect steam condensate," which is purified from organics and inorganics, is sent to steam condensate storage tanks. The package that purifies suspect type of steam condensate, called the deoiling package, had problems with control logic at startup, and due to the requirement at initial startup of refinery, the startup operation team could not wait for the package to be ready.

Therefore, the steam production unit with a capacity of 700 tons per hour of medium pressure steam was started up, and after two months the deoiling package was ready to be put into service. One day after the service of this package, the concentration of DO in the return condensate to the main header (inlet to the storage tanks) increased sharply and the pH dropped by 1.5 units. Studies showed that this package removed some of the neutralizing and organic oxygen scavenger amines, and the quality control expert should have considered the effect of reducing their concentration due to the servicing of this package. Decreasing the pH of the steam condensate system and increasing the concentration of DO are two very important factors in corrosion associated with steam condensate lines and boilers [18, 19, 23].

5.10.2 Operation History 20

Defining Specs of Performance (Performance Characteristics Definition): A utility processing site would draw seawater from the sea to supply the industrial water it needed and desalinated it through the MED–TVC process. The seawater reached the utility processing site in a chlorinated and filtered type, and sodium bisulfite was used to remove free residual chlorine up to 0.15 ppm to prevent possible chlorine damage to the desalination plant material (stainless steel and copper alloys). An inadequate definition of the performance of the dechlorination chemical caused an analysis called "after bisulfite feed" placed in the quality control regime of analysis. In this analysis, the sample person had to take a sample from the sampling point after bisulfite feeding every four hours and analyze the free residual chlorine remaining in it. The low range of the measurement, as well as the numerous factors for systematic error, and errors in this analysis, sometimes caused the amount of dechlorinated chemical feeding to reach three times the required amount. Studies have shown that during the period when the system was controlled by this quality method, in many cases the equipment was at risk of corrosion (level of Cl_2) and quality control was performed in an undesirable manner.

Later, with the establishment of EMPA, the proposed standard method for controlling the chlorinated seawater inlet to the desalination unit was a combination of routine sampling of a sodium bisulfite dosing tank (dechlorination chemical), and routine logging for the feeding rate of the pumps into the inlet stream. Then, applying of a stoichiometric dose for sodium bisulfite, finally a safe dose was proposed to ensure complete dechlorination [19]. The free residual chlorine analysis called "before bisulfite feed" was removed from the quality control regime of analysis because all the problems for chlorine analysis in the previous paragraph were also present here, and instead the EMPA document related to this section was changed to include controlling the quality of chlorine feed into seawater in SWI should be done by sampling the sodium hypochlorite reservoir and determining its concentration, and then regulating the amount of feeding into the seawater inlet to the utility processing site.

It is clear that in this schematic, chemistry control, which was incorrectly established in the system, in addition to poor quality control and the lack of meeting the operational needs, such as shock dose of both sodium hypochlorite and sodium bisulfite, also imposed a very large volume of useless analysis on the laboratory. This ultimately lead to the omission of more important analyses and increased corrosion risk (due to poor quality control) in other processing sections.

5.10.3 Operation History 21

A utility processing site received the raw water it needed from the sea, and the chlorinated and filtered seawater was then fed into a series of plate-type heat exchangers to be preheated to enter five MED–TVC desalination packages. Improper chlorination caused the marine animals to grow in some exchangers and three heat exchangers were temporarily eliminated (by-pass arrangement). Therefore, the designed flowrate caused premature destruction of the washers and leakage of the other in-service heat exchangers (due to over flowing of inflow stream).

To solve this problem, the operating team decided to leave the seawater inlet valve that lead to the heat exchanger in a semi-open position to allow less flowrate into the heat exchanger (a common mistake that increases turbulence and therefore cavitation). The angle of attack of the seawater flow was changed with this valve and severe corrosion was created in the plug of this butterfly valve. Opening the butterfly valve increases the conditions for cavitation by 20–60% [52, 53]. Heat exchanger insulation valve malfunction caused many problems for the utility site.

Studies showed that the amount of free residual chlorine entering the utility site was mistakenly estimated, and the low concentration of free residual chlorine (spec of performance) caused a corrosion in the plug of the insulation (on/off) valve and failure in three heat exchangers (Figure 5.41).

Figure 5.41 A corroded plug of seawater feed valve to series of heat exchangers.

5.10.4 Operation History 22

Preparation of an analysis regime: An ACF was used at a utility processing site for the final purification of drinking water. This ACF was designed as a corrective action to increase water quality and was installed on-site (before chlorination and in the last stage of treatment process) [27]. The analysis regime was not considered for this filter and only the periodic analysis of the activated carbon adsorption capacity (by decommissioning and servicing the spare ACF) was mentioned. After some time, the preceding activated carbon manufacturer refused to sell to the utility processing site, so the required activated carbon was purchased from another reputable manufacturer and charged in the filter. After two months, the concentration of organic pollutants in the drinking water distribution system increased and the results of carbon adsorption tests, such as the iodine number and several other factors, were assumed to be good enough.

Studies have shown that due to changes in the manufacturer of activated carbon, analyses performed from the bed cannot be generalized to drinking water treatment conditions. This meant the right thing to do was to routinely perform a series of analyses of the treated water from the activated carbon bed; the quality of the effluent water is always monitored with a set of characteristic parameters of the ACF product. Thus, ignorance of the need for a routine analysis regime of the filter effluent caused problems with the quality of drinking water due to the entry of organic pollutants into the system when changing the brand of activated carbon. In addition to improper feeding of chlorine into the distribution system, these contaminants greatly increased the risk of biofilm formation and MIC [9, 54].

5.10.5 Operation History 23

Some analysis regimes are misleading, for example, at a utility processing site a process engineer presented special conductivity of a closed cooling water cycle to the laboratory as a characteristic parameter. Daily analysis and fluctuation in the results of this parameter wasted quite a bit of time and it was thought there was a source of contamination in the system; and due to its small fluctuation, caused a small change in specific conductivity and had no impact on other parameters such as pH and alkalinity. The inclusion of iron analysis in BW (continuous blowdown) routine schedule is another example of such unnecessary and misleading analysis.

Determining the required dose: For chemicals that are dosed into the system, it must be clearly specified how much of the chemical must be fed into the stream of the treating process. This dose can be fed in two ways depending on the type of chemical:

1) For conventional chemicals whose chemical formula is well known, and the equilibrium data and reaction constants are known completely, the required dose can be based on stoichiometric calculations. Sometimes it's better to consider a safety factor for compensating the non-idealities of the system, such as inadequate mixing. In this case, only the concentration or assay of pure chemical would be required, which is determined by known standard laboratory methods. The chemical purchased in this case was fed into the process stream in pure form or dissolved in water, and its impact was monitored downstream. If monitoring its impact on the treating stream was erroneous, then routine dose monitoring can be used.
2) Feeding for licensed chemicals should only be done at the dose recommended by the manufacturer/supplier. In the event of a change in process conditions, the chemical after-sales service representative must be notified immediately and the necessary advice sought. If there is a change in the storage conditions of chemicals or inconsistency of the lot number with documents delivered to the warehouse, the quality control department must immediately contact the after-sales service department and refrain from taking any action. If any deviation is observed in the results of the analysis or the efficiency of the chemical is reduced, all process parameters must be checked first, and if no deviation is observed in the processes, the after-sales service engineer must be notified. Dosage of these chemicals can be done as either "as chemical" or "as pure." However, if a chemical includes after-sales service, it is recommended to use "as chemical" type of calculations.

5.10.6 Operation History 24

A gas plant used a film-forming CI to protect the sour water pipeline; sour water was the product of the sour gas processing and glycol regeneration. More than two years had passed since the expiration date of the CI, and due to the unpredictability of its purchase and the time-consuming steps of the purchase-tender (at least six months), an analysis was performed on the site and it was finally determined that if the chemical dose was increased to three times (meaning the residual CI at the end of the line) it could compensate for the effect of chemical decay. Therefore, the feeding was performed accordingly and the residual was increased from 5–10 ppm to 15–30 ppm.

The result was that, despite the proper corrosion prevention of the sour water pipeline, the increase in the film-forming CI dose exacerbated the foaming tendency and hydrocarbon entrainment in the aqueous phase, so that the possibility of separation of water and other contaminants was missed at the inlet of the sour water stripping unit. Therefore, organic pollutants along with some very high volume mineral compounds entered the stripping column and deposited on the outer surface of the reboiler tubes. Figure 5.42 represents organic matter (and also inorganics) precipitations on a tube bundle in a train of a sour water stripping unit.

These deposits formed localized corrosion on the tubes, causing the two trains of sour water stripping to become out of service and their tubes bundles to be replaced four times (two times for each train, see Figure 5.43). In this example, it can be seen that the improper use of a licensed CI without consideration the principles of correspondence in the purchase contract, as well as the arbitrary conclusion of a few simple corrosion tests in the processing site laboratory led to localized corrosion and eventually stopped the operation of two trains of a processing unit.

Measuring the strength of a chemical: If a chemical is purely charged in a dosing tank and fed into a process stream, it is a very simple procedure, but sometimes the situation is such that it is not possible to feed pure chemical into the system. These

Figure 5.42 Precipitation on a tube bundle of a kettle type reboiler in a sour water stripping unit.

Figure 5.43 Same bundle after cleaning (localized corrosion, pits, are clearly represented).

conditions can include such things as feeding pump limitations, storage limitations, and fluctuations in treatment operations. Therefore, a chemical charge instruction is issued for preparing a solution in dosing package with water or any other solvent. This solution preparation will have error for a variety of reasons:

- Change in the concentration of pure chemical and error in its measurement;
- Error in instrument for measuring the height of the chemical or the solution already in the solution preparing/dosing tank;
- Ingression of the atmosphere into the tank and partial deactivation of the solution in the tank;
- Error in measuring the concentration of residual solution (from preceding batches) in the solution preparing/dosing tank;
- Calculation error of dilution and execution error of instructions (or guidelines) issued;
- Problems in the mixer and improper mixing (there are many dosing tanks that have some problems with their mixers, like corroded propellers, unaligned shafts, and broken down motors); or
- Converting prepared solution into a nonhomogeneous one (formation of two phases due to heat, adsorption of humidity, and effect of dilution on solution stability).

Therefore, in order to know the concentration of the chemical in the tank, sampling and determination of the chemical concentration in the dosing tank should be included in the routine analysis regimes.

Knowing the strength of a chemical is also mandatory for some chemicals that are consumed in pure form. For example, although the amount of sodium nitrite, an anodic inhibitor [18, 55] feeding for the closed cooling system is provided to the site by the after-sales service department of the manufacturer/supplier, knowing the chemical factor or chemical strength is required to regulate the concentration of residual nitrite in the system in subsequent charges and when for any reason a large amount of MU water entered the system.

Identification of production batch: There are many examples where a new shipment has different conditions and properties from the previous shipment(s). This has happened to both a new manufacturer/supplier and a previous seller. Therefore, when delivering a chemical from the site warehouse, and transferring it to the processing site to start consuming, it is necessary to pay attention to the lot number listed on the packaging of the chemical (bag, barrel, tons, etc.), and check its specifications with technical documents delivered to the commercial department. All safety calculations and considerations must be based on the same chemical available. Sometimes chemical sellers think that the purchasing department at a processing site wants to buy 200 tons of trisodium phosphate (TSP), for example, so it makes no difference whether they are delivered in 200-L barrels with a concentration of 2% or less barrels with 5% of TSP. The purchasing department experts and quality control personnel (mainly process engineers) on a processing site should pay special attention to these issues.

Calibration of feeding packages: A liquid feeding package consists of several parts as shown in Figure 5.44. Depending on the type of chemical, a feeding package feeds chemical into the treating process stream, and more details are added to this basic form. Periodic calibration of each feeding package is a basic requirement of setting up an EMPA. In general, the only equipment that needs to be calibrated is the feed pump. Calibration tools are very simple and for feeding of liquids (which are mostly used in refineries, gas plants, petrochemicals, and utility processing sites of the energy industries) they consist of a calibration pot and a stopwatch.

For calibration, a systematic method must be created on the utility processing site, gas plant, and any other processing industries. Usually, the output of a feeding instruction should be subject to update of the calibration curves of feeding pumps. In general, there are two types of algorithms for feeding a chemical into a process stream, which are shown in Figure 5.45.

Some feeding systems depend on the flowrate of the process streams, which are called a flow pace system [47], and this dependency is placed in the computing block as a default function. Increasing the flowrate of a process stream increases the speed of rotation of the motor in the feeding pump and subsequently feeds more chemical. Therefore, the calibration graph for such a feeding system will be three-dimensional with constant velocity or stroke curves, but if feeding of the chemical is not dependent on the flowrate of the process, then the curve will have only two dimensions (Figure 5.46). There are different types of feed pumps that all need to be calibrated [47].

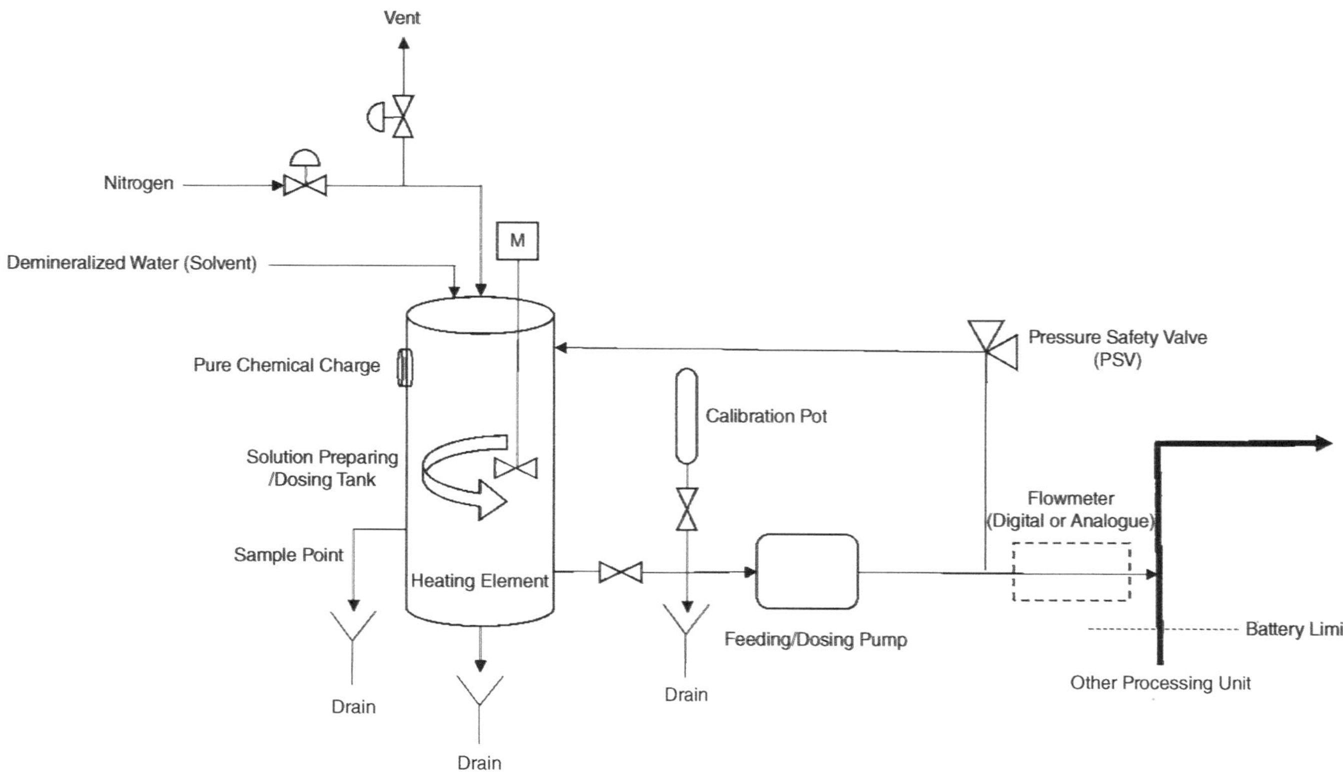

Figure 5.44 A liquid chemical feed package and its minimum requirement.

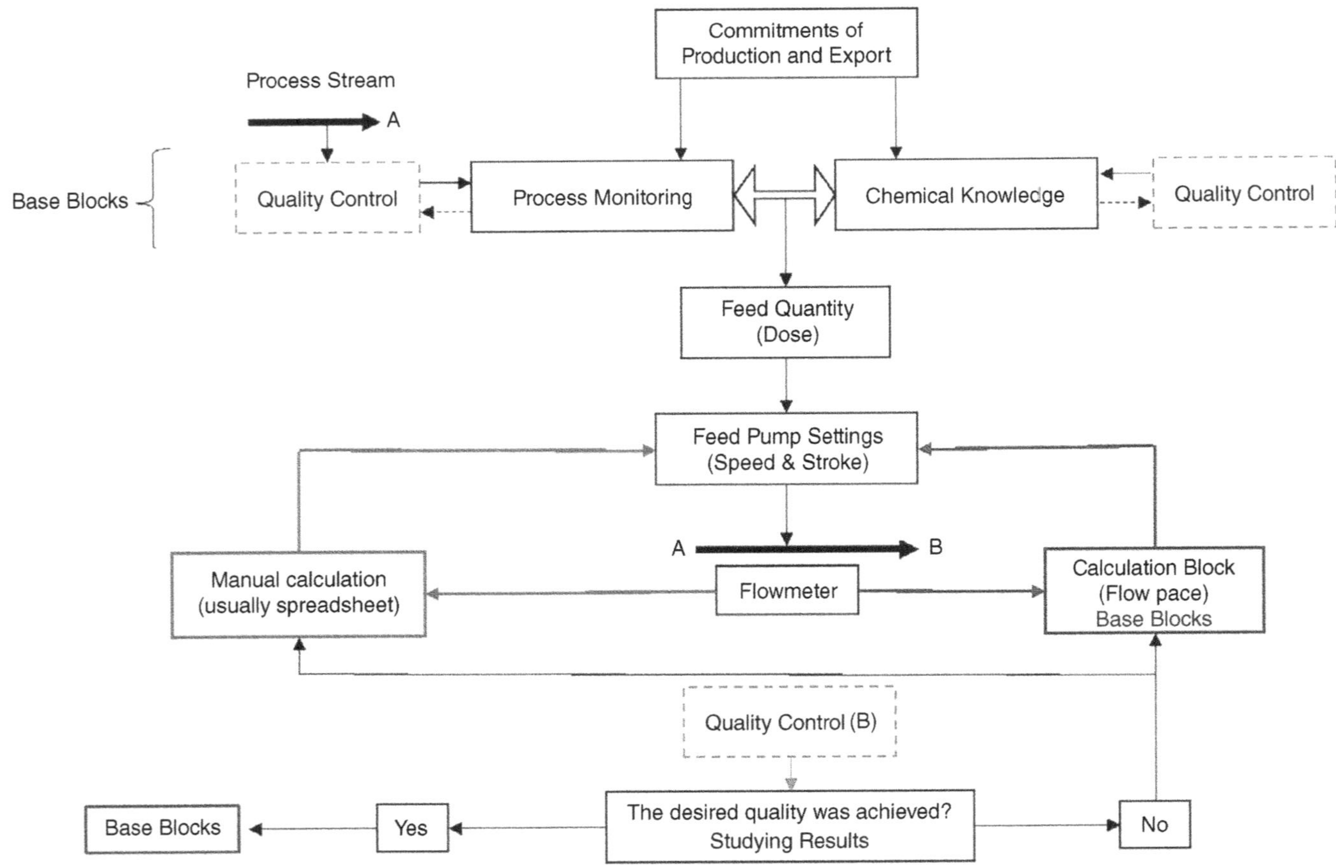

Figure 5.45 Two algorithms of feeding a chemical into a process stream.

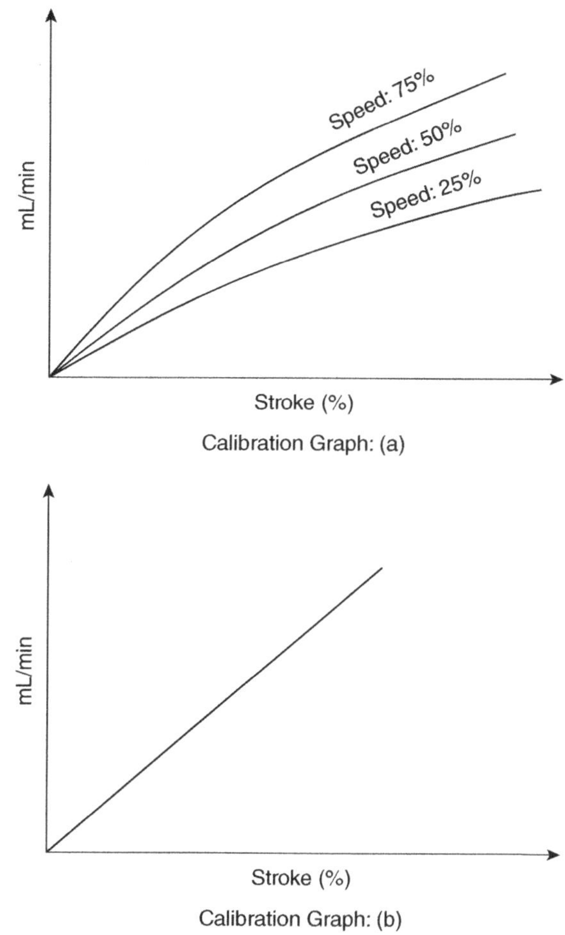

Figure 5.46 Calibration graphs; (a) variable speed and variable stroke feed pump, (b) variable stroke feed pump.

One of the requirements for updating (or generating) calibration graphs is that over time the feeding package will have conditions for which the use of calibration graphs may not be valid even for a few days past the update being issued. The most important reasons for changing the calibrated (feeding) values for a designed chemical feeding package for a liquid can be one (or more) of the following;

- Change in the concentration of the chemical and the consequent change in the physical properties of the fluid (chemical), such as viscosity and density.
- Undesired pressure drop in the suction and feeding (discharge) section, which can be due to the corrosion and precipitation (the material of some feed packages is not suitable to the chemical and severe corrosion occurs in them).

Therefore, corrosion products can accumulate as particles in the suction strainer of the feed pump and disrupt the feeding).

- Repair and replacement of feed pump parts such as diaphragm or pump course adjustment,
- Pump out of adjustment (so that it does not feed in the same setting as before).
- Leakage from the system, especially in the discharge path of feeding pumps.
- Passing pumped liquid from pressure safety valve (PSV) so that a small (or considerable) amount of the feed pump discharge stream returns to the dosing tank.

5.10.7 Operation History 25

A utility processing site used a combination of Metoxypropyleamine, MOPA, and Monoethanolamine, MEA, to both adjust and raise the pH of its return condensate. The chemical is charged in barrels in the dosing tank, and after dilution with demineralized water, is fed into eight points of the process in the steam production unit. According to the manufacturer's documentation, the recommended dose of 1.5 ppm of the chemical was to be fed into the system as pure, to reach a pH in the range of 8.8–9.2 [19]. After some time, the pH dropped and was barely controlled at 8.5. Investigations showed that a leak in the isolation valve related to the calibration pot caused the operator to make an error when updating the calibration curves for the pumps, and to misread the pumped volume of liquid in the pot. The isolation valve leak was quickly remedied, but within one week the system experienced a poor quality control reading. As we have seen, not paying attention to the basic points in the use of chemicals (here, the calibration of the feeding package) will cause the system to be at risk of corrosion due to the lack of adjustment of the pH of the steam condensate system in the predetermined amount [12, 18, 19].

Study of results: As shown in Figure 5.44, after feeding a chemical into a process stream, the expected results must be studied. For example, if a chemical is to produce a certain amount of zinc and phosphate in a process stream to inhibit the corrosion [18, 54], then the concentration of these two quality control parameters must be analyzed both up- and downstream (points A and B), where the mixing is done, and the treatment efficiency must be determined and recorded. Comparison of the values obtained from the analysis (after performing the error analysis in the laboratory [29]) with the expected values obtained through stoichiometry or the manufacturer's recommended dose (after performing the error and uncertainty analysis by analytical techniques in the feeding calculations, logics or manual [56]) will determine the level at which the EMPA is running. Obtaining a reasonable approximation between the expected values and the measured values for some chemical parameters is an integral part of the chemical consumption activity at the processing sites.

5.10.8 Operation History 26

One steam-generating site used a combination of Morpholine and two other amines to adjust the pH of return steam condensate. According to the tables in the chemical documentation, 1.2 ppm of this chemical was sufficient to bring the pH to 9 (which, of course, is also influenced by the return of the steam condensate containing the amine). Considering the process stream flowrate, 2.6 tons of this neutralizing amine were required annually, but after three years of operation, the consumption of this chemical had gradually reached 66 tons per year. After much correspondence with the manufacturer and additional laboratory tests, it was found that the utility processing site had a major problem with pH measurement and did not follow pH measurement instructions from ultra-purified dosed water [57].

Unfortunately, this utility processing site did not have an EMPA for its chemical handling activities and did not calculate and report as pure amine doses. However, the lost calculations could be recovered by considering the chemical charged in the dosing tank over three months, and the result was the dose of neutralizing amine (which was a combination of three amines) in some cases reached 30 ppm, as chemical. Therefore, a fast stabilizing pH sensor that is suitable for purified dosed water was immediately purchased for this site. By improving the measurements, the required amount of amine started to decrease and after two months reached 110% of the predicted amount (197 kg/month, or 2.86 tons per year).

As can be seen in this example, a simple mistake and ignoring the principles of working with chemicals (routine dose monitoring) has caused an inappropriate event in terms of the use of the neutralizing amine chemical. Some amines fail molecularly to form organic acid at high operating temperatures in the boiler [18]. Therefore, considering the production of organic acids in the system at an operating temperature of 435 °C, it can be expected that the system also had a risk of acidic corrosion; however, this is difficult to prove due to the adsorption capacity in some types of CPP (that contain ACF) as well as the neutralizing capacity of the newly fed amine into the system. Nevertheless, increasing the required amine concentration by 25 times (30 ppm/1.2 ppm equals 25) was a serious issue to increase the rate of amine degradation in the system.

5.10.9 Operation History 27

A gas plant had a closed cooling water system. This system consisted of a volume of approximately 20 000 m^3 along with two expansion tanks, each with a volume of 1350 m^3. The chemical, sodium nitrite, was fed into the system as a batch of anodic CI. The decrease in nitrite concentration made personnel think that there was an active corrosion in the system, because they recognized nitrite as a "dangerous corrosion inhibitor" [18].

Examinations showed that the field operator did not follow an important guideline in EMPA; in addition to (i) monthly controls of nitrite concentration at seven points of the system (for better monitoring of residual CI), to be applied as diluted CI (ii) for the volume of MU water received. The field operator ignored part ii above. Part of the closed cooling system leaked and the water level in both expansion tanks reached a critical level once every 20 days. The field operator opened the MU water hand valves, and at the same time did not start the diluted anodic CI feeding pump. Hence, pure water entered the system and as a result, the nitrite concentration was reduced without any destructive events. However, it was obvious the reduction of CI level could place the system at risk of corrosion [18].

It is therefore recommended that the process engineer, or any other person responsible for quality control resulting from chemical-related activities, receive feedback from all its activities and carefully study the unexpected results that can be considered as process deviations. Figure 5.44 shows a simple diagram of the "studying results" position.

Documentation: The largest workload in any of the EMPA activities is related to the consumption stage. All steps in this stage must be documented with appropriate quality. The availability of all documents is an important aid in determining the reasons for deviations from the desired chemical control (which leads the processing site to the optimal control of all operating parameters, including corrosion). It can be considered that recording and classifying all the steps along with the necessary details can increase the level of awareness of quality control officials about the conditions of the final and intermediate products and help them to reach an EMPA.

However, it should be noted that there are many process sites that, despite documenting and classifying almost all the steps in the consumption stage, suffer from damage (and other operating parameters) due to the non-implementation of an EMPA. Therefore, it is obvious that in order to increase the impact factor of the documentation in the consumption stage, a proper configuration must be used. The experience of more than 10 large processing sites in the field of oil, gas, and related utilities indicates a serious shortcoming when not using an effective configuration to document the steps in this area.

Tracing industrial cases in which misuse of chemicals has led to corrosion of equipment, and subsequently tracing the root causes of these misuses in consumption, have all proved to the author that following a histogram configuration works best. To better understand histogram configuration, the concept of *treatment histogram* must first be introduced. The performance of a sand filter, chlorine feeding for disinfection, and finally the use of a demulsifier to separate oil from water are all routinely measured, reported, and interpreted. The preparation of these results and interpretations in the form of a graduated tape based on operating days (or hours) implicitly gives us a treatment histogram for a system.

5.10.10 Operation History 28

A proper treatment histogram helped the author realize, after three years, that at a gas plant in early November, the feeding of neutralizing amine and organic oxygen scavenger (volatile) for the steam and its condensate system would always increase by 50% or more to keep the quality of steam condensate and BFW. The reason was the increase in heat transfer in the steam condensers and the consequent increase in the condensation rate, which caused the oxygen ingress into the condensate collection system and also reduced the performance of the steam traps (due to the decrease in condensate temperature by more than 70 °C). Therefore, the atmosphere penetrated the steam condensate collection system, DO increased in the steam condensate system, and pH decreased in the boiler system. The gas plant was located in the tropics, with average temperatures and wind speeds changing each year in early November.

5.11 Reporting

All activities in EMPA (Figure 5.32) must be reported, clearly and effectively. A report should include the necessary details at each step of the process, as well as feedback on each separate activity. Because the process and its needs are changing continuously, reporting should be done routinely at proper intervals. This will allow a processing site to review events and make decisions at the right time. Existence of such reports, in addition to the possibility of analyzing operating conditions and predicting future needs, helps the process engineer and other people in charge on the site to better understand the root(s) of many deviations, problems, and consequently to be able to plan for eliminating them.

Knowing that most activities in this cycle (EMPA) have a humane approach and are not controlled by logics, reporting (which is known to be an important part of the documentation) must be able to link the activities performed in EMPA with the data obtained from the DCS of a processing site. A purposeful set of reports over a period of time, such as one year, that have properly analyzed the processes can be a great feed for engineering management (or quality control) system designers who decide to make a cross-link between manual documentation and the histograms obtained from the DCS data.

In utility and gas processing industries where a significant number of chemicals are used on processing sites, separate reports are required for each chemical. Sometimes the same chemical is used in several processing units; in this case, a separate analysis should be considered for each point of use on the processing site, as well as completely independent feedback.

Sometimes the emphasis on preparing a separate report for each chemical, and then preparing a separate report for each point of use (for each chemical) may distress the head of process engineering, but it should be known that this is a systematic approach to reduce errors, to achieve maximum quality and placement of the process at its best performance point. There are examples that a processing site, after six months of strict adherence to this approach, has been able to recognize all the parameters of its process sensitivity, and by implementing a quality control system (in interaction with DCS), were able to assign a large part of the full implementation of EMPA to the custom software.

One of the important points in these routine reports is the monitoring that ultimately leads to the announcement of the inventory and expiration date of a particular chemical. It may be interesting to note that inadequate monitoring of the inventory of a chemical can eventually cost engineers time to study and replace an improper chemical, and due to the urgency of supplying the chemical, the same chemical can be purchased again to avoid unplanned shutdowns. Inventory monitoring also has positive impacts on procurement, delivery, and storage activities.

Depending on the type of chemical, its storage restrictions, and consumption at the processing site, inventory can be announced at specific intervals. A simple application is to be able to estimate the average consumption rate in the coming days and reorder according to inventory [47]. To determine the end point and the order point, one can refer to the extrapolation of consumption rate, and refer to the "inventory strategy document" of the same chemical, respectively.

Some processing sites have developed an inventory strategy for their sensitive chemicals. For example, they always want to have at least two activated carbon filters in stock. It should be noted that this strategic inventory is reduced from the total need of the site for that chemical, and after receiving the new purchased chemical, that strategic inventory should be put in the consumption queue first. Corrosion never stops and the fight against it should not stop; hence, one of the most important applications for "strategic inventory" is to think of measures to prevent corrosion. In the meantime, the plant must have a clear definition of their strategic inventory of all kinds of chemicals that affect corrosion in some way.

5.12 Documentation

The last two paragraphs of Section 5.7 and other sections of the book reiterate the importance of documentation when working with chemicals along with some of the differences in documentation in other sections of a process site. Here are the features of effective documentation:

- Each document should help us to visualize ourselves in the past by studying it, so, we have to be good and patient writers.

- Have a set of predefined templates to take the least amount of time to refer to past actions, as well as describing past treatment conditions. Therefore, all your documentation should be consistent.
- An improvement cycle should be established to update the prepared documents, as well as identify all the documents needed to cover the details of the activities. Years of working on a processing site can tell process engineers and other quality control officials what documents should be routinely or non-routinely prepared to help effectively review system history. Having a set of notebooks for each system covering all the site events related to chemicals can be a very effective means to create a comprehensive coverage for all the required documents.
- Considering that each of these documents must contain specific details and that many of their contents may be site specific; the details of each document are beyond the scope of this chapter. Table 5.6 is an attempt to identify and provide all the documentation required for each of the EMPA activities.

Table 5.6 Identified documents for each activity in EMPA.

Title	Type of generating	Frequency of issuing	Main purpose(s)
		Selection	
Quality control plan	Non-routine, flowchart	–	To present targets of quality control by chemicals
Impact evaluation	Routine, technical report	Monthly	To present chemical impacts on entire plant operation (Figure 5.11)
Pre-evaluation of project chemicals	Non-routine, technical report	–	To ensure proper chemical selection in project prior to consumption
Chemical selection	Non-routine, technical report	–	To review and modify initial chemical selection document
Management of changes	Non-routine, technical report	–	To present all reactions to changes in process conditions related chemical(s) feeding
Chemical performance	Routine, technical report	Weekly	To present the performance of a chemical
Feasibility of manufacturing/supply	Non-routine, technical report	–	To inform the process engineering department regarding replacement or substitution

(*Continued*)

Table 5.6 (Continued)

Title	Type of generating	Frequency of issuing	Main purpose(s)
		Procurement	
Purchasing contract	Non-routine, legal format	–	To clear legal aspects and commerce
Procurement feedback	Routine, inter-departmental format	Weekly	To modify related actions extracted from feedback
		Delivery	
Delivery log sheets	Routine, pre-defined logs	Weekly	To inform process supervisors and the process engineering department about re-calculating the required dose
Delivery events	Routine, pre-defined format	Weekly	To inform process engineers about documentation and use it for further troubleshooting
		Storage	
Point of reordering	Routine, pre-defined format	Monthly	To cover the required inventory
		Consumption	
Process sensitivity analysis	Routine, technical report	Weekly	To clear all sensitive parameters impacted by chemical(s)
Chemical impact	Routine, technical report	Monthly	To present chemical impacts on entire plant operation (Figure 5.11)
Plant performance	Routine, technical report	Monthly	To present plant performance related to chemical(s)
Analytical regime	Non-routine, table	–	To cover testing regime by laboratory
Dosing calculation notes	Routine, spreadsheet	Once per shift	To document the method of calculating required dose
Calibration graphs	Routine, graph (table)	Weekly	To present a new calibration graph
Solution-preparing guidelines	Non-routine, guideline format	–	To instruct the method of solution preparation
Manufacturer/supplier correspondence	Non-routine, letter classification format	–	To record documents for legal aspects

Table 5.6 (Continued)

Title	Type of generating	Frequency of issuing	Main purpose(s)
After-sale services	Routine, letter classification format	Daily	To follow product specialist guidelines (or procedures) for licensed chemicals
Troubleshooting	Non-routine, technical report	–	To present all process deviations, problems, failures, and issues, as well as the way(s) of modifying and returning the process condition to normal

5.13 Summary

Every processing site in the oil, gas, and petrochemical industries needs chemicals to continue their operations. A serious look at these chemicals and their impacts on the main parameters of the operation makes their importance even more clear. These chemicals have important impacts on the economy, corrosion, energy, quality control, and the environment. These chemicals can eventually lead to reduced production as well as the production of off-spec products. Following a proposed work frame called Effective Management of Process Additives (EMPA), can help optimize the operation of a processing site to minimize or eliminate the undesired impacts of chemical consumption. A look at all the industrial cases given in this chapter shows that analyzing the impacts of the use of these chemicals on the overall operation of a process site is not possible without considering the process analysis skills from an engineering perspective. Although all of these industrial cases have tried to explain how equipment corrosion is affected by the six activities available in EMPA, it is clear that not following and not knowing any of EMPA activities can impact other management parameters in a processing site, this is stated in the several operation histories.

Abbreviations

ACF Activated carbon filter
AI Aggressiveness index
API American petroleum institute
BFW Boiler feed water

BIOX	Biological oxidation
BW	Boiler water
$C5^+$	Pentane plus (hydrocarbon heavier than pentane)
CCPP	Calcium carbonate precipitation potential
COD	Chemical oxygen demand
CSC	Cold steam condensate
DAP	Di-ammonium phosphate
DCS	Distributed control system
DO	Dissolved oxygen
EMPA	Effective management of process additives (chemicals)
FG	Fuel gas
GHG	Greenhouse gas
IGF	Induced gas flotation
IM	Insoluble matters
IOX	Ion exchange
LIMS	Laboratory information and management system
LSI	Langelier stability index
LPG	Liquefied petroleum gas
MEA	Monoethanolamine
MED–TVC	Multiple effect distillation–thermo vapor compression
MIC	Microbiologically influenced corrosion
MOC	Management of change
MOPA	Metoxypropyleamine
MU	Make up
NGL	Natural gas liquids
OWIS	oily water inlet sump
OWS	Oily water sewer
OWTP	Oily water treatment package
PCV	Pressure control valve
POS	Potentially oily sewer
ROS	Recovered oil sump
RVP	Reid vapor pressure
SBR	Sequencing batch reactor
SCNP	Spent caustic neutralization package
SRB	Sulfate reducing bacteria
SWI	Seawater intake
TDS	Total dissolved solid
TSP	Trisodium phosphate
TSS	Total suspended solids
WWTP	Wastewater treatment plant

References

1 Guo, B. and Ghalambor, A. (2005). *Natural Gas Engineering Handbook*. Houston, TX: Gulf Publishing Company.

2 Kidnay, A.J. and Parrish, W.R. (2006). *Fundamentals of Natural Gas Processing*. Boca Raton, Florida: CRC Press.

3 Stewart, M. (2014). *Surface Production Operations, Volume 2: Design of Gas-Handling Systems and Facilities*, 3e. Houston, TX: Gulf Publishing Company.

4 Stewart, M. and Arnold, K. (2011). *Gas Dehydration Field Manual*. Elsevier: Gulf Professional Publishing.

5 Metcalf & Eddy / Aecom (2003). *Wastewater Engineering, Treatment and Resource Recovery*, 5e, vol. 1. McGraw-Hill.

6 Davis, M.L. (2010). *Water and Wastewater Engineering: Design Principles and Practice*. McGraw-Hill.

7 Beychok, M.R. (1967). *Aqueous Wastes from Petroleum and Petrochemical Plants*. London: John Wiley & Sons.

8 American Water Work Association (AWWA). (2011). *Desalination of Seawater*, AWWA Manual M61, Denver, CO.

9 Lahav, O., Voutchkov, N., and Birnhack, L. (2012). *Post Treatment of Desalinated Water*. Hopkinton: Balaban Desalination Publications.

10 WHO. (2004). *Guideline for Drinking-Water Quality*, 3rd ed., Chapter 8, http://www.wh-o.int/water_sanitation_health/dwq/GDWQ2004web.pdf.

11 Buecker, B. (2000). *Fundamentals of Steam Generation Chemistry*. PenWell.

12 Nalco Company (2011). *The Nalco Guide to Boiler Failure Analysis*, 2e. McGraw-Hill.

13 Saji, S.V., Sorour, A.A., and Meroufel, A.A. (2020). *Corrosion and Fouling Control in Desalination Industry*. Switzerland: Springer.

14 Duranceau, S.J., Pfeiffer-Wiler, R.J., Douglas, S.A. et al. (2011). *Post-Treatment Stabilization of Desalinated Water*, Water Research Foundation.

15 Cotruvo, J., Voutchkov, N., Fawell, J. et al. (2010). *Desalination Technology: Health and Environmental Impacts*. Boca Raton, FL: CRC Press.

16 USEPA. (1980). *Volatile Corrosion Inhibitors and Boiler Water Additives: Potential for Nitrosamine Formation, Toxic Substances*, EPA. 560/11-80-023.

17 Javaherdashti, R. and Nikraz, H. (2010). *A Global Warning on Corrosion and Environment: A New Look at Existing Technical and Managerial Strategies and Tactics*. VDM Verlag.

18 Nalco Company (2009). *The Nalco Water Handbook*, 3e. McGraw-Hill.

19 Association of Water Technologies (AWT), Inc. (2001). *Technical Reference and Training Manual*, 2e. Rockville, MD.

20 Stewart, M. and Arnold, K. (2009). *Emulsions and Oil Treating Equipment: Selection, Sizing and Troubleshooting*. Gulf Professional Publishing, Elsevier.

21 Rossmoore, H.W. (ed.) (1995). *Handbook of Biocide and Preservative Use*. Springer.
22 The American Society of Mechanical Engineers (ASME). (2011). *High-Purity Water Treatment Systems*. PTC, 31, (Performance Test Codes).
23 BetzDearborn, Inc. (1991). *Betz Handbook of Industrial Water Conditioning*, 9e. Trevose, Pennsylvania.
24 Crits, G.J. (2003). *Condensate Polishing: Purification Technology for Steam Power*. Tall Oaks Publishing.
25 The American Society of Mechanical Engineers (ASME). (1998). *Consensus on Operating Practices for he Control of Feedwater and Boiler Water Chemistry in Modern Industrial Boilers*, CRTD, 34.
26 American Water Work Association. (2005). *Granular Activated Carbon*, ANSI/AWWA B604-05.
27 Chowdhury, Z.K., Summers, R.S., Westerhoff, G.P. et al. (2013). *Activated Carbon: Solutions for Improving Water Quality*. Denver, CO: American Water Works Association.
28 The American Society of Mechanical Engineers (ASME). (2002). *Consensus for the Lay-Up of Boilers, Turbines, Turbine Condensers, and Auxiliary Equipment*, CRTD, 66.
29 *Ullmann's Encyclopedia of Industrial Chemistry*. 2002. Wiley VCH-Verlag.
30 Kurita Water Industries Ltd. (1999). *Kurita Handbook of Water Treatment, 2nd ed.*, Japan.
31 Crits, G.J. (2012). *Crits Notes on Water and Ion Exchange*. Chemical Publishing Company.
32 The American Society of Mechanical Engineers (ASME). (2014). *Performance Test Code on Deaerators*, PTC 12(3).
33 Stewart, M. and Arnold, K. (2011). *Produced Water Treatment Field Manual*. Gulf Professional Publishing, Elsevier.
34 Lazarova, V., Choo, K., and Cornel, P. (2012). *Water-Energy Interactions in Water Reuse*. IWA Publishing.
35 American Petroleum Institute (API) (1990). *Monographs on Refinery Environmental Control -Management of Water Discharges*, 1e. API Publication 421.
36 Berne, F. and Cordonnier, J. (1996). *Industrial Water Treatment: Refining, Petrochemicals and Gas Processing Techniques*. France: Gulf Publishing Company.
37 Water Environment Federation. (2002). Activated Sludge, 2nd ed. *Manual of Practice OM-9*, USA.
38 Lawrence, K.W., Yang-Tse, H., Howard, H.L. et al. (2006). *Waste Treatment in the Process Industries*. Boca Raton, FL: CRC Press.
39 Javaherdashti, R. (2017). *Microbiologically Influenced Corrosion: An Engineering Insight*, 2e. Switzerland: Springer International Publishing.
40 Connell, G.F. (1996). *The Chlorination/Chloramination Handbook*. Denver, CO: American Water Works Association.

41 Todd, B. (1967). *The corrosion of materials in desalination plants. Desalination* 3: 106–117.
42 Silverman, D.C. (2003). Aqueous corrosion. In: *Corrosion: Fundamentals, Testing, and Protection*, vol. 13A (eds. S.D. Cramer and B.S. Covino), 190–195. *ASM Handbook, ASM International.*
43 American Water Work Association. (2005). Sodium Metabisulfite. *ANSI/AWWA* B601–05.
44 The American Society of Mechanical Engineers (ASME). (2008). *Steam and Water Sampling, Conditioning, and Analysis in the Power Cycle.* PTC, 19(11), (Performance Test Codes).
45 The American Society of Mechanical Engineers (ASME). (2006). Consensus on Operating Practices for the Sampling and Monitoring of Feedwater and Boiler Water Chemistry in Modern Industrial Boilers, *CRTD*, 81.
46 Treybal, R.E. (1981). *Mass Transfer Operations.* McGraw-Hill Publications.
47 Lauer, W.C., Barsotti, M.G., and Hardy, D.K. (2009). *Chemical Feed Field Guide for Water Treatment Plant Operators.* Denver, CO: American Water Works Association.
48 American Water Work Association (AWWA) (2006). *Water Chlorination/ Chloramination Practices and Principles.* Denver, CO: *AWWA Manual M20.*
49 Black and Veatch Corporation (2010). *White's Handbook of Chlorination and Alternative Disinfectants*, 5e, vol. 287. John Wiley & Sons.
50 American Water Work Association. (2004). *Hypochlorites.* ANSI/AWWA B300-04.
51 American Water Work Association. (2010). *Calcium Chloride.* ANSI/AWWA B550-10.
52 Yanga, B.S., Hwanga, W.W., Myung-Han, K. et al. (2005). Cavitation detection of butterfly valve using support vector machines. *Journal of Sound and Vibration* 287: 25–43.
53 American Water Work Association (AWWA) (2012). *Butterfly Valves: Torque, Head Loss, and cavitation Analysis.* Denver, CO: *AWWA Manual M49.*
54 American Water Work Association (AWWA) (2011). *Internal Corrosion Control in Water Distribution Systems.* Denver, CO: *AWWA Manual M58.*
55 Nalco Company (2015). *The Nalco Guide to Cooling Water Systems Failure Analysis*, 2e. McGraw Hill.
56 The American Society of Mechanical Engineers (ASME). (2005). *Test Uncertainty*, PTC, 19(1), (Performance Test Codes).
57 ASTM D1293-18 (2018). *Standard Test Methods for pH of Water.* West Conshohocken, PA: *ASTM International.*

6

Application of TRIZ for Corrosion Management

Reza Javaherdashti[1] *and Mehdi Basirzadeh*[2]

[1] *Eninco Engineering B.V., Netherlands*
[2] *Kavosh Meyar Tabasum Pasargad Company, 81, Nami st. Ahwaz, Iran*

6.1 Introduction

Creativity is an essential part of innovative product development and the origin of successful products. Several methods support activities in developing innovative products, one of them is the Theory of the Inventive Problem Solving (TRIZ).

TRIZ (pronounced TREEZ) is a Russian acronym: "теория решения изобретательских задач"

T	**T**eorija
R	**R**eshenija
I	**I**zobretatelskih
Z	**Z**adach

Literally: TRIZ is "a problem-solving, analysis and forecasting tool derived from the study of patterns of invention in the global patent literature" [1].

TRIZ is a theory for solving inventive problems. It supports the idea that unresolved problems are the result of conflicting goals (constraints) and unproductive thinking. It suggests breaking out of a nonproductive thinking mold by reframing the contradicting and competing goals in such a way that the contradictions disappear [2].

TRIZ is a state-of-the-art technology for innovation that can be used in many industries and sciences. TRIZ is a systematic process that develops critical thinking skills and encourages creativity and innovation. TRIZ elements can be used effectively by a wide range of people – from children to adults. The origin of TRIZ is derived from experimental data, patents, and the documentation of how innovative people solved inventory problems [3].

Corrosion Policy Decision Making: Science, Engineering, Management, and Economy, First Edition.
Edited by Reza Javaherdashti.
© 2022 John Wiley & Sons, Inc. Published 2022 by John Wiley & Sons, Inc.

It was developed by the Soviet inventor and science-fiction author Genrich Altshuller (1926–1998) and his colleagues, beginning in 1946. In English the name is typically rendered as "the theory of inventive problem solving" and occasionally abbreviated in English as TIPS [4].

"TRIZ is a systematic approach for understanding and solving problems which allows clear thinking and the generation of innovative ideas" [5].

TRIZ is a problem solving and brainstorming technique that that is especially popular among design engineers [6].

TRIZ provides a wide range of domestic products and services for the chemical, petrochemical, gas transportation, oil-producing, energy, and other industries.

TRIZ is one of the leading companies in the machine-building industry with a complete production cycle from design to service, and specializes in the production of compressors, pumps, and auxiliary equipment. The company also provides services for the repair, reconstruction, increase of reliability, productivity, dynamic stability, resource, and service life of dynamic equipment from various manufacturers.

TRIZ's intellectual and production capabilities enable the implementation of projects for the reconstruction and modernization of equipment without the involvement of contractors, which makes it possible to continuously monitor all production processes.

TRIZ systematically expands production facilities using modern equipment in order to fully and timely meet customer needs.

TRIZ is a comprehensive source of resources and services for the largest companies in Ukraine, Russia, Belarus, Uzbekistan, Poland, and other countries in Europe and Asia [7].

TRIZ is a young but evolving science. TRIZ is evolving not only deeper, but also broadly (expanding), creating for the exact sciences. Today, TRIZ begins to penetrate scientific and artistic systems. There was a time when trying to formulate the principles for solving a creative technical problem seemed timid and hesitant [8].

TRIZ is almost certain to produce a simpler, less expensive, and more effective solution than what would otherwise be obtained [9].

One can think of TRIZ as another way of Lateral Thinking.

TRIZ is based on two basic principles:

1) Someone, sometime, somewhere has already solved your problem or one similar to it. Creativity means finding that solution and adapting it to the current problem
2) Do not accept contradictions. Resolve them [10].

TRIZ is unique because of its "problematic" approach. In particular, TRIZ faces all possible problems as a conflict of two different situations that are not usually applicable to each other. As a tool developed by engineers, there may be

40 industry-oriented TRIZ principles, however, they can apply to business sector or even everyday problems.

TRIZ can be done in three main steps;

1) Define the problem and the expression of the contradiction in a clear and simple way.
2) As a system, approach the problem and control all interactions and stakeholders at the time of zooming in and out of the system over time. Nine Window Technique is the most appropriate method for such an analysis.
3) Identify the most powerful principles that can be used for the respective problem and easily come up with creative ideas to improve the strategy [11].

Altshuller concluded that similar approaches, used in many different areas, have created very effective solutions in his research. The creators have repeatedly reviewed these solutions.

During his research, Altshuller felt the need for creativity theory with the following conditions;

1) Being regular and systematic.
2) Take the lead to place the ideal solution in a very wide solution space.
3) Reproducibility and reliability, but it does not depend on psychological tools.
4) Access to creative information.
5) Add creative acknowledgment.

The benefit of TRIZ is that conflicts and contradictions can be resolved using innovative solutions. The three basic principles of TRIZ are as follows;

1) The ideal design is the goal.
2) Contradictions help to solve problems.
3) The innovation process can be systematically configured.

6.2 Basic Structure of TRIZ

TRIZ is a structure with innovative philosophy, methods, and tools (Figure 6.1). The TRIZ philosophy of excellence displays resources and contradictions. The most important tool of TRIZ is ARIZ. ARIZ is a creative algorithm for problem solving. The most widely used TRIZ tool is the Contradiction Matrix (Appendix 6.A).

Altshuller describes TRIZ as "a methodology that disciplines thinking to stimulate the inspiration that leads to daring solutions to problems" [12].

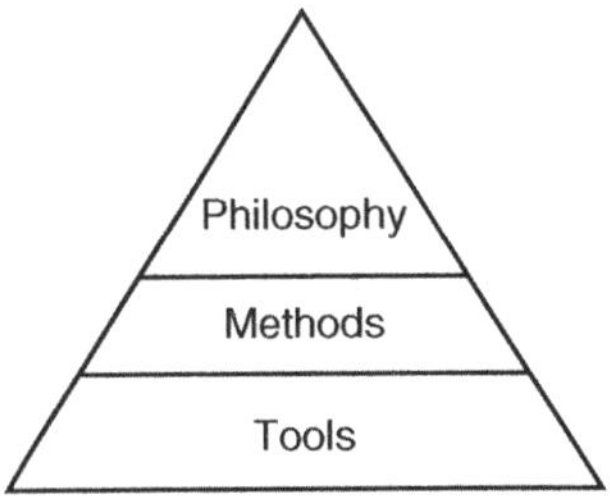

Figure 6.1 Basic structure of TRIZ.

The thinking about a problem's solution must begin with the "what is" *of the problem space and move toward the* "what will be" of the solution space [13].

Application of the TRIZ methodology provides the bridge between these two; it is the problem space which is characterized and resolved. TRIZ has the reputation as the only systematic problem solving and innovation tool-kit available.

The TRIZ toolkit consists of several tools; each of them is most effective against a separate type of problem [14], but they can all be used in relation to each other.

TRIZ is not so much a theory as it is a global practice used by some of the world's most innovative companies. Some of these include Proctor & Gamble, Boeing, Siemens, 3M, Hewlett-Packard, Eli Lilly, Honeywell, NASA, Toyota, Intel, Johnson & Johnson, Motorola, and many more. Although TRIZ is used in a wide range of industries and organizations, it is still a relatively undiscovered method. Part of this is due to its abstract nature, even if it is something that gives TRIZ the innovative problem-solving power. The other part is that the entire business is nowhere in the S-curve where it has embraced innovation as a systematic driving force – although it is close and some companies have taken the path [15].

TRIZ is a large and accurate toolkit, but many tools overlap, because TRIZ is designed to fit a variety of problem-solving techniques. Part of TRIZ's ingenuity is that it allows people to create their own TRIZ personalized tool that best fits their problem-solving style. It's a bit like having a gym with a wide range of equipment – people choose and use only a fraction of that equipment, depending on what suits them and what fitness problem they choose. The TRIZ toolbox is almost identical – there are tools that have specific goals and tools that one uses extensively [16].

The tools that make up this method include Inventive Principles, Evolution Laws, Smart Little People, Ideality, Substance-Field Analysis (SFA), Flow Analysis, Feature Transfer, Standard Solutions, Separation Principles, Multiscreen (9-windows), Trimming, and Contradiction. It must be noted that Contradiction is on the most used tool. This process is very powerful for breaking down existing design paradigms and moving on to new and exciting ones [17].

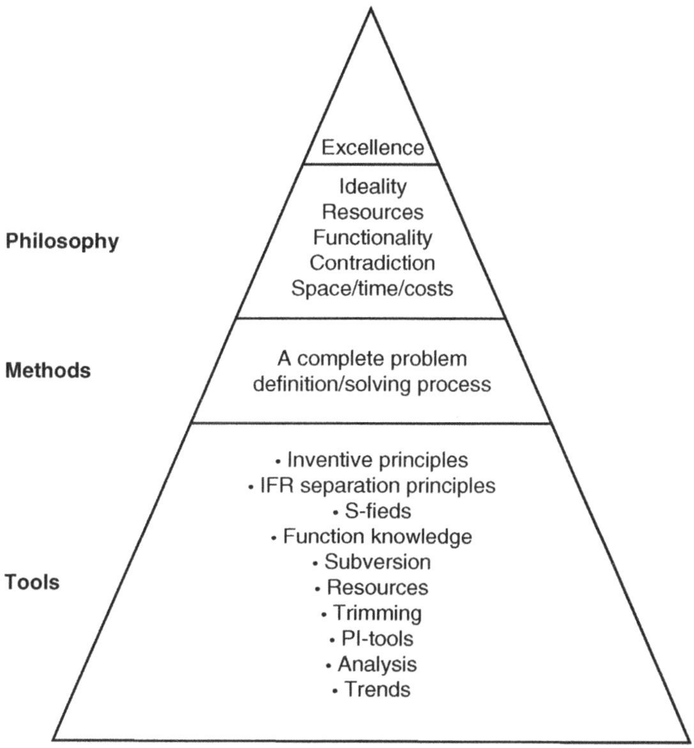

Figure 6.2 Tool, Method, & Philosophical Levels of Systematic Innovation.

Systematic Innovation works on several levels: (Figure 6.2) first, a set of tools; second, a complete process that links different tools together for any given innovation situation; and third, a set of philosophical ideas [18].

The philosophy of TRIZ: TRIZ is a powerful problem-solving philosophy based on logic and data. You can solve specific problems by applying generic solutions to similar issues.

It is designed to estimate creative problem solving and product design.

It is designed to establish principles that are common to all fields of technology.

It is designed to eliminate contradictions.

It is designed to use materials, energy, and knowledge effectively to create beneficial effects [19].

6.2.1 The Essence of TRIZ in 50 Words

An answer to the question "What is the essence of TRIZ?" stated in 50 words.

Essence of TRIZ:
Recognition that
technical systems evolve
towards the increase of ideality
by overcoming contradictions
mostly with minimal introduction of resources.
Thus, for creative problem solving,
TRIZ provides a dialectic way of thinking,
i.e.
to understand the problem as a system,
to image the ideal solution first, and
to solve contradictions. [20]

6.3 Level of Invention

Through the analysis of many inventions in the development of TRIZ, Genrich Altshuller discovered that different inventions involve different levels of creativity. Therefore, it seems that different tools and techniques are needed to create all types of inventions. (Obviously, the invention of the pencil with an internal eraser is very different from the invention of the steam engine.) In the late 1960s, Altshuller defined various levels for inventive problems related to;

1) The number of trial-and-error attempts required to guarantee a solution of a certain level.
2) The scale of change imposed on the original system. According to Altshuller, inventive problems can be divided into five levels, as follows:
   LEVEL 1 – Apparent (no invention). ~1–10 solutions are considered.
   Established solutions; well-known, and easily accessible.
   LEVEL 2 – Improvement. ~10–100 solutions considered.
   Existing system improved, usually with some compromise (example; bifocals).
   LEVEL 3 – Invention within paradigm. ~100–1000 solutions considered.
   A concept for a new generation of an existing system (example; automatic transmission).
   LEVEL 4 – Invention outside paradigm. ~1000–100 000 solutions considered.
   A new concept for performing the primary function of an existing system (example; jet aircraft, integrated circuit).
   LEVEL 5 – Discovery. More than 100 000 solutions considered.
   Pioneering invention of an essentially new system (example; laser, radio) [21].

6.4 History of TRIZ

In 1946, Genrich Altshuller, a 20-year-old patent investigator, recognized the patterns of ordinary thinking when inventing and developed the basic idea of TRIZ theory. He sent a proposal to Joseph Stalin, but was sent to a camp in Siberia, where he continued to develop his ideas. He was released after five years, published his work from TRIZ, and opened a number of TRIZ schools in various parts of the Soviet Union. His activities were banned in 1974, but he was allowed to return during the Perestroika years. During these 50-plus years of development, thousands of researchers/engineers have been involved in TRIZ research and development to claim the world's patents, i.e. 2.5 million patents in total, as is sometimes being claimed, in its technological semantics and to establish the system of the TRIZ methodology.

Since the 1980s, especially after the end of the Cold War, a number of ex-USSR TRIZ specialists immigrated to the western countries and brought TRIZ ideas with them. Sweden, the US, and Israel were active in receiving them. Particularly in the United States, some companies have started developing software tools of TRIZ. In Japan, introduction/promotion activities have been started in a significant scale only since 1997 [22].

For a period from 1970 to 1980, many TRIZ schools were opened throughout the Soviet Union, training hundreds of students, and during this time, Altshuller traveled there to conduct seminars. This phase ended in 1980 when the first TRIZ special conference was held in Petrozavodsk, Russia. In the next period, from 1980 to 1985, TRIZ received quite a bit of publicity in the Soviet Union. Many people became familiar with TRIZ and Altshuller veterans, and the first TRIZ specialists and semiprofessionals appeared. Altshuller was highly efficient at developing his TRIZ model due to the large number of seminars he conducted, the various TRIZ schools that were founded, and the followers of those who joined the ranks, allowing for the rapid testing of ideas and tools. TRIZ schools in St. Petersburg, Kishinev, Minsk, Novosibirsk, and other cities became very active under Altshuller's leadership.

In 1986, the situation changed dramatically. Altshuller's deteriorating health limited his ability to work on TRIZ and control its growth, and for the first time in the history of TRIZ, perestroika allowed it to be used commercially. The Russian TRIZ Association was founded in 1989 with Altshuller as president.

The economic situation in the former Soviet Union worsened from 1991 onwards, and many capable TRIZ professionals, most of whom had established their own businesses, were forced to emigrate. Many TRIZ professionals immigrated to the United States and began promoting TRIZ individually. Others found international partners and established TRIZ companies.

There are many consultants and companies today that offer training, consulting, and software tools. http://Trizjournal.com is a source for the dissemination of knowledge and information about the development and application of TRIZ worldwide [23].

6.5 About the Founder of TRIZ

6.5.1 Genrich Saulovich Altshuller

The life of Genrich Altshuller, the father of TRIZ as we know him, inspires all TRIZ lovers. His career as a patent secretary, the fierce struggles in the Verkota and Glag prisons, and the great inventions of TRIZ are fascinating to anyone who studies it. This section is a collection of different facts and events in his life from different sources. I thank all the websites hosting Altshuller's biography, which helped me to know him better.

Genrich Altshuller was born on October 15, 1926 in Tashkent, Uzbekistan (former Soviet Union) to a family of journalists. A few years later, the family moved to Baku, Azerbaijan (Soviet Union). He was Jewish. He studied in Baku and spent most of his life there. After finishing high school in Baku, he studied at the Azerbaijan Industrial Institute.

After graduating, he joined the Russian Navy. He was trained as a pilot on a World War II fighter plane, but he was not much involved on the battlefield. He worked at the Russian Naval Innovation Center where his job was to screen patents. This was the ideal place for his creative thinking to flourish. Altshuller began his career in 1946 when he was only 20 years old. He studied thousands and thousands of patents and discovered the logic of innovation, later known as TRIZ.

He was brilliant and inventive from his childhood. He was awarded his first author's certificate (equivalent to a patent) when he was only 14. He invented diving gear and a jet engine boat while he was studying in school.

During 1946–1948 he discovered the key TRIZ techniques. He used TRIZ to survive World War II devastation.

In 1948, Altshuller and his friend made fundamental proposals to the Russian government, but the result was negative and he was imprisoned for a long time. He was forced to stay in a labor camp in Vorkuta, in the terrible cold above the Arctic Circle. He went through a very difficult time in GULAG (Soviet prison system).

Interestingly, Altshuller was not the only prison elite, many other elites and academics were imprisoned under the dictatorship of Stalin.

He took this unique opportunity to learn many fields, including mathematics, logic, science, foreign languages, and more. This knowledge helped Altshuller to understand different systems from a generic perspective. After his arrest, Altshuller began writing stories and articles on science and fiction. His first article on TRIZ was published in 1956.

His first story, Icarus and Daedalus, was published in 1957. He was writing under the pen-name Altov. He wrote many other books, such as "Ballad of the Stars," "Donkey and Axiom," etc., which ware later translated into many other languages.

All his writings were full of brilliant and creative ideas. Altshuller wrote the book "And Suddenly the Inventor Appeared," published it in 1984, it was a view at the young generation's decisive response. This was his best-selling book.

Altshuller spent most of hi life studying patents. He displayed more than 200 000 patents to see how those problems were solved. He found that very few of them were real inventions; the majority are just direct advanced modifications of previous submissions. He found that all of those inventions used a specific set of rules to solve problems.

Basically, the same set of rules has been used over and over again to solve all types of innovative problems. He mentioned 40 such rules as his Inventive Principles, which is considered to be the main technique of TRIZ. Instead of categorizing patents into common industry classes, Altshuller classified these patents into five different levels according to the novelty of their patents.

They are; (i) common design problems (solved by knowledge within the specialty), (ii) minor improvements in the existing system (use of knowledge in industry), (iii) basic/fundamental improvements in an existing system (use of knowledge inside and outside the industry system), (iv) new inventions (using knowledge about technology), and (v) rare discoveries (using knowledge of science and beyond). 99% of patents were in the first four groups, and less than 1% were found to be new discoveries.

Altshuller distinguishes between a general problem and an Inventive problem. He stated: "An innovative problem is an innovation in which the solution causes another problem." For example, "Increasing the strength of a metal plate makes it heavier." We need strength of the plate but not its weight.

Inventors have traditionally compromised between strength (improving feature) and weight (worsening feature) and resorted to some trade-off, such that the plate is neither too thick nor too weak. However, these solutions did not bring the maximum desired (ideal) result, because the plate was neither too strong nor too light. An inventive solution does not compromise, but eliminates the contradiction, that is, it increases the benefits of improving the feature (strength) and reduces the effects of worsening the feature (weight).

Altshuller not only wrote articles and books, but also organized several workshops and seminars on TRIZ. Some workshops lasted as long as intensive training programs (for several weeks). In 1971, he founded the "Public Institute for

Innovative Creativity;" the world's first TRIZ Institute. He helped organize local TRIZ schools through students and TRIZ enthusiasts. There were more than 500 such TRIZ schools throughout the former Soviet Union.

He wrote approximately 20 books and 400 articles on TRIZ, and he held some 65 seminars, taught more than one thousand students, and was the author of several patents. Altshuller believed that a healthy society needed more creative people, because creative people pursue their noble goals. Without these people, the problems of human society would become acute and lead to destruction. Philosophically, he was a great rationalist and believed that the power of reasoning could solve all problems.

TRIZ is the method of his rational approach to strong thinking and problem solving. In 1989, Altshuller became president of the "International TRIZ Association," founded by his friends and students. In 1990, he and his family moved to Petrozavodsk, Russia, where he lived for the rest of his life. His presence in Petrozavodsk turned this town into a TRIZ research and communication center.

This great soul passed away on September 24, 1998, of complications from Parkinson's disease. Altshuller's wife, Valentina Zhuravleva, lived until 2004. He was also co-author of many science and fiction books. Altshuller left behind a revolutionary science, the TRIZ, which will keep him alive in the memory of thousands of people around the world. His great discoveries and his contribution to humanity will immortalize him forever into the future [24].

6.6 Contradiction as a Means to Formulate an Inventive Problem

In TRIZ matrix terminology, a problem is often called a contradiction. This is because the Russian scientist realized that improving the product on the one hand would make it worse on the other.

There are two types of system contradictions.

Technical Contradictions (TCs), where the improvement of one characteristic degrades a different characteristic (power vs. weight, speed vs. size, etc.). Traditionally, this contradiction has led to system compromise. TRIZ seeks to eliminate the contradiction and prevent compromise.

Physical Contradictions (PC), where the characteristics of the system contradicts itself (i.e. they must be both higher and lower, present and absent, etc.). TRIZ tries to turn PCs into TCs [25].

Here are examples of some common TCs that are often encountered:

- As the product gets stronger, its weight begins to increase.
- Large engines improve acceleration, but also increase vehicle costs.

- As the service is customized for each customer, the service delivery system becomes complex.
- Training is inclusive, but it distracts employees from their regular assignments [26].
- As the car accelerates (improves), both the fuel and the emissions get worse.
- The braking system is very responsive, but the rotors break down quickly.
- These parts are easy to assemble, but difficult to disassemble.
- The product becomes stronger (improved), but its weight increases (bad).
- We improve the ability of our workers (good), then they find other jobs (bad).
- Exercise is good for you, but it is relatively uncomfortable and time consuming (bad) [27].

Here are some classic examples of PCs:

- Coffee should be warm enough to drink, but cold enough to prevent burning the mouth.
- The software should be sophisticated enough to accommodate interesting features, but it should be simple enough to be user-friendly.
- Training should be complete enough, but it should be quick enough not to waste more time.
- The temperature must be high enough to melt a mixture quickly, but it must be low enough to reach a homogeneous mixture in the cooling.
- The wing of the plane should have a large area (to provide an elevator for takeoff and landing), but the same wing should have a small area (to reduce drag and achieve greater speed).
- The car airbag should be deployed quickly (to protect the adult passenger), but should also be deployed slowly (to avoid injuring a child passenger).
- To draw fine lines, the tip of the pen should be both sharp and smooth to prevent the paper from tearing.
- The office temperature should not be too hot to be comfortable for Sue or too cold for Sam to be comfortable.
- The surgeon has to spend a lot of time cleaning to prevent infection, but he has to scrub quickly because everyone is waiting.
- Aircraft landing gear must be present for take-off and landing, and absent (to prevent dragging) during flight.

Probably the strongest point of TRIZ's approach is that its goal is to solve so-called contradictions. While the traditional engineering approach is to find compromise, or trade-off, for contradictions, TRIZ tries to resolve contradictions on a regular basis. It does this by identifying the reason why progress is impossible. Using 40 innovative principles offers a solution to contradictions [28].

Problem solving often involves understanding and resolving conflicting requirements – improvement in one area is the detriment of something else (TCs), or we

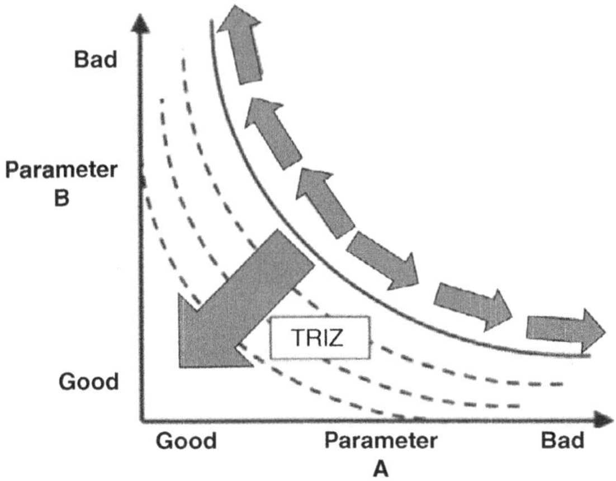

Figure 6.3 Reducing the contradiction.

may want to do the same in opposite situations but at different times or places. An umbrella should be small and large (physical contrasts).

Once we understand the conflict in the required conditions, we can use the TRIZ processes to discover the contradictions and the tools to resolve them [29].

TC means that when parameter "A" changes, another "B" changes at the same time in one system. For example, power versus fuel consumption and weight versus power in the car system. Problems related to technical conflict can be solved through the contradiction matrix and related inventive principles. After interpreting these principles, solutions were found using the TRIZ method. As shown in Figure 6.3, by reducing the contrast, one parameter "A" improves without the other parameter "B" becoming worse [30].

6.7 Procedure of Inventive Design

The 40 inventive principles and contradiction elimination of TRIZ techniques are good ways to solve the problem of innovative engineering design with system contradictions. TRIZ does not create successful solutions by "better brainstorming" or teaching people to "think creatively." In dealing with contradictions, TRIZ provides breakthrough solutions by providing tools to find the problem behind the problem and eliminate it. The general method for inventive design with TRIZ is shown in Figure 6.4 [31].

There are the 39 domains (technical parameters) defined in TRIZ as the source of the contradiction. Most inventions are related to contradiction resolution. Differences between the two items occur; one is what you are looking for and the other is what you would expect.

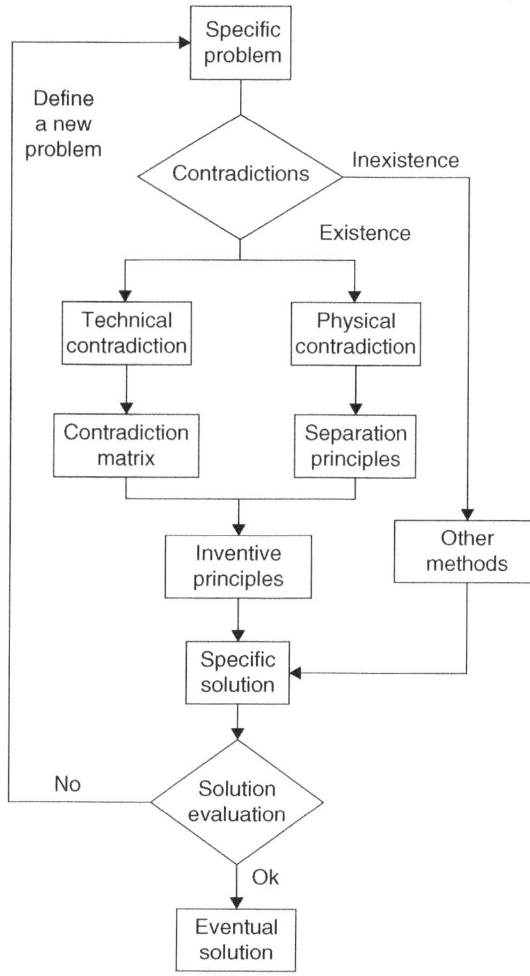

Figure 6.4 General procedure for inventive design with TRIZ.

For example, to make the car accelerate faster, you may put a larger engine in it, but the weight of the engine reduces its performance. This is the contrast between strength (you want) and weight (which you do not want) [32].

6.8 Concept Development Using TRIZ

The patterns and lines of evolution of the technology system are useful in identifying new ideas, but they are not specific concepts of next-generation technologies. Concept development for next generation technologies should be completed after generating ideas. There are two types of problems for developing the concept, as follows:

- Transition from one stage of evolution to a higher stage if there is no technical or PC. The next generation of inventions cannot be fully developed unless the contradictions of these system are resolved.
- The direction or ideas of high potential for the next generation of inventions are known, but the physics to realize that direction are missing.

The TRIZ Tools which are useful for use in these situations are inventive principles, standard solutions, and effects; Figure 6.5 shows this. The ideas are first analyzed and then decisions are made. If there are contradictions, inventive principles or standard solutions are chosen to resolve them. If there is no contradiction, meaning that physics is lost for the realization of ideas, effects must be chosen. New concepts developed are evaluated. If one or two concepts are accepted they are input into the design and manufacture process, otherwise they should be returned to the idea analysis.

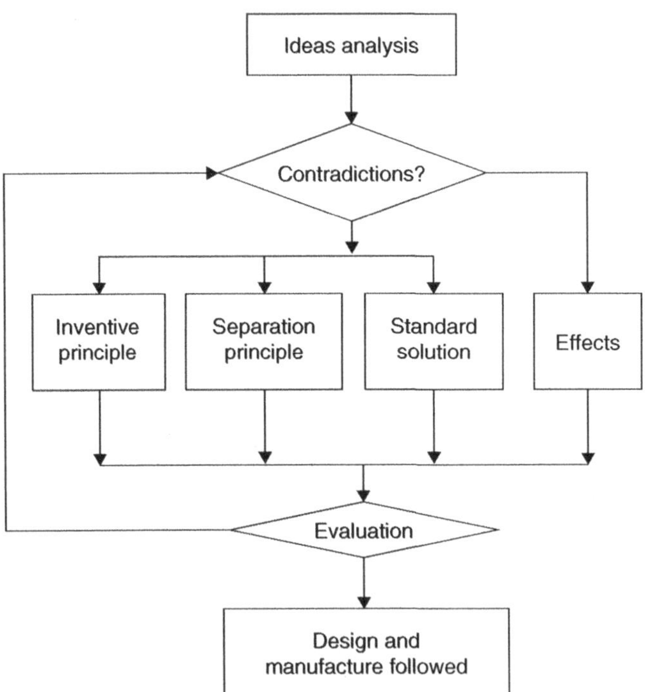

Figure 6.5 Process of concept development using TRIZ.

There are two types of inventive principles for resolving technical and physical conflicts. 39 engineering parameters and a matrix are used to select a few of the 40 principles, which are related to resolving a particular conflict. There are four principles of separation to resolve PCs; they are selected according to space, time, conditions, or parts of the whole.

In TRIZ there are 76 standard solutions that are divided into 5 classes. First, Substance-Field models are developed to solve the problem. Then, a standard solution from class 1 to class 4 is selected by the Su-Field model and a solution is obtained from that. Class 5 is last to be used to correct the pre-solution, and the final solution is obtained.

Impacts in TRIZ are useful for creating physics or working principles to realize the ideas generated in Fuzzy Front End (FFE). About 10 000 effects have been described in the sciences, including physics, mathematics, chemistry, and so on. Some of them may be used to determine the principles of work in product design.

An ordinary engineer usually knows about 100 effects, so a data base of effects is a type of assistant tool for product designers. Some computer-aided innovation (CAI) software is needed as a tool or assistance. These include databases for evolution, principles, standard solutions, and effects. Many companies in the developed world have used them, for example GM, MADMAX, DARPA, REVEO, and partners who they took advantage of this program [33].

6.9 Contradiction Matrix (39 × 39)

The paradox and contradiction appear when the search to improve one desirable feature makes another desirable worse! Solving a typical problem usually leads to a compromise solution. As mentioned earlier, the most innovative solution is achieved when a technical problem containing a contradiction is solved by the complete elimination of the contradiction.

Altshuller, from his research on more than 40 000 of the most patented inventions, found that there were only "39 features" that could either be improved or destroyed. Therefore, any problem can be described as a conflict between a pair of parameters (2 of 39 parameters).

In the past, many patents have resolved these individual conflicts in a number of different contexts. Conflicts were resolved over and over again, sometimes several years apart. He concluded that the "40 Principles of Invention" were used to fully resolve these contradictions, and not as a trade-off or compromise. He added: "If the second researchers had known these preliminary results, they would have solved their problems more easily."

Altshuller therefore seeks to extract and organize common contradictions and principles for resolving these contradictions. He uses it in the form of a matrix of 39 improving parameters and 39 worsening parameters (39 × 39 matrix) with each cell entering the most innovative principles (maximum 4). This matrix is known as the "CONTRADICTION MATRIX" and is still the simplest and easiest TRIZ tool. (See Contradiction Matrix) [34].

Assuming we improve the speed of the car by installing a larger engine, the mileage will be affected; this creates a "contradiction." The challenge that TRIZ has addressed is how to improve one or more features without compromising the performance of the other features. The Russian scientist found this issue at the core of all patents that were being filed. He also found that most of the "contradictions" faced by different industries are the same. For example, cost versus speed, accuracy versus speed, and so on.

Because the contradictions are more or less the same across industries, so are their solutions. The creators of the TRIZ matrix recorded their solutions in what is called the 40 problem-solving inventive principles [35].

6.9.1 List of the 39 Features

1) Weight of moving object
2) Weight of stationary object
3) Length of moving object
4) Length of stationary object
5) Area of moving object
6) Area of stationary object
7) Volume of moving object
8) Volume of stationary object
9) Speed
10) Force
11) Stress or pressure
12) Shape
13) Stability of the object's composition
14) Strength
15) Duration of action by a moving object
16) Duration of action by a stationary object
17) Temperature
18) Illumination intensity (jargon)
19) Use of energy by moving object
20) Use of energy by stationary object
21) Power (jargon)
22) Loss of energy
23) Loss of substance
24) Loss of information

25) Loss of time
26) Quantity of substance/the matter
27) Reliability
28) Measurement accuracy
29) Manufacturing precision
30) External harm affects the object
31) Object-generated harmful factors
32) Ease of manufacture
33) Ease of operation
34) Ease of repair
35) Adaptability or versatility
36) Device complexity
37) Difficulty of detecting and measuring
38) Extent of automation
39) Productivity

In search of the patent, Altshuller identified 39 technical parameters, including mass, volume, and stability. According to Altshuller's early publications, it is incorrect production that improves one parameter and leads to worsening another. These so-called TCs can be resolved with one of the elements of TRIZ – the contradiction matrix – which suggests which of the 40 identified "inventive principles" should be used to overcome the contradiction. Improved parameters are listed vertically, and worse parameters are listed horizontally, see Figure 6.6 [36].

The 40 Principles and Matrix of Contradiction are two of the innovative problem-solving tools in TRIZ. We learn which of the 40 Principles and Matrix of Contradictions is more important and significant as we try to discover the contradiction in the matrix with the help of their frequency, that is, how many times they appear [37]. All 40 principles are ranked according to the frequency of use in the contradiction matrix. This study found that only 20 principles are able to address more than 75% of the contradictions [38].

6.9.2 List of the 40 Principles

One of the tools used to overcome TCs is called "principles." TRIZ offers 40 inventive principles to resolve contradictions in the system. The 40 principles for implementing a technical system and within are a general suggestion. Altshuller discovered these principles in the research and synthesis of thousands of patents. These were some of the keys to how innovative people could solve problems that were pioneering and independent of industry or science. These principles are general enough to be applied to various problems, products, and industries to create innovative solutions. TRIZ has laid the groundwork for systematic innovation and provided a constantly evolving discipline for organized learning. 40 TRIZ

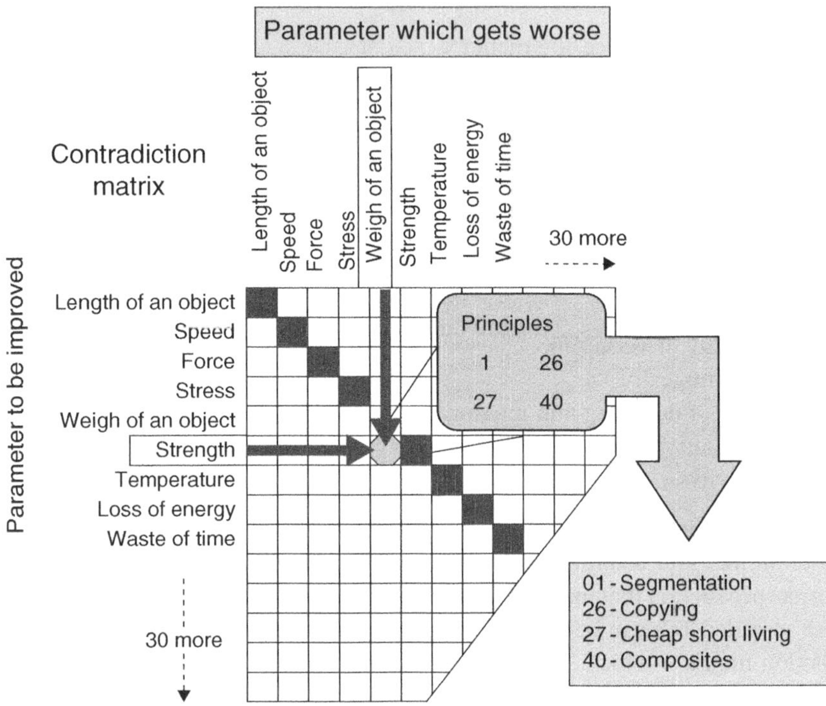

Figure 6.6 Altshuller's contradiction matrix.

Principles are a list of known solutions. Studying these existing solutions can inspire you to solve new problems and come up with innovative solutions [39] Table 6.1.

6.10 Using the TRIZ Matrix

The TRIZ matrix is at the heart of the TRIZ problem solving methodology. Therefore before we begin understanding the specific steps that apply to the TRIZ method, it is essential that we understand the logic behind those steps, i.e. the TRIZ problem solving methodology.

6.10.1 TRIZ Problem Solving Methodology

The TRIZ method is based on the assumption that all technological innovations that have taken place in various industries are based on a set of "inventive principles." The TRIZ method states that any particular problem that an organization

Table 6.1 List of TRIZ 40 inventive principles and their opposites.

#	Principle	Opposite
1	segmentation	merging (#5), integration, agglomeration
2	taking out	merging (#5); adding in
3	local quality	universality (#6); global quality
4	asymmetry	symmetry; balance
5	merging	segmentation (#1); separating
6	universality	local quality (#3); locality
7	nested doll	mutual exclusivity or mismatch
8	anti-weight	weight
9	preliminary anti-action	preliminary action (#10); afterward anti-action
10	preliminary action	preliminary anti-action (#9); afterward action
11	beforehand cushioning	afterward cushioning
12	equipotentiality	increase potentiality
13	the other way around	internally contains opposites
14	spheroidality – curvature	linearity
15	dynamics	statics
16	partial or excessive actions	this is its own opposite
17	another dimension	increase or decrease dimensionality
18	mechanical vibration	remove vibration
19	periodic action	periodic inaction; continuous action (#20)
20	continuity of useful action	periodic action (#19); continuity of useful inaction
21	skipping	do at low speed to get combination of actions
22	blessing in disguise; turn lemons into lemonade	curse in disguise; turn sugar into vinegar
23	feedback	lack of feedback; uncontrolled; positive feedback (feed forward)
24	intermediary	remove intermediary; simplify; self-service (#25)
25	self-service	single purpose device; intermediary (#24)
26	copying	avoid copies; use original
27	use cheap replacement objects	use expensive replacements, use expensive original
28	substitution for mechanical means	substitution by mechanical means
29	pneumatics and hydraulics	mechanical

(Continued)

Table 6.1 (Continued)

#	Principle	Opposite
30	flexible shells and thin films	rigid shells and thick slabs
31	porous materials	impermeable materials
32	color changes	use monochrome systems; use negative images
33	homogeneity	heterogeneity
34	discarding and recovering	this is its own opposite
35	parameter changes	parameter constancy
36	phase transitions	phase stability
37	thermal expansion	dimensional stability
38	strong oxidants	strong reducers
39	inert atmosphere	active atmosphere; presence of atmosphere; remove neutral parts; add active parts
40	composite materials	monolithic materials

faces can be reduced to a general problem that it has already encountered. This general problem has a general solution based on one of the 40 inventive principles of the invention in which the matrix exists. Therefore, the use of matrices suggests a general solution. This general solution can then be used to provide a specific solution to the specific problem that the organization is facing. Terms can be confusing and therefore an example in this scenario would be useful.

6.10.2 Reality of the "Four-Box Scheme" Theory

As the fundamental general scheme of problem solving, the Four-Box Scheme has been recommended not only in TRIZ, but also more widely in science and technology. For solving a user's specific problem, the scheme describes converting the problem into a generalized problem at an abstract level, finding a generalized solution to it with reference to some known models, and then interpreting it back into a specific solution in the user's real situation. To assist problem solvers in this scheme, TRIZ, science, and technology in general have developed a variety of knowledge bases and theories. The current general situation is shown in Figure 6.7

The accumulated models and knowledge bases are presented to the users in parallel as a wide range of different alternative advices, suggestions, or hints.

A guidance which is either irrelevant to the case or is proved to be wrong and inappropriate is neither effective nor reliable. There is no clear general way to select the models; the ways of abstracting the problem into the model are often

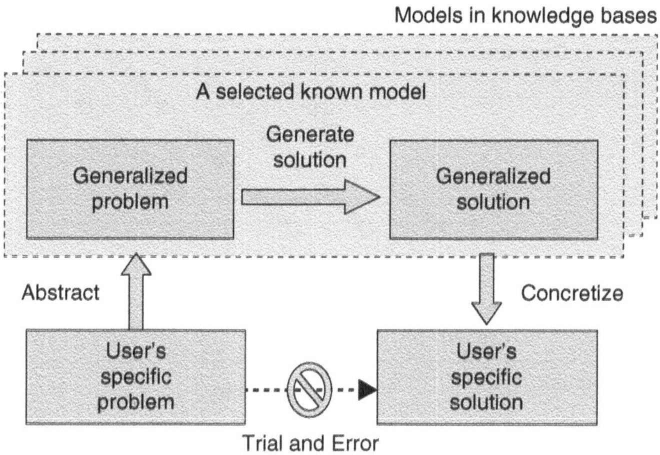

Figure 6.7 The Four-Box Scheme of Problem Solving Supported by Knowledge Base.

vague, and the ways of concretizing the model solution to specific solutions rely on intuition.

Thus the "Four-Box Scheme" in theory places problem-solvers in an ambiguous (obscure) world. The contents of the "Four Boxes" have yet to be described in a meaningful, yet general, way to cover the field of creative problem solving in technology [40].

TRIZ method for solving problems:

1) Identify the problem of the current project;
2) Compare the problem to an existing TRIZ general problem, as discussed later;
3) Identify the TRIZ solution for the general problem;
4) Use the suggested solution to determine the problem of the project.

For example: Intersection of external damaging factors and functional time of stationary object in the contradiction matrix (Figure 6.8): 17 – Transition into another dimension, 1 – Segmentation, 40 – Composite materials, 33 – Homogeneity [41].

6.11 Physical Contradiction Resolution

A systematic creativity process may consist of four major steps. As shown in Figure 6.9, definition is the first step. The second step is to select the most appropriate tools. The third step is to try to generate solutions. The fourth step is to evaluate and down-select.

Gets worse, Degrades →

↓ Gets better, Improves

		29 Precision of manufacture	30 External damaging factors	31 Object-generated harmful factors	32 Ease of manufacture
15	Functional time of moving object	3, 27, 16, 40	22, 15, 33, 28,	21, 39, 16, 22,	27, 1, 4
16	Functional time of stationary object		17, 1, 40, 33	22	35, 10
17	Temperature	24	22, 33, 35, 2	22, 35, 2, 24	26, 27

Figure 6.8 A portion of the contradiction matrix.

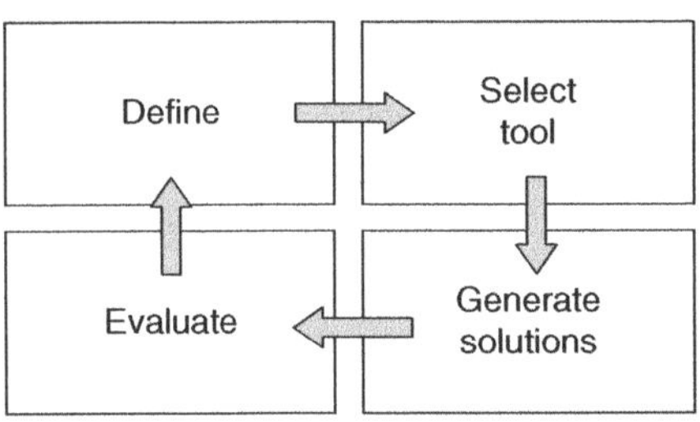

Figure 6.9 Four major "Systematic Creativity" steps.

PC: An action should be provided to achieve useful results and not provided to avoid harmful result. PC describes opposite requirements to the same parameter.

Separation principles help when some PC stands between you and an innovation, and you need to resolve the conflict with minimal or no trade-off. For example, you need the water in the system to be hot for some functions but cold for others.

The separation principles come from the TRIZ, and they are defined a little differently by different experts. For simplicity, we characterize the separation principles by separating contradictory properties in time, space, scale, and condition.

TCs are typically related to properties of the entire technical system, but PCs relate to physical properties of one characteristic of an element of the system. The Tables 6.2 and 6.3 give some examples for these types of contradictions.

Table 6.2 Technical contradiction examples.

Technical Contradictions (examples)	
Improving Parameter	**Worsening Parameter**
Power	Weight
Complexity	Functionality
Adaptability	Reliability
Productivity	Precision
Convenience of Use	Manufacturability

Table 6.3 Physical contradictions examples.

Physical Contradictions (examples)	
Characteristic "A"	**Characteristic "Non-A"**
Electro conductivity	Dielectric
Liquid	Solid
Hard	Soft
Fast	Slow
Strong	Weak

Contradictions arise in this process when technical requests are made to improve the existing system. An important concept to note is that based on any TC, we can find the physical cause of the contradiction. Almost all TCs can be turned into a corresponding PC. When we turn our TC into a PC, we define a specific physical problem that can be easily solved using "physical" principles and physical, chemical, and geometric effects along with other phenomena [42].

A PC exists when solving a problem in one parameter of a system is improved and the same parameter of that system is also degraded.

Algorithm for the resolution of a PC:

1) Establish a specific problem;
2) Through analogy, establish a generic problem by determining the bi-polarity within one parameter;

3) Apply the separation principles to generate generic solutions;
4) By analogy, determine specific solutions from the set of generic solutions;
5) First, create a particular problem. You want to experience high impact responses from a supportive audience, and I want to experience less impact responses from a critical audience;
6) Next, determine if it is bipolar;
7) High impact and low impact responses;
8) Then we use the principles of separation to form general solutions and produce specific solutions by analogy [43].

When dealing with a known PC, one can use one of the following four principles for overcoming this type of contradiction.

1) **Separation in time:** We use the separation principles to form generic solutions, and by analogy, generate specific solutions changing a property, response, or behavior vs. time.

   By time separation, inconsistencies can be resolved by identifying a time period in which one function is performed and another time period in which another function is performed. It resolves these contradictions as long as the separations do not overlap in time.

   Example: Water faucet
   The contradicting requirements for a water faucet are; (i) the water must be BLOCKED, and (ii) the water must FLOW. These contradicting requirements can be separated in time; (i) the water should be BLOCKED during the time the faucet is shut, and (ii) the water should FLOW during the time the faucet is open (Figure 6.10).

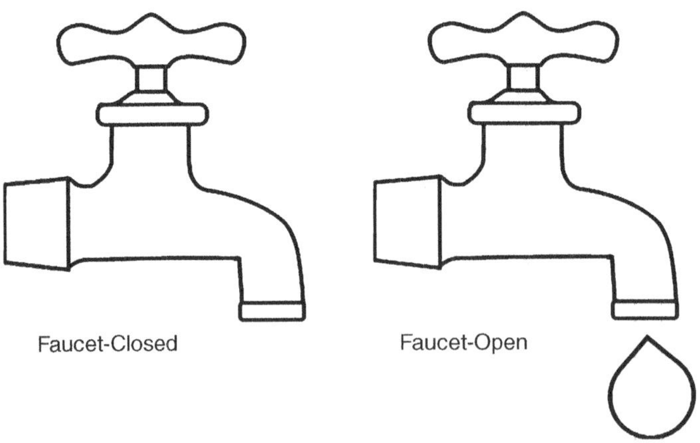

Figure 6.10 Example; Water faucets.

2) **Separation in space:** Changing the property, response, or behavior based on a special location.

   Separation in space can resolve contradictions in which two or more contradictory systems requirements must be fulfilled, at the same time. That is, if an object is required to have property "A" and not have property "A," then one can separate the object into two objects, each with its own properties. Conversely, the contradiction may be resolved if one part of the object has property "A" and the other part does not have it.

   Example: Hot and cold Water faucet

   The contradicting requirements for a water faucet are; (i) the water must be hot, and (ii) the water must be cold. These contradicting requirements can be separated in space. There is one faucet and two handles; one for hot water and another for cold water (Figure 6.11).

Figure 6.11 Example: Hot and cold water handles.

3) **Separation between parts and the whole:** Changing a property so as to make it different in the sub-system/system/supersystem.

   Resolving contradictions by way of separation between the parts and the whole may be employed when contradictory requirements state that a system exhibits specific properties, and at the same time and space, one or more of its parts exhibits opposing properties.

   Example: Motorcycle chain

   The contradicting requirements for a mechanical force transmission system between a motorcycle engine sprocket and a rear-wheel sprocket are; (i) the transmission system must be RIGID in order to transmit force and match properly with the two sprockets, and (ii) the transmission system must be FLEXIBLE in order to wrap around the two rotating sprockets (Figure 6.12).

The property expressed at the micro level (the chains links) is rigid when made to interface with the sprockets' teeth as well as each other by way of hinging pins. However, the overall system behavior expressed at the macro level (i.e. the motorcycle chain) is flexible [44].

Figure 6.12 Example; Motorcycle chain.

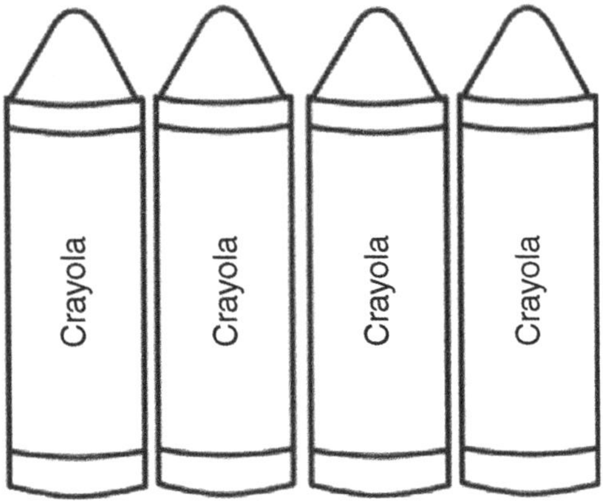

Figure 6.13 Example; The Crayola kids crayons.

4) **Separation upon condition:** Where a physical property of a system changes in response to an external condition

   Example: An interesting consumer example of separation upon condition is the new Crayola® product for kids to color. It uses a special crayon and paper design, which only enables coloring on a specialty paper that you must buy from Crayola. The crayons will not write on walls! (Figure 6.13).

The inventive principles most applicable to separation in space include segmentation, taking out/trimming, local quality, asymmetry, nested doll, other way around, curvature, another dimension, intermediary, copying, and flexible shells/thin films.

The inventive principles most applicable to separation in time include preliminary anti-action, preliminary action, beforehand cushioning, dynamism, partial or excessive action, mechanical vibration, periodic action, continuity of useful action, skipping, copying, discarding and recovering, and thermal expansion.

The inventive principles that can be used for condition-based separation – the physical properties of a system in response to an external condition – are mechanics substitution, pneumatics and hydraulics, porous materials, color changes, parameter changes, phase transition, and strong oxidants [45].

6.12 Ideality and the Ideal Final Result (IFR)

The TRIZ concept of Ideal Final Result (IFR) suggests that we determine what is ideally expected in a problem situation, irrespective of whether it is possible or not. The ideality of a product, process, or situation is achieved when all the functions of the product or process are achieved without facing any problems or cost [46].

Ideality is one of the building blocks of the TRIZ logic. Defining ideality is helpful to understanding where you currently are. You can use it to describe where you want to get to, and it is essential when you are comparing options. Ideality is the ratio between all the good things you want (benefits) and any downsides (costs and harms).

Benefits divided by all costs, Formula (6.1).

$$\text{Ideality} = \frac{\uparrow \text{Benefits}}{\downarrow \text{Costs} + \text{Harms} \downarrow} \tag{6.1}$$

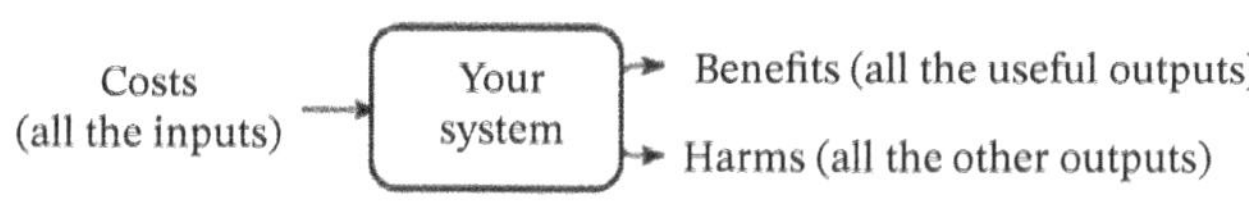

Benefits are all good outcomes. A list of benefits is a list of all the things that you want, without any detail of how you get them (benefits cannot contain solutions).

Costs are all inputs required to create your system, not only money, but also time, resources, and energy.

Harms are all outcomes you don't want. Nothing is neutral in TRIZ; an outcome is either wanted (and a benefit) or not (and therefore a harm) [47].

According to TRIZ, the Ideal Final Goal is what we ultimately wish to achieve. The Ideal Final Product is no product; only the results.

The IFR wants to achieve all the useful functions without any resources. In other words, the ideal product should need no space, no time, no cost, and no maintenance. The ideality is generally measured by the following function, Formula (6.2).

$$\text{Ideality} = \frac{\text{All good properties and useful function}}{\text{All harmful functions} + \text{Cost}} \tag{6.2}$$

According to the above function, with the increase of positive features, the degree of ideality increases. Ideally, it becomes 100% when the system contains all the positive features or achieves all the desired functions without the need for any effort or cost, and without creating harmful effects. However, it is almost impossible to achieve the above definition of the ideal. Nevertheless, when it is not possible to achieve a 100% ideal, the next ideal level may be targeted. In general, the ideality of the system can be increased by using the following methods:

- increasing the useful functions (in the numerator of the fraction);
- decreasing the harmful functions and cost (in the denominator of the fraction);
- a combination of the above.

From commonsense point of view we can see that a higher level of ideality can be achieved by obtaining maximum results and by using minimum resources. In some cases, very complicated problems can be solved by using unwanted or harmful resources to produce something useful. Inventive Principle-22: Blessings in disguise, also suggests the same; to turn harmful elements into useful resources. Let's see how to use the resources more effectively and efficiently [48].

The ideal anti-corrosive material is no corrosion in material. The ideal corrosion solution is a "No Corrosion" environment without implementing any effort (Figure 6.14).

Useful effects include all the valuable results of the system's functioning. Harmful effects include undesired inputs such as cost, footprint, energy consumed, pollution, danger, corrosion, etc. The ideal state is one where there are only benefits and no harmful effects. It is to this state that product systems will evolve. From a design point of view, engineers must continue to pursue greater benefits and reduce cost of labor, materials, energy, and harmful side effects. Normally, when improving a benefit results in increased harmful effects, a trade-off is made, but the Law of Ideality drives designs to eliminate or solve any trade-offs or design contradictions. The IFR will eventually be a product where the beneficial function exists but the machine itself does not. The evolution of the mechanical spring-driven watch into the electronic quartz crystal watch is an example of moving toward ideality.

Ideality is a powerful concept since it requires defining an ultimate system; an "ideal" system. An ideal system is a system which does not exist, but its function is delivered. Altshuller noted that increasing the degree of ideality is a trend which governs evolution of almost each technical system. The same happens with business systems; the more we can deliver with less, the more effective and efficient the

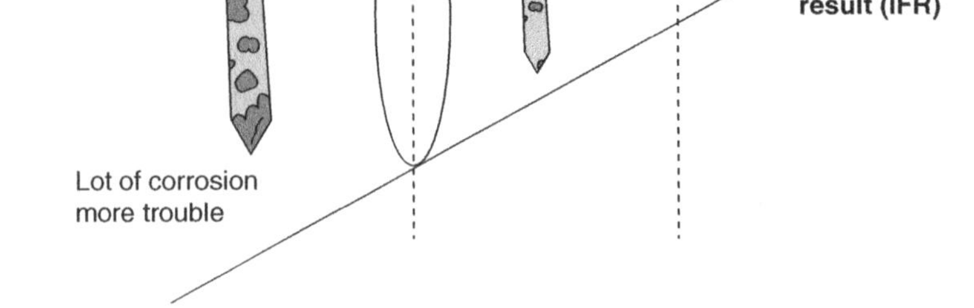

Figure 6.14 Moving from corrosion problem to IFR.

system will be. For instance, introducing IT support helps businesses to greatly reduce expenses by automating business processes. Using web-based marketing through social networks helps entrepreneurs reach millions of potential customers around the globe without leaving the house. Of course, a completely ideal system may not exist due to the law of energy preservation, but keeping the concept of ideality in mind when solving problems or designing new systems provides a platform for the "right thinking."

Although new management methods, such as Lean and Six Sigma, also increase the degree of idealism, they only do so within a certain range, while TRIZ techniques help to make disruptive changes to the ideal degree. Existence of systems will increase sharply; that's why many Six Sigma professionals learn TRIZ and integrate TRIZ with Six Sigma practices [49].

An ideal system does not exist, but all of its functions are fulfilled at the right time and at the right place; without energy, substance, or other resources, and without any ill effects.

$$I = \alpha \frac{\sum F_i Q_i}{\sum C_j + \sum \beta_k H_k} \Rightarrow \infty \tag{6.3}$$

Here:
I – idealization level (dimensionless performance);
F – useful function (effect);
Q – quality of useful function;
C – time and mean cost for useful function implementation;
H – nuisances;
α, β – accommodation coefficients.

Common sense suggests that the value of useful functions should increase, and the cost and nuisances should decrease, and then the ideality increases. When the numerator approaches infinity or the denominator approaches zero, this will occur.

It is often assumed that growth is the ideal attribute of progress, and therefore the traditional method of increasing numerator and reducing denominators seems justified. A. Seredinski has suggested expanding this concept [50].

If one looks at the ideality formula from a mathematical point of view, it becomes necessary to analyze all the other possibilities. We present the formula in a simple way (6.4).

The idealization level Formula (6.3) can be shown as:

$$I = \frac{\text{Numerator}}{\text{Denominator}} \tag{6.4}$$

Values of numerator and denominator can change; they can decrease, remain constant, or increase. We shall consider all of the possibilities and construct

Table 6.4 Idealization level behavior.

Numerator (F)	Denominator (P)		
	↓	=	↑
↓	1 ?	2 ↓	3 ⇊
=	4 ↑	5 =	6 ↓
↑	7 ⇈	8 ↑	9 ?

Table 6.4 where the rows will specify the types of "behavior" of a numerator, and the columns specify a denominator. The arrow, when pointing upwards, means growth, and when pointing downwards, means reduction. In cells where a line and column cross, we shall mark the "behavior" of ideality. Double arrows mean strong change.

Let's look at all nine cells.

The increase in degree of ideality is traditional, and is considered to be a simultaneous increase of the numerator and decrease of the denominator – as shown in cell #7. Cells #4 and #8 also characterize growth of ideality, though it is as not as rapidly as in cell #7.

Cell #3 shows the worst alternative. If the sum of useful functions decrease, and the harmful functions and costs increase, it leads to a sharp decline in ideality. Ideality also decreases in cells #2 and #6, though not as fast as in cell #3.

A constant ideality is shown in cell #5.

What is happening in the cells located on the ends of a diagonal, cells #1 and #9? The answer is not obvious. What occurs if the numerator and denominator either increase or decrease simultaneously?

For the answer to this question we will use simple reasoning.

Let us assume that the numerator has increased by four times, and the denominator by two times. Naturally, the factor of ideality will increase by two times, and vice-versa. Therefore, it seems that ideality can also grow when both the numerator and the denominator change "in the same direction," i.e. both either increase or decrease [51].

6.13 TRIZ Crossover QMS

Today, quality practitioners are placing considerable emphasis on achieving compliance with Quality Management Standards (QMS) (i.e. ISO 9001, EN 15224, ISO/TS16949, etc.) which are essential to ensure compatibility in the supply chain. TRIZ is a tool to improve creativity, which is needed to help solve difficult problems or predict the future development of systems. Most often it is helpful in situations that present a serious threat to the organization's survival (i.e. a competitor has better products or services at lower prices, the cost of materials is increasing faster than the prices of our products, or we could improve our market position if we could improve xxx). Predictions about the future form the basis upon which all decisions are made, and any improvement requires the prediction of evolution. To find the best solution the cooperation of interdisciplinary teams is required, which is why we decided to develop a tool based on merging Deming's System of Profound Knowledge with Altshuller's Theory of Inventive Problem-Solving. Such a tool will enable us to fulfill Fignebaum's foundation for business success; Innovate in product, service leadership, and cycle-time management.

To use TRIZ tools for improving your QMS, consider your organization as a system generating value. The concept of value can be used as a quality measurement indicator.

$$Q_{\text{indicator}} = \frac{\text{Value}_{\text{for customer}}}{\text{Cost}_{\text{for organization}}}$$

In the modern management approach, the strategic focus is on the ratio of value and costs. The difference between value and costs creates a variety of strategic options for setting the competitive price.

In order to compete in the new global economy, we must invent products and services that are of high value to customers and are produced by less expensive processes and systems. TRIZ crossover QMS is a tool that can help quality managers think outside the box to find the right solutions for clients and organizational needs, and be implemented with available resources [52].

6.14 The Evolutionary S-Curve

The evolutionary s-curve governs the evolvement of all systems. Research on the dynamics of evolution has also shown that all successful innovations are stimulated by an ideal end state. That end-state – defined as an Ideal Final Result (IFR) – is that the system delivers the functions and benefits that a customer requires, without any cost or negative harms. While this end-state might often

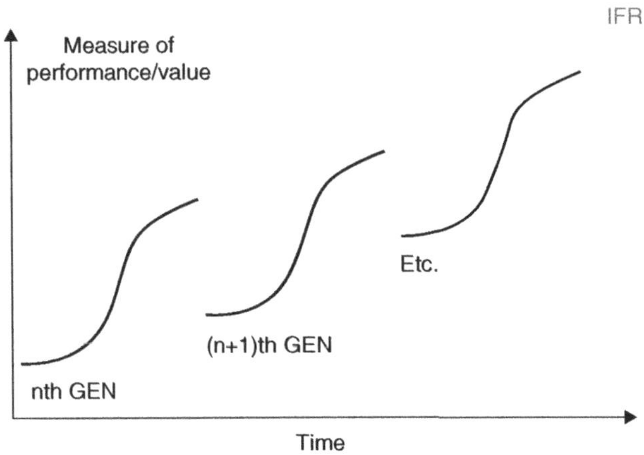

Figure 6.15 The fundamental dynamic of system evolution.

sound somewhat theoretical, there are many examples of systems and components that have evolved to such a state. What Figure 6.15 shows is that the dynamic of evolution toward this end-state occurs through a succession of s-curves. Key to the understanding of the overall dynamic is the recognition that all systems hit fundamental limits; the flattened profile at the top of an s-curve is not an indication that the market or engineers cease to be interested in improving a system, rather that something emerges to prevent the improvement from taking place. In other words, a conflict or contradiction arises and a system thus breaks a fundamental limit. Therefore, the only way to cross this fundamental limit is to find a new s-curve. Finding a new s-curve means resolving the contradiction [53].

The concept of an ideal final result is related to the concept of an "Ideal System," i.e. a system that does not exist (does not require time, effort, use of any source or place), while at the same time performing its functions fully. For the issue chosen at the previous step an IFR is formulated.

Possible options (in ideality decreasing order, with corresponding shifts from reasons to consequences):

- The problem goes away on its own;
- The required action happens by itself;
- The undesirable effect disappears by itself;
- The undesirable effect is compensated by itself;
- Attitude toward the undesired effect becomes negative.

The IFR in terms of the functional analysis is shown below.
("Reason-Consequence" and "System to Supersystem" scale-wise):

- The need for the function of this element disappears;
- The necessity of the function of this element disappears;
- The function is performed by the object itself;
- The function is performed by other elements of the given system;
- Operation is performed by the nearest system or a subsystem [54].

6.15 Nine Windows

Nine Windows is a tool commonly used in TRIZ. This tool is based on the concept that we usually see the world through a window. This tool forces us to observe and evaluate the world through nine windows; considering the past, present, and future in combination with the system, subsystem, and supersystem level. This is presented as a 3 × 3 matrix, as shown in Figure 6.16; by looking at the past, present, and future, we can get a historical perspective and context of the existing problem. In addition, by examining the problem of a system, system and environment (supersystem), and subsystem, one can understand the overall system and its structure/relationships [55].

- **Super-system (or Macro system):** External environment and components that the problem or system interacts or may interact with.
- **System:** The problem or system that has been created.
- **Subsystem (or Micro system):** A component or parts of the problem or system.

To complete the System row, list what started the problem in the Past/System cell, and then list the target – where the project will ideally end up after solving – in the Future/System cell.

	Past	Present	Future
Super-system			
System			
Subsystem			

Figure 6.16 Nine windows matrix.

To complete the Super-system row, first list all the things a person can do to prevent the current problem (in the environment where the system works) in the Past/Super-system cell. Next, list all the items that can be listed in the Future/Super-system cell correct the problem.

Once the team has reviewed both the problem and the system in the present, move on to the past and future. To do this, list all the things that can be done in the past to prevent problems in the Past/Subsystem cell. Then list all the things one can do in the future (if the problem still exists) in the Future/Subsystem cell. Explore all nine windows by asking:

- Can the company or team do something at the subsystem, system, or supersystem level **in advance** to correct or prevent the problem or improve the system?
- Can the company or team do something at the subsystem, system, or supersystem level **in the future** to correct or prevent the problem or improve the system?
- Can the company or team do something at the subsystem, system, or supersystem level **in the present** to correct or prevent the problem or improve the system?

6.16 Trends of Engineering System Evolution

These trends are a set of "empirical directions derived from the development of the engineering system that describes the natural transfer of engineering systems from one state to another." This process was initially achieved by analyzing thousands of patents. They have been validated by careful study of the history of technology, and today it is said that these trends are true for all categories of engineering systems.

Figure 6.17 shows the hierarchical nature of these trends and high-level categories. For each of these trends, there are usually several sub-trends that describe slightly different directions. The bottom line, according to those who use it, is that the way your product may grow over time is predictable; if you understand these trends, you can predict how the next generation of your product will take shape. Therefore, with the availability of technology to realize (or create) this next generation, you should be able to be one step ahead of your competition.

As an example, one of the most important trends in product evolution is the "Trend of Increasing Value." For this purpose, value is defined as being equal to total functionality, F/total cost, C. The trend states that an engineering system evolves over time so that its value always increases.

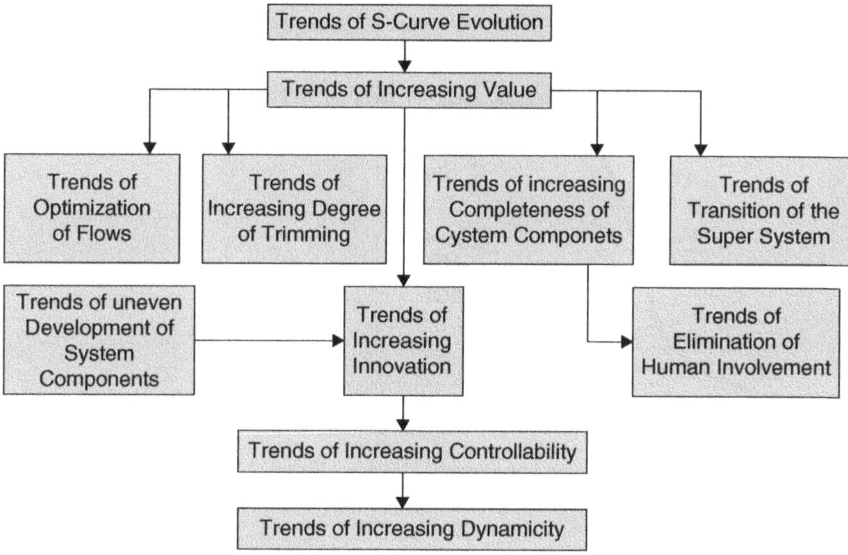

Figure 6.17 Hierarchy of trends.

Think of this as an analogy to Darwinian evolution; just as plants and animals continuously compete in their environments for territory, food, and reproductive success, your products are competing in the marketplace for sales, shelf space, and market share. Over time, you will succeed by bringing the best overall value to the market, in comparison to your competitors.

Defining value as F/C is important and suggests many different strategies for increasing value. Which approach you take should be tied to where your product lies on its S-curve of evolution? The Trends of Increasing Value and S-Curve Evolution in (figure 6.18) illustrates a variety of strategies. For example, an appropriate strategy for the early part of Stage 2 is to significantly increase overall functionality while allowing costs to increase at a slower rate. Conversely, a good strategy for stage 3 is to keep functionality constant while reducing product costs. The timing of the change from one strategy to another is not always clear, as it is sometimes difficult to decide on the need to jump to a new s-curve. Knowing that these changes in product development and pricing strategy are almost inevitable should encourage you to think about them regularly [56].

The evolution trend toward increasing ideality also applies in both technical and non-technical contexts. In the technical context, this trend has a strong influence on many of the other evolutionary trends observed by TRIZ researchers.

Some of these trends may be seen to possess direct relevance in the non-technical, business and, organizational contexts. Segmentation of substance and objects – a process that shows the transfer of objects from the macro to micro scale

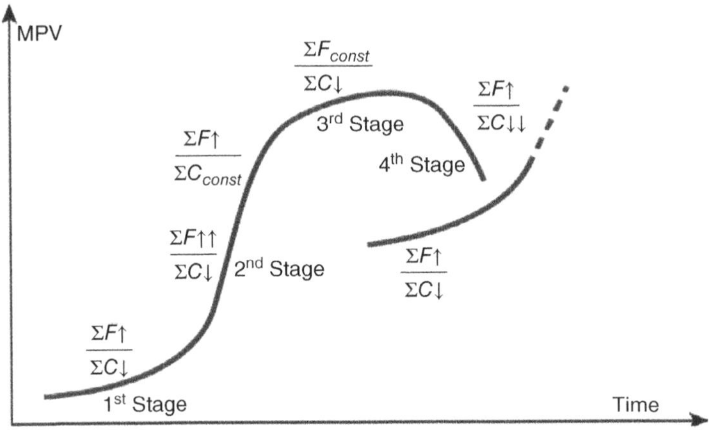

Figure 6.18 Trend of increasing value and s-curve evolution.

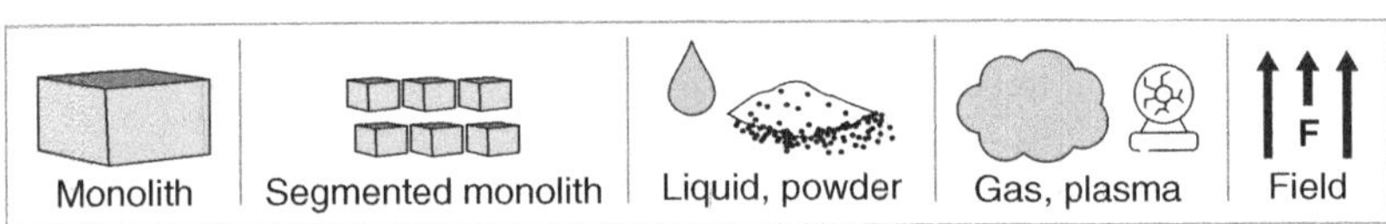

Figure 6.19 "Substance and Object Segmentation" trend.

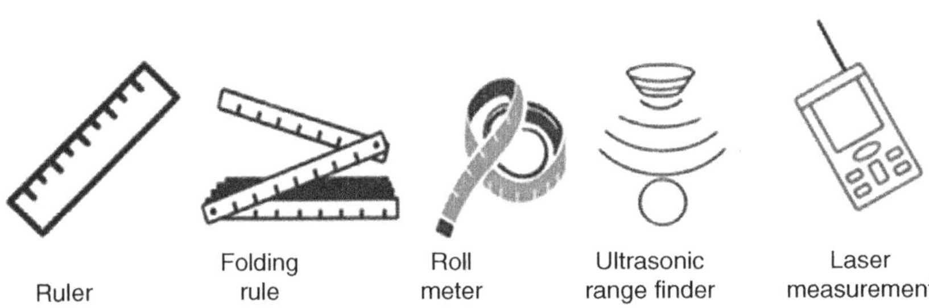

Figure 6.20 "Substance and Object Segmentation" trend example for measurement device.

(Figure 6.19) is applied to the evolution of business from the perspective of both customers ("mass customization") and organizations (evolution from "blue collar"). For example, "machinist" to "work team" to "worker" to "person").

The evolution toward "fields" has relevance in the non-technical context if the term is considered analogous to "emotions" or "feelings." Figure 6.20 For example, many products are now designed to respond not only to customers, but also to customer spirit; e.g. hotel rooms which allow the occupant to alter the feel of a room through use of variable color lights.

6.17 Geometric Evolution of Linear Constructions

Figure 6.21 shows another trend with direct non-technical corollaries. The trend may be seen to apply to a number of contexts in connection with both customers and internal organization and communication structures. For example, the evolution from individual artisans to 1D hierarchical organizations, to 2D matrix-management structures, to the emerging 3D 'spherical organizations, to – if 'time' is interpreted as a fourth dimension – the idea of time-variant organization structures. The evolution in straws (Figure 6.22) is another good example for geometric evolution of linear constructions trend.

Mono-Bi-Poly is another trend with direct applicability in a non-technical system evolution context. The mono-bi-poly trend shown in Figure 6.23 is particularly evident in symbiotic marketing applications such as the integration of film [57],

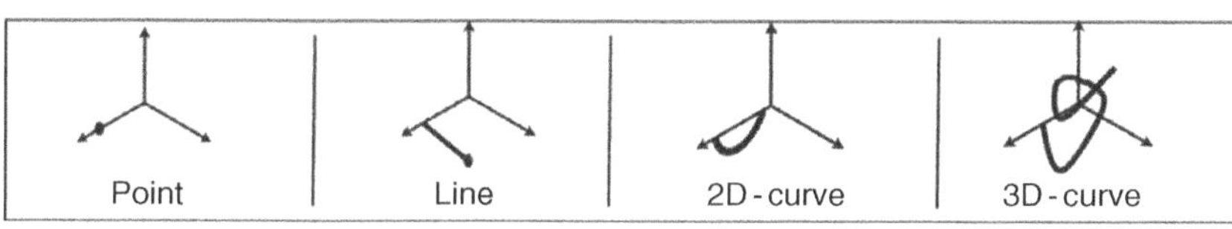

Figure 6.21 "Geometric Evolution of Linear Constructions" trend.

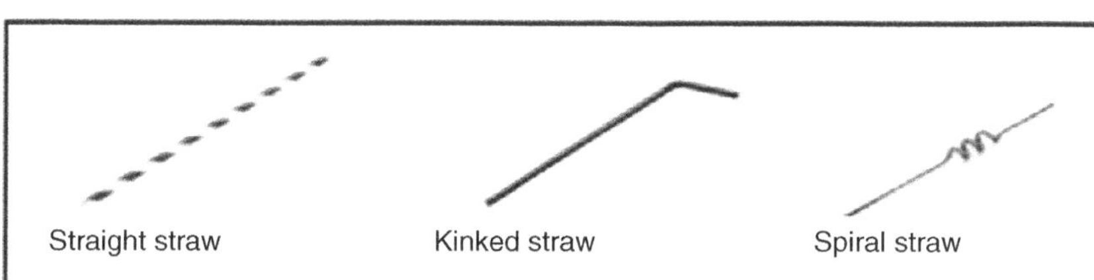

Figure 6.22 Example for straws' geometric evolution of linear constructions' trend.

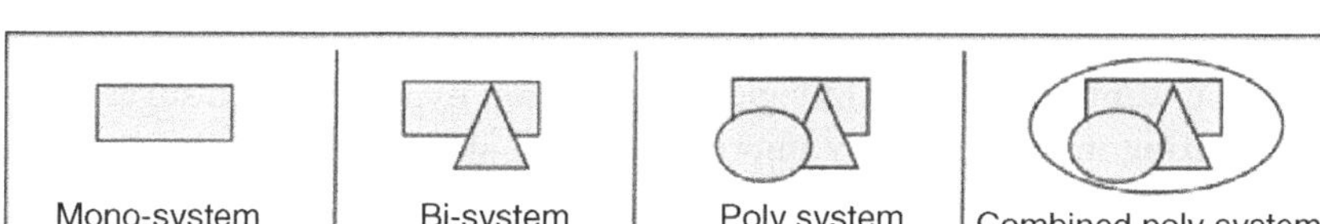

Figure 6.23 Mono-Bi-Poly trend.

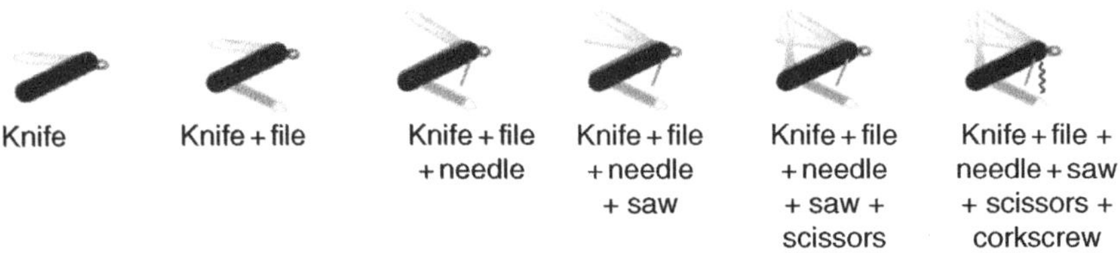

Figure 6.24 Mono-Bi-Poly trend for knife.

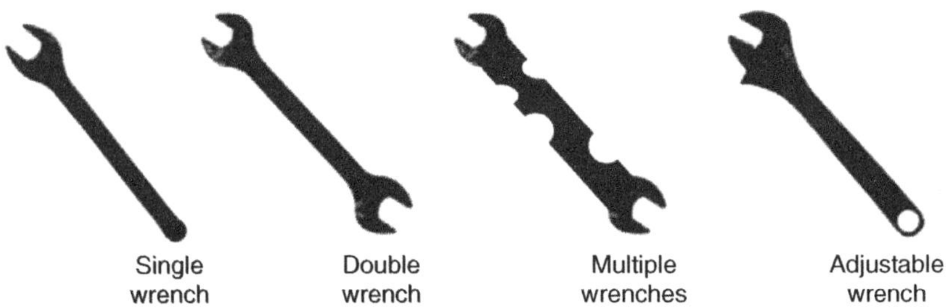

Figure 6.25 Mono-Bi-Poly trend for wrenches.

soundtrack, and merchandising in the entertainment industry, or in a number of multi-media applications [58].

The evolution in Swiss knives (Figure 6.24) and wrenches (Figure 6.25) are good examples for the mono-bi-poly trend.

A British Innovation Expert, Darrell Mann collected 31 different types of trends of evolution in his paper [59]. Twenty-five of them have been chosen as the useful evolution rules to link with biological cases, as shown in the left side of Table 6.5.

6.18 Trimming

6.18.1 Making Things Better and Less Expensive

Trimming is radically different to traditional problem solving; you start with the idea that you're going to improve your system by taking things away, and get more of what you want with less. TRIZ logic has important hypotheses and suggestions for minimizing inputs to ideally improve systems, and this is done by applying intelligence about resources. At TRIZ, we aim to minimize costly input sources and use free or inexpensive input sources, as well as any resources within the system. An important element of TRIZ is examining harmful sources of input, and if

Table 6.5 Inventive principles related to trends of evolution.

Trends of evolution	TRIZ inventive principles
Smart materials	31, 40
Space segmentation	2, 31
Surface segmentation	1, 31
Object segmentation	1, 2
Evolution macro to nano scale	1
Webs and fibers	31
Decreasing density	8, 35
Increasing asymmetry	4, 5
Boundary breakdown	5
Geometric evolution (Linear)	14,17
Geometric evolution (Volumetric)	17
Dynamization	15, 28
Action co-ordination	19, 2
Rhythm co-ordination	19, 20
Rhythm co-ordination	19, 20
Matching to external	15
Mono-Bi-Poly	5, 8
Mono-Bi-Poly (increasing difference)	5, 8, 7
Reduced damping	11, 29
Increasing use of senses	23
Increasing use of color	32
Increasing transparency	32
Degrees of freedom	15, 17
Trimming	2, 8, 22
Controllability	22, 23, 24
Reducing number of energy conversions (trending to zero)	35

they are inevitable and unavoidable, follow TRIZ's simple to-do list on how to turn harms into benefits.

TRIZ also has a very powerful tool called TRIMMING, which is part of the TRIZ system and function analysis process; this is a completely pure method that examines each component to see if it can be removed or trimmed without affecting the required outputs of the system. Its purpose is to maintain all the benefits while reducing costs, damages, and complications. This can be true for both technical and managerial problems. Recently, we have been working with companies to modify and trim out unnecessary process steps in management systems. A simple success by removing or correcting barriers to direct communication between teams reduced many communication barriers.

For the Trimming, TRIZ gives four strategies. It is assumed that the component to be trimmed is A, and the action object is component B; the following strategies are adopted:

- **Strategy 1:** If the value of component B is low, it can be considered to be trimmed. Therefore, the system does not need the function of component A, so it can be modified and trimmed.
- **Strategy 2:** If component B can perform the functions of component A by itself, component A can be trimmed.
- **Strategy 3:** If the function of component A can be performed by other components in the system, component A can be trimmed.
- **Strategy 4:** If the functions of component A can be performed by new, lower-cost components, then component A can be trimmed.

The priority of the above four strategies is reduced one after the other and the strategy with a highest priority is approved [60].

6.19 Input–Output–Trimming Operator (I–O–T)

The input–output–trimming operator is one of the best tools to express the problem correctly, if we need to create new concepts for existing machines, devices, or components.

We can think of any machine as a chain of energy conversion from input to output (according to TRIZ language we use the word "field" instead of "energy").

For example, let's use the usual kitchen mixer. Here we have the chain of transformation of the following context:

Electric field (input) → Electromagnetic field → Mechanical field of motor rotation → Mechanical field of tool rotation → Mechanical field of mixture motion (output).

According to the I–O–T operator, we have to trim the chain and express the problem of the correct conversion of input to output without intermediate links. For example, we want to convert electrical energy (field) into components of mixture motion without intermediate links. Of course, we can modify and trim only part of our chain instead of the entire chain, then we have to express the trimming problem for the input and output of this part instead of the entire chain.

The next step is to find the physical effects that solve the problem of converting this input directly to this output and create a new concept(s) to implement this transformation. For example, for our mixer; the mixer can be made on the piezo-electric effect, or if we trim only part of the chain, we can use electromagnetic vibrators. An extraordinary solution is to use electrical discharges in liquids, for example, electro-hydraulic shock, but I do not think so [61].

Like many TRIZ tools, trimming rules give you a specific set of steps that must be followed in a specific order. However, once you start generating ideas, you may find that you are moving in interesting directions that do not seem to be relevant to the task at hand.

The purpose of trimming rules is to make you think of other ways to achieve your goals. So, if you discover a rich window of new ideas that will lead you to interesting and useful tips, definitely follow them, provided you are thinking of new solutions and generating interesting suggestions and points for discussion.

This seam will naturally be less efficient and you will eventually run out of steam; at this point you need to get back to work. Following the Trimming Rules steps ensures that you cover all possible solutions. If you are thinking of solutions that do not seem to be related to the TRIZ task, that is fine. Do not waste energy and put them in the reverse process. Just go ahead and go back to following the steps.

If you come up with something out of the ordinary, then you're usefully coming up with new solutions. However, always make sure you do not stop early, get back on track. Just because you had good ideas in the past does not mean that you will not think of better ideas; let yourself be surprised! Following the trend means that you can be sure that you have been looking for any possible solution.

Trimming Infinity and Beyond – Trim and trim again if you are looking for completely innovative concepts.

When do you stop trimming? When you've gone too far. Knowing how much to measure is difficult, so a good rule of thumb is to keep trimming until you get something out of it. Suddenly your useful function is lost and your system no longer works. At that point, take a step back.

Trimming begins as a thinking exercise; as you trim, you see new ways your system can work and you develop conceptual solutions. This time of thinking is (relatively) inexpensive, and in a few minutes you will realize that deleting another component will make everything wrong. At that point you will find that you've pushed yourself – and your system – as far as it can go. You keep going because the more trimming you do, the more innovative your solutions will be. Figure 6.26 graphically shows the relationship between your trimming and the final system innovation you create.

When you get to the very deep stage of trimming – for example, you have already modified a part that you used to present your prime function, or re-trimmed a system that has already been trimmed – you can ask completely different questions

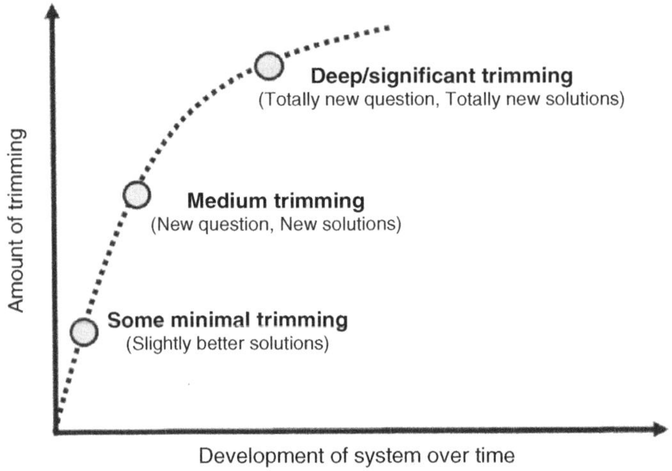

Figure 6.26 Trimming and system development.

about how it works and possibly consider new technology offering new opportunities. Eventually, you will build a much better system.

An example of deep trimming is online grocery shopping that shortens and trims the actual time you spend in a physical store. The functions of grocery stores are now separated in a timely manner; you create your list online and choose the food you want while you are still at home – the actual delivery and putting away of the food is much later. This provides extra benefits that you can see very early; both for holidays (which often require a lot of other work) and for your comfort (in the middle of the night, while at home on the sofa watching TV). The cost of your time is also reduced because you do not need to travel to and around the store, and your time and energy to shop is reduced because you do not need to pick up items, put them in your trolley, then on the conveyor, pack them, load them in the car, and take them home; all these steps are done by someone else. The dangers and stresses of walking around a crowded supermarket (especially with small children!) are gone. Other additional benefits are available, such as the potential for a wider range of products. Some household items can be easily ordered automatically every week. Plus, you also get to buy exactly what you need, without being tracked by attractive supermarket displays [62].

6.20 Resource Analysis

Resources play a key role in TRIZ while solving a problem. Proper use of existing resources helps to achieve cost-effective and ideal solutions without the complexity of the system and the introduction of expensive components and new materials.

Resources are available at both system and subsystem levels, and can be tangible (e.g. substances, fields) and immaterial (e.g. information). Although resource analysis was originally part of ARIZ, today resource analysis is used alongside other TRIZ techniques [63].

Terninko, Zusman, and Zlotin (1998) and Pannenbaecker (2001) divide resources into six categories; (i) substances, (ii) fields, (iii) functional, (iv) informational, (v) time, and (vi) spatial.

1) **Substances (material)** are any materials of which the system and its surroundings are composed. Readily available resources include raw materials or semi-finished products, as well as waste or lack of material.
2) **Fields (energy)** are any type of energy in or around the system, e.g. gravity, light, or electromagnetic radiation.
3) **Functional resources** are any type of effects. They include the capability of a system or its surroundings to perform additional functions. Another feature is an additional (unexpected) benefit that arises as a result of innovation, e.g. producing heat from cow dung (methane).
4) **Informational resources** are any tangible information. Additional information about the system could be extracted from existing fields in a system. For example, the crankshaft carries not only strength, but also information; the speed of pressing together.
5) **Time resources** are any kind of time including time intervals before, during, and after a process, e.g. use of online access to a computer.
6) **Spatial resources (space)** are free, unused space in a system or its environment [64], e.g. use of the interior of spare wheel space in a car [65].

After selecting the contradiction to solve, we must prepare a list of the available resources within the systemic context of the contradiction (Table 6.6), which is done in accordance with classical TRIZ methods [66]:

6.21 Function Analysis

Altshuller understood the importance of a functional approach to problem solving since the early days of TRIZ. For example, his concept of the ideal system states that the ideal system performs its function but does not exist, meaning that the ideal system performs its function free of charge and with no harm. However, the need to integrate performance analysis into TRIZ was identified after developing methods to solve generic problems in innovation. Function analysis plays a major role in problem formulation. Here we describe an advanced development called the Tool-Object-Product (TOP) Function Analysis and its benefits.

Table 6.6 Resource analysis.

#	Resource	Explanation
1	Time	Time before sales; time during sales; time after sales
2	Space	Customer space, company space, car space, bag space, external possibilities
3	System	Company people, sales people, engineers, workers, managers, analysts, software developers, IT infrastructure of the company, product, domain expertise, business expertise, communication capabilities inside company
4	Super-system: environment	Customers, product environment, suppliers, investors, independent analysts, Internet, independent experts, retailers, research and academic facilities, communication capabilities among customers, suppliers, experts, analysts
5	Super-system: similar/identical/inverse	Companies that produce similar combinations of device-software
6	Information	Information about existing customers, competitors, suppliers, information about domain, information about short- and long-term benefits, information about persons

6.22 Substance-Field Analysis

Altshuller developed a method and a set of symbols to describe generic types of problems and their solutions; the method was called SFA. Altshuller's model of the simplest useful system is composed of three elements; the two substances and the field. We can see examples of this in Figures 6.27–6.29.

Althsuller's SFA enables you to describe models of systems to be improved and models of improved systems. These are a set of the 76 most effective generic transitions from models to models of improved systems, which he called Standard Solutions to Inventive Problems. Substance-Field models describe system models rather than functions. However, to support function analysis, you must describe models of functions.

6.23 Tool-Object-Product (TOP) Function Analysis

TOP Analysis, the next generation of material-field analysis, was developed by Zinovy Royzen in 1989. The simplest useful function has four components: a

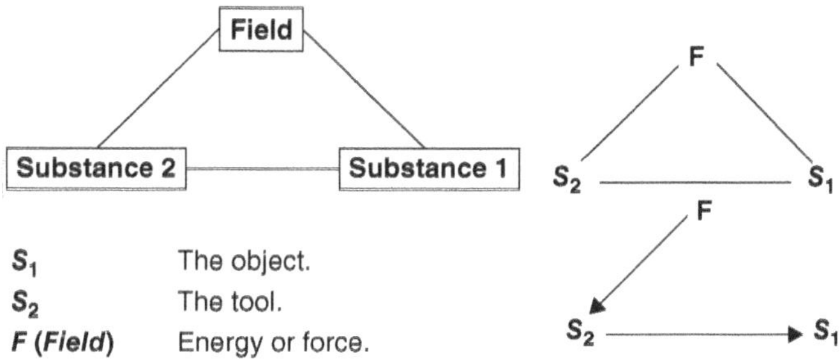

Figure 6.27 Models of the simplest useful system.

Figure 6.28 Models of an incomplete useful system.

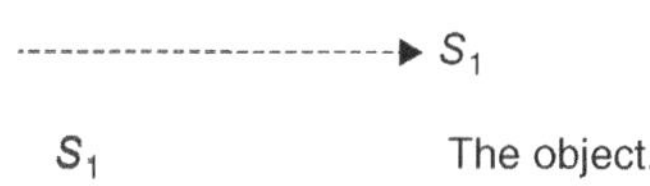

Figure 6.29 Model of the simplest system having a harmful action.

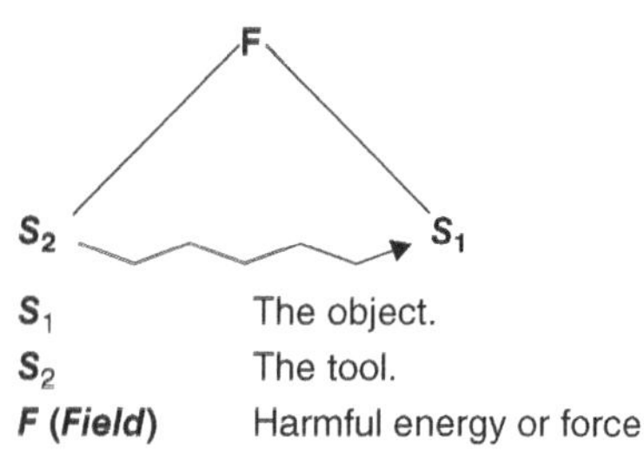

function (or function provider); a function target (or instrument function receiver); a tool function in the object; and the function product. Useful function is a tool to obtain the product of the function of the object. The action is described with an arrow, which simplifies the models in Figures 6.30 and 6.31.

Very often a useful action also causes an unwanted effect, or an attempt to improve a function leads to deterioration in another function of the system. Conflicts are the most difficult type of problem in innovation, and TRIZ offers models to describe any type of conflict (Figure 6.32).

Modeling a function by describing all four components – the tool, the object, the action, and the product – improves understanding of both the function and the

O	The object of the useful action
T	The tool of the useful action
F (Field)	Energy or force, or description of the useful action
U.P.	A useful product.

Figure 6.30 Model of a useful function.

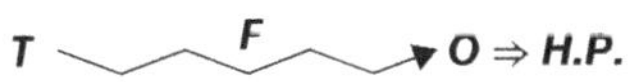

O	The object of the harmful action
T	The tool of the harmful action
F (Field)	Energy or force, or description of the harmful action
H.P.	A harmful (unwanted) product or products

Figure 6.31 Model of a harmful function.

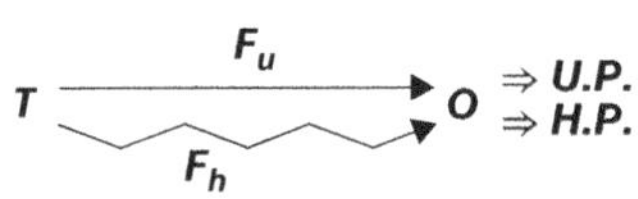

O	The object of the useful and the harmful function
T	The tool of the useful function
F_u	Energy or force, or description of the useful function
F_h	Energy or force, or description of the harmful function
U.P.	A useful product.
H.P.	A harmful (unwanted) product or products

Figure 6.32 Model of a conflict.

best methods for its improvement. The following sections describe some of the advantages of TOP Function Modeling.

6.24 Generic Model of a Function

Neither the tool of the function nor the object of the function should matter. TOP Function Modeling allows you to model any function in any system. It is a more generic method of modeling a function than Substance-Field Modeling.

6.24.1 Precise Description of a Function

Desired and unwanted products of the functions of a modeled system improve understanding of the system and simplify system resource analysis.

6.24.2 Link between Functions

Introducing the product of a function into its model creates a simple and understandable link between functions. For example, a product from the first function could be a tool or an object from the next function.

The link between functions is important in understanding not only the optimal performance of a product, but also the chain of unwanted functions. Links between functions simplify cause-and-effect analysis and improve the process of revealing the cause of a potential or current failure of a product.

6.24.3 Increasing Effectiveness of Function Analysis

Function analysis guides you in analyzing your product performance into single functions; both useful and unwanted. The system approach guides you in describing the function of the supersystem of your product and interactions between the product and its supersystem. It also guides you in analyzing and describing interactions between the product and its surroundings that are not part of the supersystem. A single function can then be considered separately if needed to be improved.

6.25 TRIZ Offers Five Basic Function Models

1) Adequate useful function (may require technology forecasting).
2) Insufficient useful function (needs to be improved).
3) Absent useful function (needs to be introduced).
4) Harmful or undesired function (needs to be removed).
5) Unknown undesired function (need to reveal the cause) [67].

6.26 Psychological Inertia

The psychological meaning of the word "inertia" means a reluctance to change; a certain bafflement due to human programming. It shows the inevitability of behaving in a certain way; the way that has been indelibly inscribed somewhere in the brain. It also indicates the impossibility – as long as a person is guided by their habits – of ever behaving in a better way [68].

Gordon Cameron identifies eight routine causes of psychological inertia:

1) Having a fixed view (model) of the solution or the root cause.
2) False assumptions (trust in data).
3) Specific terminology in a language that is a strong carrier of psychological inertia.
4) Experience, expertise, and trust in previous results.
5) Limited knowledge, hidden resources, or mechanisms.
6) Inflexibility (model worship), trying to prove a particular theory, stubbornness.
7) Reusing the same strategy.
8) Rushing to a solution, incomplete thinking [69].

Psychological stillness reflects many barriers to personal creativity and problem-solving ability. In problem solving, it is the inner, automatic voice of psychological inertia that whispers, "You are not allowed to do this" or "Tradition wants to do this" [70].

In the past, many methods and tools have been developed to deal with uncertainty by describing the conceptual design phase of a new product. In particular, TRIZ has been shown in several industrial fields to be a powerful tool in guiding designers to define innovative features for mechanical products.

This structural method of inventive problem solving aims to overcome the psychological inertia that can hinder the achievement of an optimum design, replaces the non-systematic trial and error approach, and helps engineers "find the right way" in searching for a solution (Figure 6.33).

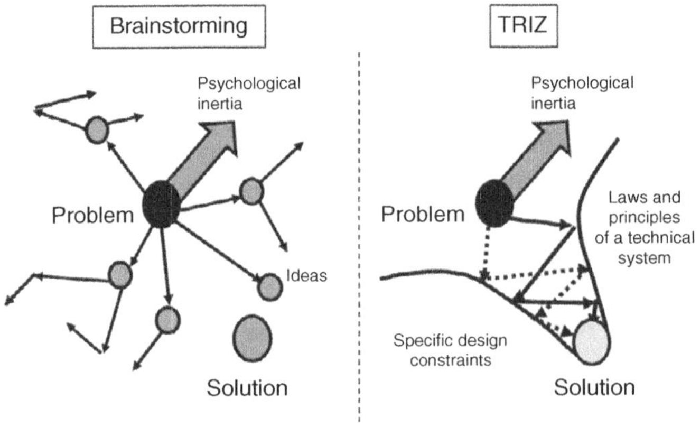

Figure 6.33 The TRIZ approach to overcome the psychological inertia in design.

TRIZ is based on the hypothesis that a few universal principles of invention are the basis of all creative innovation. Therefore, after identification and coding, these principles may be applied to make the invention process more predictable. This TRIZ knowledge-base may support design teams to deal with the poor nature of the PSS design problem, in which one or more steps are often either unknown or incoherent, there is insufficient information in the initial state, or the properties of the goals are not fully defined in advance [71].

6.27 Size-Time-Cost Operator

The Size–Time–Cost (STC) Operator heuristic engages a user in redesigning a problematic situation. The STC operator suggests six conditions for situation improvement. A user is expected to think of her/his actions when each of the three parameters (size, time, and cost) sequentially reaches two limits; zero and infinity (e.g. what would I do if I had ZERO budget for progress?) [72]. If any of these features are exaggerated, how can I solve the problem?

Size: the size of the Operational Zone (OZ); **Where** does conflict occur?

Time: the Operational Time (OT); **When** does conflict occur?

Cost: the cost of the known solution or assigned to the operation cost.

The STC operator can be used to overcome psychological barriers. It is in the form of a simple matrix (Table 6.7) in which the three parameters (size, time, and cost) can assume two opposite values; zero and infinite [16].

STC, is the three-dimensional thinking between 0 and ∞, and exaggerated thinking is a tool to challenge perceptions of constraint (Figure 6.34).

For example, considering cell number one in Table 6.7, the questions to be formulated are: "If the system size increases infinitely, what can be done to solve the problem?" Similarly, in cell number six, the question that is formulated is: "If system costs are to be zero, what can be done to solve the problem?"

First, the STC operator (size, time, and cost) was used to change the parameters during the problem analysis process. The TRIZ developers then tried to go beyond

Table 6.7 Size–Time–Cost Operator.

	Infinite	Zero
Size	1	2
Time	3	4
Cost	5	6

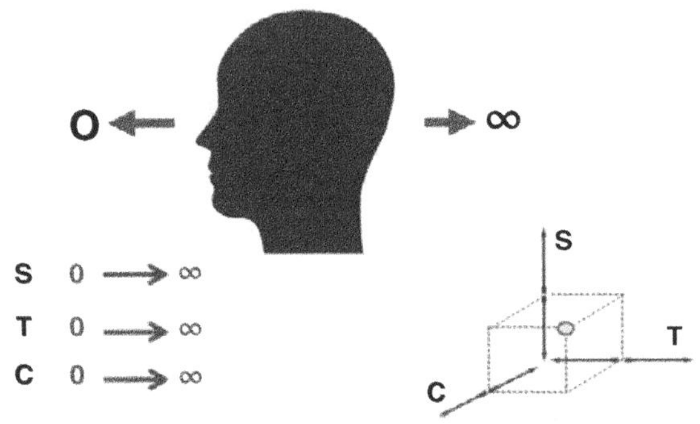

Figure 6.34 STC is a three-dimensional thinking between 0 and ∞.

such parameters and the performance of the resulting model was satisfactory. For example, if a system has a pressure parameter, the value of the analysis is the result of reducing this parameter to a minimum or, conversely, increasing it infinitely. Thus the STC operator became a scaling operator; one of the properties of the selected object and its variations along the axis (from zero to infinity) are considered.

However, some parameters cannot be changed along a line. Variability in the broadest sense presupposes selecting and changing a feature of the object. At the same time, changes in the entire multi-screen scheme caused by changes in a single parameter must be analyzed [73].

6.28 Applying the 40 Inventive Principles in Corrosion Management

1) **Segmentation**
   a) Divide an object into independent parts.
   b) Make an object sectional (for easy assembly or disassembly).
   c) Increase the degree of an object's segmentation.

Examples:

- Cast iron radiator in even number of sections, up to 44 sections. Each section can be exchanged after corrosion.
- Divide car components into different parts and components.
- Use powdered welding metal instead of foil or rod to get better penetration of the joint.
- The corrosion area segmentation.
- Multi-pin connectors.
- Corrosion can be divided into three main groups: 1; Wet corrosion, 2; Corrosion in other fluids, 3; Dry corrosion.

2) **Extraction** (Extracting, Retrieving, Removing)
   a) Extract the "disturbing" part or property from an object.
   b) Extract only the necessary part or property from an object.

   Examples:

   - Air conditioning in the room where you want it, with the noise of the system outside the room.
   - Using anti-corrosion gutters to remove rainwater from the building.
   - Using anti-corrosion chimney design so as not to disturb different parts of the building.
   - Clean off any corrosion before applying the paint.
   - Often it is possible to chemically remove the products of corrosion. For example, phosphoric acid in the form of naval jelly is often applied to ferrous tools or surfaces to remove rust.
   - Several standard methods for corrosion removal are available. The methods normally used to remove corrosion are mechanical and chemical. Mechanical methods include hand sanding using abrasive mat, abrasive paper, or metal wool; and powered mechanical sanding, grinding, and buffing, using abrasive mat, grinding wheels, sanding discs, and abrasive rubber mats. However, the method used depends upon the metal and the degree of corrosion.
   - Aircraft must be kept clean and corrosion-free at all times.

3) **Local Quality**
   a) Transition from homogeneous to heterogeneous structure of an object or outside environment (action).
   b) Different parts of an object should carry out different functions.
   c) Each part of an object should be placed under conditions that are most favorable for its operation.

   Examples:

   - Material surface treatments/coatings; plating.
   - Erosion/corrosion protection, case hardening, non-stick, etc.
   - Crevice corrosion is a localized form of corrosion occurring in confined spaces (crevices), to which the access of the working fluid from the environment is limited.
   - Corrosion can be concentrated locally to form a pit or crack, or it can extend across a wide area more or less uniformly corroding the surface.
   - There is a lot of corrosion on the bottom of the car.
   - Use a gradient instead of constant temperature, density, or pressure.

- Localized Corrosion: The most common, and most detrimental, form of localized corrosion is pitting. Pitting means when the attack happens in one single location on the surface and creates a pit, or small cavity, in the metal. This type of corrosion attack is hard to prevent, engineer against, and often times difficult to detect before structural failure is met due to cracking. Pipes are often compromised due to pitting.

4) **Asymmetry**
   a) Replace symmetrical form(s) with asymmetrical form(s).
   b) If an object is already asymmetrical, increase its degree of asymmetry.

   Examples:

   - Eccentric drive.
   - Wing design compensated for asymmetric flow produced by propeller.
   - Asymmetrical electrodes (AEs) of mild steel can lead to increase of the current noise detection. This phenomenon can occur if an oxide layer is formed on the alloy surface.
   - An electrical furnace has electrodes asymmetrically placed permitting the continuous loading of ore and discharge of molten metal.
   - To improve mixing, use asymmetrical tank or asymmetric blade in a symmetric tank.
   - Both the corrosion pattern and the corrosion intensity play an important role in the ductile properties. The asymmetrical distribution of the corrosion around the surface is a decisive factor, which can influence the ultimate strain at maximum force more seriously.

5) **Consolidation**
   a) Consolidate in space homogeneous objects, or objects destined for contiguous operations.
   b) Consolidate in time homogeneous or contiguous operations.

   Examples:

   - Combine harvester.
   - Manufacture cells.
   - Combined hot and cold water taps.
   - Bundle of tubes in a heat exchanger in order to increase transfer and decrease the size.

6) **Universality**
   a) An object can perform several different functions; therefore, other elements can be removed.
   b) Use standardized features.

Examples:

- Generalized corrosion: Typically never happens, aside from in acidic conditions. This uniform corrosion over the entire surface of the metal is rare and leads to overall thinning, which has little effect outside of fatigue and stress conditions.
- Uniform corrosion: The most common variety of corrosion, such as rusting of iron. Other metals and alloys also corrode uniformly such as aluminum, copper, brass, and magnesium.
- In a crusher with several cylinders, the cylinders turn for grinding and simultaneously they are cooling to avoid overheating (of cylinders and product).
- The American Society for Testing and Materials (ASTM) is a recognized world leader in the field of development and supply of international standards. Rust inhibitors are put through a standard industry test known as the ASTM D665 Rusting Test.

7) **Nesting**
   a) One object is placed inside another, that object is placed inside a third one, and so on.
   b) An object passes through a cavity in another object.

Examples:

- Nested dolls (Russian)
- Ant-nest corrosion is a type of premature and localized failure, which is observed in Cu tubes used mainly in air-conditioning and heat exchanger units, induced by the presence of organic matter.
- Heat exchanger principle; both cold stream and hot stream pass through tubes (one inside the other outside) to exchange heat with the other stream.

8) **Counterweight**
   a) Compensate for the weight of an object by combining it with another object that provides a lifting force.
   b) Compensate for the weight of an object with aerodynamic or hydrodynamic forces influenced by the outside environment.

Examples:

- As the percentage of aluminum rises, and all strongly resisting corrosion in air or seawater.
- Zinc is used to protect other metals from corrosion.
- Hydrofoils lift ships out of the water to reduce drag.
- Forklift counterweights are commonly made of metals.

- Iron and cast iron counterweights are often used in elevators. Steel counterweights are suitable for scaffolding and may weigh as much as 7000 lbs. Tungsten counterweights are designed for aircraft control surfaces, aircraft rotor blades, guidance platforms, and vibration-dampening governors.

9) **Prior Counteraction**
   a) Preload counter tension to an object to compensate excessive and undesirable stress.

Examples:

- Prepare a system with an inhibitive shape for avoiding corrosion.
- Corrosion can be prevented through using multiple products and techniques including painting, sacrificial anodes, cathodic protection (electroplating), and natural products of corrosion itself.
- Galvanizing: Coating iron or steel with zinc to prevent corrosion is known as galvanizing.
- Car underbody is sprayed with anti-corrosion paint to prevent rusting.
- Corrosion engineering is the field dedicated to controlling and preventing corrosion.
- Pre-cleaning before epitaxy, oxidation, diffusion, metallization, etc.

10) **Prior Action**
   a) Perform required changes to an object completely or partially, in advance.
   b) Place objects in advance so that they can be put into action immediately from the most convenient location.

Examples:

- Create beforehand stresses in an object that will oppose known undesirable working stresses later on, and prepare a system with an inhibitive shape for avoiding corrosion.
- Handling of a particularly corrosive wastewater would be reason for planning in advance the procedure to be used in the event that tubing failure during operation was detected. Such a procedure might be to immediately begin injection of a non-corrosive liquid into the well until the well bore was completely cleared, then to shut the well in until the reservoir pressure had died away to a level that would allow removal of the damaged tubing without backflow of the corrosive wastewater. Such a procedure would help to prevent damage to the casing, packer, etc.

11) **Cushion in Advance**
   a) Compensate for the relatively low reliability of an object with emergency measures prepared in advance.

Examples:

- In a heat conducting system with salts, the pipe must be preheated to avoid crystallization.
- Put an emergency stop system on a process.
- Clean water for emergency use should be available in the immediate work area before starting work for removing rust and corrosion from objects.
- Materials splashed in the eyes should be promptly flushed out with water, and medical aid obtained for the injured person.
- Suitable fire extinguishing equipment should be available to the cleaning/treating area.
- If materials (acid, alkali, paint remover, or conversion coatings) are spilled on equipment and/or tools, treat immediately by rinsing with clean water, if possible, and/or neutralizing acids with baking soda and alkalis with a weak (5%) solution of acetic acid in water.
- Corrosion management includes planning actions for corrosion mitigation and prevention. This is achieved by use of anti-corrosion measures, corrosion monitoring, regular inspection, and study of each accident, implementation of meetings, publications of minutes, education, and knowledge transfer.

12) **Equipotentiality**
    a) Change the condition of the work in such a way that it will not require lifting or lowering an object.

Examples:

- Any efforts directed at making corrosion control easier for operating personnel are highly relevant. In attempting to improve corrosion ergonomics, interactions among several human sub-factors such as design, procurement, manufacturers, engineers, and users need to be considered.
- Equipotential of corroded surfaces.
- Equipotentiality of the workstation.
- In general, corrosion threats should be mitigated to a point where the expenditure of resources is balanced against the benefits gained. One outcome of this is that a financial analysis might conclude that a technically sound corrosion mitigation action is unjustified. To determine whether a corrosion management investment is appropriate, it can be compared to the potential corrosion consequence through a return on investment (ROI) analysis. ROI is a benefit (or return) of an investment divided by its cost.

13) **Do It in Reverse**
   a) Instead of the direct action dictated by a problem, implement an opposite action (i.e. cooling instead of heating).
   b) Make the movable part of an object, or outside environment, stationary – and a stationary part moveable.
   c) Turn an object upside-down.

Examples:

- Iron ore is a combination of minerals from which metallic iron can be extracted.
- Corrosion is the reverse process of metallurgy. In other words, the energy used to transform ore into a metal is reversed as the metal is exposed to oxygen and water. As the metal is exposed to these elements, the corrosion process begins, oxides are formed on the surface, and in some cases combine with sulfides and carbonates.
- Anodization is the process of converting an anode into a cathode by bringing a more active anode in contact with it.
- Rotate the part instead of the tool.
- The major anti-corrosion strategies are:
  - The selection of appropriate materials;
  - The design of the product;
  - Protective methods (coatings, anodes, etc.);
  - Correct installation and maintenance;
  - Research, development, and testing.

14) **Spheroidality**
   a) Replace linear parts with curved parts, flat surfaces with spherical surfaces, and cube shapes with ball shapes.
   b) Use rollers, balls, spirals.
   c) Replace linear motion with rotational motion; utilize centrifugal force.

Examples:

- Corrosion tests in a rotary drum furnace.
- Rotary blower with corrosion-resistant abradable coating.
- The swirler is formed from a powder-sintered compact of martensitic stainless steel having corrosion resistance and wear resistance. They show greater resistance to corrosion, have better bendability, and are fire-resistant.
- Corrosive products inside the spherical vessel also cause internal corrosion.
- Rust or corrosion can lead to significant damage to bearings.

15) **Dynamicity**
   a) Characteristics of an object or outdoor environment must be altered to provide optimal performance at each stage of an operation.
   b) If an object is immobile, make it mobile. Make it interchangeable.
   c) Divide an object into elements capable of changing their position relative to each other.

   Examples:

   - Adjustable steering wheel (or seat, back support, or mirror position).
   - Static and dynamic corrosion tests in liquid Pb–Bi eutectic.
   - Dynamic corrosion is defined as the corrosion that occurs under mechanical actions such as sliding friction, which exposes fresh surfaces and accelerates wear.
   - Corrosion dynamic test is still a key step for carmakers.

16) **Partial or Excessive Action**
   a) If it is difficult to obtain 100% of a desired effect, achieve more or less of the desired effect.

   Examples:

   - Some corrosion mechanisms are less visible and less predictable.
   - Partial or excessive corrosion.
   - Over-spray when painting, then remove excess.
   - Good corrosion management aims to maintain, at a minimum life cycle cost, the levels of corrosion within predetermined acceptable limits. This requires that, where appropriate, corrosion control measures be introduced and their effectiveness ensured by judicious, and not excessive, corrosion monitoring and inspection.

17) **Transition into a New Dimension**
   a) Transition one-dimensional movement, or placement, of objects into two-dimensional; two-dimensional to three-dimensional, etc.
   b) Utilize multi-level composition of objects.
   c) Incline an object, or place it on its side.
   d) Utilize the opposite side of a given surface.
   e) Project optical lines onto neighboring areas, or onto the reverse side, of an object.

   Examples:
   - Dump truck.
   - Heat exchanger with a bundle of tubes (instead of one tube).
   - Corrosion caused by magnesium salts, often presents in large amount in underground and seawater.

- Corrosion protection coating of three-dimensional metal structure by electrophoretic deposition of graphene oxide.

18) **Mechanical Vibration**
    a) Utilize oscillation.
    b) If oscillation exists, increase its frequency to ultrasonic.
    c) Use the frequency of resonance.
    d) Replace mechanical vibrations with piezovibrations.
    e) Use ultrasonic vibrations in conjunction with an electromagnetic field.

    Examples:

    - It is clear that ultrasonic vibrations can have profound effects on electrochemical reactions, but there is little evidence of vibration-induced changes for frequencies less than ~10 kc/s.
    - Vibration effect on corrosion wear.
    - Mixing alloys in an induction furnace.
    - Ultrasonic testing to monitor corrosion.
    - Modeling and simulation of pipeline corrosion in the oil and gas industries.

19) **Periodic Action**
    a) Replace a continuous action with a periodic one (impulse).
    b) If the action is already periodic, change its frequency.
    c) Use pauses between impulses to provide additional action.

    Examples:

    - Hitting something repeatedly with a hammer.
    - In a batch production, the periods of apparatus inactivity are used for cleaning or to realize a new production.
    - Periodic corrosion testing is the best preventive measure to avoid component failure or complete shutdown.
    - Periodic corrosion control of aircraft will reduce aircraft fires and accidents, and at the same time increase the quality of maintenance, thereby increasing the operational readiness of the deterrent force.

20) **Continuity of Useful Action**
    a) Carry out an action without a break. All parts of the object should constantly operate at full capacity.
    b) Remove idle and intermediate motion.
    c) Replace “back-and-forth” motion with a rotating one.

Examples:

- Electrical continuity testing corrosion.
- Feed unit operation continuously (reactor, distillation column) instead of batch operating conditions.
- A Corrosion Management System (CMS) is the documented set of processes and procedures required for planning, executing, and continually improving the ability of an organization to manage the threat of corrosion for existing and future assets and asset systems.

21) **Rushing Through**
   a) Perform harmful and hazardous operations at a very high speed.

Examples:

- Use catalyst to accelerate chemical reaction to avoid sub-product formation.
- Alkali metals like sodium need to be stored in oil as they corrode quickly.
- Metals usually corrode faster if they are wet.
- Oxy-fuel cutting is a chemical reaction between pure oxygen and steel to form iron oxide. It can be described as rapid, controlled rusting.
- Corrosion and wear-protection with ultra-high-speed laser material deposition process.

22) **Convert Harm Into Benefit**
   a) Utilize harmful factors, especially environmental, to obtain a positive effect.
   b) Remove one harmful factor by combining it with another harmful factor.
   c) Increase the degree of harmful action to such an extent that it ceases to be harmful.

Examples:

- The formation of oxides on stainless steels can provide a protective layer preventing further atmospheric attack, allowing for a material to be used for sustained periods at both room and high temperatures in hostile conditions.
- Passivation: Some corrosion processes will create solid metal compounds that will coat the initial site of corrosion and prevent further corrosion at that site.
- Biofilm coatings: A new form of protection has been developed by applying certain species of bacterial films to the surface of metals in highly corrosive environments. This process increases the corrosion resistance substantially.

23) **Feedback**
   a) Introduce feedback.
   b) If feedback already exists, change it.

Examples:

- Look for signs of corrosion.
- If left unchecked, corrosion will eventually weaken the pipeline.
- All planes are being inspected for possible cracking and corrosion.
- Use corrosion control regulations.
- Corrosion Monitoring Tracer (CMT) can indicate rate of metal loss.
- Monitoring corrosion of metals is essential for material evaluation studies, inhibitor selection, and for adopting protective measures to combat corrosion problems in operating plants. Measurements of direct weight loss are time consuming as compared to electrical or electrochemical testing.
- Rust spots inside pipes can discolor water or pepper it with rust flakes. Rusty water can also taste or smell metallic, even when it's clear. If you notice discolored or metallic tasting water, contact a professional plumber for a more formal evaluation.

24) **Mediator**
   a) Use an intermediary object to transfer or carry out an action.
   b) Temporarily connect the original object to one that is easily removed.

Examples:

- Pot holder to carry hot dishes to the table.
- A variety of instruments are used for testing and monitoring corrosion of metals.
- The presence of an intermediate conversion layer is presumed to improve the corrosion resistance of the system.
- Electron transfer mediators accelerated the microbiologically influenced corrosion against carbon steel by nitrate reducing *Pseudomonas aeruginosa* biofilm.

25) **Self-service**
   a) An object must service itself and carry out supplementary and repair operations.
   b) Make use of waste material and energy.

Examples:

- Principle of recycling in chemical engineering process.
- Heat recovery in process.
- Interestingly, aluminum doesn't corrode, unlike other metals, even though it is reactive. This is because aluminum is covered by a layer of aluminum oxide already. This layer of aluminum oxide protects it from further corrosion.
- Some metals acquire a natural passivity, or resistance, to corrosion. This occurs when the metal reacts with, or corrodes in, the oxygen in air. The result is a thin oxide film that blocks the metal's tendency to undergo further reaction. The patina that forms on copper and the weathering of certain sculpture materials are examples of this.

26) **Copying**
   a) A simplified and inexpensive copy should be used in place of a fragile original or an object that is inconvenient to operate.
   b) If a visible optical copy is used, replace it with an infrared or ultraviolet copy.
   c) Replace an object (or system of objects) with their optical image. The image can then be reduced or enlarged.

   Examples:

   - The use of X-ray photoelectron spectroscopy in corrosion science.
   - X-ray diffraction (XRD) analyzer is used to identify the scaling or corrosion that causes blockages.
   - Using image-analysis techniques for investigating localized corrosion processes.
   - Using statistical modeling and computer simulation of corrosion growth in aluminum alloy.
   - Modeling and simulation of corrosion and cathodic protection systems can be used to optimize the protection systems and reduce costs by orders of magnitude.

27) **Dispose**
   a) Replace an expensive object with an inexpensive one, compromising other properties (i.e. longevity).

   Examples:

   - Development of anti-corrosion coating systems using disposable waste materials.
   - Improve corrosion resistance with low-cost coating.
   - Paints provide a relatively inexpensive method of increasing corrosion resistance. However, the processes involved are highly inefficient; during application, up to 50% of the coating can evaporate and oven curing produces harmful byproducts that are both hazardous and expensive to dispose of at high volume.

28) **Replacement of Mechanical System**
   a) Replace a mechanical system with an optical, acoustical, thermal, or olfactory system.
   b) Use an electric, magnetic, or electromagnetic field to interact with an object.
   c) Replace fields that are:
      1) stationary with mobile;

2) fixed with changing in time;
3) random with structure.

d) Use fields in conjunction with ferromagnetic materials.

Examples:

- Laser cleaning rust.
- Electroplating is a process where a metal is coated by electrolytic deposition with chromium, silver, or another metal.
- Impressed current cathodic protection (ICCP) systems use anodes connected to a DC power source (such as a cathodic protection rectifier). Anodes for ICCP systems are tubular and solid rod shapes of various specialized materials; these include high-silicon cast iron, graphite, mixed metal oxide, or platinum coated titanium or niobium coated rod and wires.

29) **Pneumatic or Hydraulic Constructions**

a) Replace solid parts of an object with a gas or liquid. These parts can now use air or water for inflation, or use pneumatic or hydrostatic cushions.

Examples:

- They are sprayed with oil to prevent corrosion.
- "Corrosion in other fluids" refers to the corrosion of metals/alloys in non-aqueous environments, such as fused salts, sometimes referred to as molten salts.
- Using a pneumatic rust corrosion slag removing debarring tool.
- Preventing corrosion in hydraulic fittings and applications requires proper material and plating selection.
- Hydraulic fluids consist of rust inhibitors that create a protective film on metallic surfaces to prevent rust. The film cannot be negatively affected by water and will completely prevent rust once it settles throughout the hydraulic system.

30) **Flexible Membranes or Thin Films**

a) Replace customary constructions with flexible membranes or thin films.
b) Isolate an object from its outside environment with flexible membranes or thin films.

Examples:

- During galvanization, a coat of melted zinc is applied directly to the steel, protecting it from corrosion.
- The US military shrink-wraps equipment such as helicopters to protect them from corrosion, and thus save millions of dollars.
- Passivation refers to the spontaneous formation of an ultrathin film of corrosion products, known as a passive film, on the metals surface which act as a barrier to further oxidation.

- Car body is made up of thin flexible sheet metal to provide more safety at high impact.
- Electroplating is the best way to apply a protective zinc layer to fasteners for one key reason; electroplating creates a thinner, more uniform layer of zinc on the work piece than hot dip galvanization or mechanical plating.

31) **Porous Material**
    a) Make an object porous, or use supplementary porous elements (inserts, covers, etc.).
    b) If an object is already porous, fill pores in advance with some substance.

Examples:

- Pitting corrosion occurs under certain conditions, and leads to accelerated corrosion in certain areas rather than uniform corrosion throughout the piece.
- Crevice corrosion occurs in confined spaces where access of fluid from the environment is limited such as gaps and contact areas between parts, under gaskets or seals, inside cracks and seams, and spaces filled with deposits.
- Use a porous metal mesh to wick excess solder away from a joint.

32) **Changing the Color**
    a) Change the color of an object or its environment.
    b) Change the degree of translucency of an object or its environment.
    c) Use color additives to observe an object or process which is difficult to see.
    d) If such additives are already used, employ luminescent traces or trace atoms.

Examples:

- Copper (Cu) corrodes and forms a basic green carbonate, and lead corrodes to form a white lead oxide or carbonate.
- Rusting, the formation of iron oxides, is a well-known example of electrochemical corrosion. This type of damage typically produces oxide(s) or salt(s) of the original metal and results in a distinctive orange coloration.

33) **Homogeneity**
    a) Objects interacting with the main object should be made out of the same material (or material with similar properties) as the main object.

Examples:

- Homogeneous corrosion refers to the corrosion occurring at almost the same rate on the surface of the steel pipe pile as for the wind turbine.
- Uniform corrosion commonly occurs on metal surfaces having a homogeneity of chemical composition and of microstructure.
- Duplex stainless steels, the cast equivalents of alloys 2205, 2507, J92205, and J93380, have similar properties and corrosion resistance.

34) **Rejecting and Regenerating Parts**
   a) After completing its function, or becoming useless, an element of an object is rejected (discarded, dissolved, evaporated, etc.) or modified during its work process.
   b) Used-up parts of an object should be restored during its work.

Examples:

- Any batteries, pods, packs, or trays showing signs of corrosion or rust must be replaced.
- Repair of corrosion damage includes removal of all corrosion and corrosion products.
- As the steel surface "rusts" it creates a regenerating corrosion-retarding layer that protects the main structure of the steel.

35) **Transformation of Properties**
   a) Change the physical state of the system.
   b) Change the concentration or density.
   c) Change the degree of flexibility.
   d) Change the temperature or volume.

Examples:

- Gold is a vital industrial metal, as it is an excellent conductor of electricity, it is extremely resistant to corrosion, and it is one of the most chemically stable of the elements, making it critically important in electronics and other high-tech applications.
- The corrosion properties of electroplated zinc will be enhanced when nickel is codeposited with it.
- The change in appearance is a direct result of corrosion.

36) **Phase Transition**
   a) Using the phenomena of phase change (i.e. a change in volume, the liberation or absorption of heat, etc.).

Examples:

- Corrosion is defined as "an irreversible interfacial reaction of a material (metal, ceramic, and polymer) with its environment which results in consumption of the material or in dissolution into the material of a component of the environment."
- Use of heat pipes in radiator rather than metal tubes for high heat transfer at specified size.
- Testing the change in volume due to corrosion of metals.
- Phase Change Materials (PCMs) provide good potential for reducing energy consumption in thermal energy storage systems.

37) **Thermal Expansion**
   a) Use expansion or contraction of material by changing its temperature.
   b) Use various materials with different coefficients of thermal expansion.

   - The engine exhaust pipe is provided with flexible pipe elements, compensating for thermal expansion and reducing thermal stresses on rigid parts.
   - The expansion of the corrosion products (iron oxides) of carbon steel reinforcement structures may induce mechanical stress that can cause the formation of cracks and disrupt the concrete structure.
   - Lower thermal expansion reduces system stresses, higher thermal conductivity, lowers operating metal temperature, i.e. reduces corrosion.

38) **Accelerated Oxidation**
   a) Make transition from one level of oxidation to the next higher level:

   1) ambient air to oxygenate;
   2) oxygenated air to oxygen;
   3) oxygen to ionized oxygen;
   4) ionized oxygen to ozoned oxygen;
   5) ozoned oxygen to ozone;
   6) ozone to singlet oxygen.

   Examples:

   - Oxidation literally means "to combine with oxygen." Corrosion occurs fastest when the metal is in contact with air, water, and salt.
   - In the most common use of the word, corrosion means electrochemical oxidation of metal in reaction with an oxidant such as oxygen or sulfates.
   - A passivation technique using highly concentrated ozone has been developed for electropolished stainless steel tubing that supplies corrosive gases. The strong oxidizing ability of ozone enables the inner surface to form an anticorrosive passivation film with thickness up to 60 Å under room temperature and atmospheric pressure conditions.

39) **Inert Environment**
   a) Replace a normal environment with an inert one.
   b) Introduce a neutral substance or additives into an object.
   c) Carry out the process in a vacuum.

   Examples:

   - Stress-corrosion occurs when a material exists in a relatively inert environment.
   - In recent years, the application of corrosion-resistant vacuum pumps is more and more extensive. They are mainly used to transport corrosive or non-allowed polluting media.
   - The upper deck in inert gas environments in cargo oil tanks (COT), and for pitting corrosion that occurs on the inner bottom in highly acidic environments.

40) **Composite Materials**
   a) Replace homogeneous materials with composite ones.

   Examples:

   - Galvanized zinc coatings on steel combine corrosion resistance with high strength to make an outstanding composite material.
   - An alloy is a mixture of two or more metals. Alloying is a process where metals like iron or steel are mixed with a less reactive metal like chromium, magnesium, etc., for protection against corrosion and to create non-rusting alloys. For example, brass is an alloy which consists of copper and is an inexpensive and non-reactive alloy. Another example of a non-rusting alloy is stainless steel, a mixture of iron and carbon.
   - Aluminum is an important structural metal, but even aluminum goes under oxidation reactions. However, aluminum doesn't corrode or oxidize as rapidly as its reactivity suggests. An alloy of aluminum or any other metal like magnesium can make aluminum stronger, stiffer, and harder.
   - When steel is alloyed with nickel, its resistance to corrosion increases dramatically.

6.29 Conclusion

The proposed view on TRIZ helps to systematize TRIZ knowledge and brings more clarity into the issue of what TRIZ is and how to look at it.

TRIZ was invented by Genrich Altshuller in Russia by analyzing more than two million patents. He found recurring and typical patterns among high-level inventions. The essence of this database has been compiled into a generic list of 40 inventive principles known as the TRIZ 40 Principles. TRIZ is based on the fundamental premise that problems occur due to PC or TC in the system. TRIZ provides 40 inventive principles to resolve contradictions in the system. The principles are generic enough to apply across different problems, products, and industries to create innovative solutions.

In this chapter we provide basic theoretical concepts and tools of TRIZ, along with definitions of inventive principles and many new examples and approaches from Altshuller's practical experiences in applying the 40 Inventive Principles in corrosion management.

6.30 Glossary of TRIZ Terms

This glossary is not by any means exhaustive; it contains mainly basic terms of TRIZ gleaned from various English-language literature sources known to the author; books, university courses, conference papers, and on-line articles.

Assembling a comprehensive glossary of TRIZ that would adequately reflect the evolving body of knowledge of this science would require contributions from many enthusiasts. The author would greatly appreciate any future help in improving this glossary [74].

Algorithm for Inventive Problem Solving (ARIZ) The central analytical tool of TRIZ (ARIZ is a Russian abbreviation). Its basis is a sequence of logical procedures for analysis of a vaguely or ill-defined initial problem/situation which is transformed into a distinct System Conflict. Consideration of the System Conflict leads to the formulation of a PC whose elimination is provided by maximal utilization of the resources of the subject system. ARIZ puts together, in a system, the most fundamental concepts and methods of TRIZ such as Ideal Technological System (Ideal System), System Conflict, PC, SFA, Standards, and the Laws of Technological System Evolution.

Altshuller's Matrix See Conflict Matrix

Auxiliary function A function supporting the system's Primary Function.

Bi-system A system consisting of two mono-systems.

Conflict Matrix (Contradiction Matrix, Altshuller's Matrix) A 39×39 matrix linking Typical System Conflicts with the Inventive Principles.

Conflicting components System's components involved in a System Conflict.

Contradiction Matrix See Conflict Matrix.

Degree of Ideality A measure of the system's ideality, usually expressed as the ratio of the system's functionality over the system's cost.

Effect Result of interaction of fields and substances. An individual effect can be modeled as follows:

Elementary sufield A sufield containing two substances and a field.

Environment Immediate physical surroundings of a technological system or of its part.

Field The energy needed for interaction of two substances. In addition to four fundamental fields – electromagnetic, gravitational, and nuclear fields of weak and strong interactions – TRIZ deals with engineering fields, such as mechanical, thermal, electric, magnetic, and chemical. These fields manifest themselves through many groups of physical and chemical phenomena.

Function A purposeful physical interaction between two components of a technological system. Description of a function includes the names of the physical action and of the object of the action.

Generic Principles for Overcoming Physical Contradictions See Separation Principles

Harmful function/action A function/action that hinders performance of the Primary Function.

Heterogeneous bi (poly)-system A system consisting of mono-systems performing different functions.

Higher-level system See Supersystem.

Homogeneous bi (poly)-system A system consisting of mono-systems performing similar or identical functions.

Ideal Final Result An ideal solution of an engineering design problem based on the notion of an Ideal Technological System.

Ideal Technological System A system that is absent as a physical entity, but that fully performs the prescribed function.

Incomplete sufield A suffield containing fewer than three elements.

Initial Situation (Initial Problem) Original problem statement, usually a cluster of various problems.

Inventive Principles There are 50 typical techniques (operators) for overcoming System Conflicts; many of them may contain a few sub-techniques.

Inverse bi-(poly)-system A system whose sub-systems have opposite properties.

Knowledge Base of Engineering Applications of Physical, Chemical, and Geometric Effects A set of physical, chemical, and geometric effects arranged by a functional principle.

Law of Completeness This law states that an autonomous technological system must include four minimally functioning principal parts; an engine, a transmission, a working means, and a control means.

Law of Elimination of Human Involvement This law states that technological systems evolve in the direction of delegation functions performed by humans to technological systems.

Laws of Engineering System Evolution See Laws of Technological System Evolution

Law of Harmonization This law states that the necessary condition for existence of an effective technological system is coordination of periodicity of actions (or natural frequencies) of its parts.

Law of Increasing Controllability This law states that technological systems evolve in the direction of increased controllability of their components; this is often achieved by transition from elementary suffields to double and chain suffields.

Law of Increasing Degree of Ideality This is the primary law of evolution of technological systems; it states that technological systems evolve in the direction of increasing their Degree of Ideality.

Law of Increasing Flexibility (Law of Increasing Dynamism) This law states that technological systems evolve in the direction toward more flexible structures capable of adaptation to changing environmental conditions (multi-functionality) and to varying performance regimes.

Law of Non-Uniform Evolution of Sub-Systems This law states that different sub-systems of technological systems evolve at different rates (along their own s-curves); this causes the development of System Conflicts.

Law of Shortening of Energy Flow Path This law states that technological systems evolve in the direction of shortening of energy passage through the system (from the engine to the working means).

Laws of Technological System Evolution (Laws of Engineering System Evolution, Patterns of Technological System Evolution, Trends of Technological System Evolution) These laws reflect significant, stable, and repeatable interactions between elements of technological systems, and between the systems and their environments in the process of evolution.

Law of Transition to a Higher-Level System (Law of Transition to a Supersystem) This law states that technological systems evolve in the general direction from mono-systems to bi- and poly-systems.

Law of Transition to a Micro-Level This law states that technological systems evolve in the general direction of fragmentation of their components (first of all, fragmentation of working means).

Level of Invention A qualitative measure of the degree of novelty of an invention.

Lines of Evolution The Lines identify specific stages of evolution associated with particular Laws of Technological System Evolution.

Macro Physical Contradiction In ARIZ, a PC formulated at the level of the whole component (e.g. the rod must be hot and cold).

Main tool A tool performing the Primary Function.

Maxi-problem A problem associated with major modifications of a system, i.e. with changing its physical principle of functioning.

Micro Physical Contradiction In ARIZ, a PC formulated for the components' ingredients (particles), e.g. for the rod to be both hot and cold, its particles must be moving both fast and slowly.

Mini-problem A problem formulated according to the rule: "The system remains unchanged or even simplifies, but the harmful effect disappears, or a useful effect is obtained." When solving a mini-problem, the physical principle of the system's functioning is not changed.

Mono-system A system performing one function.

Object (Article, Product) A component of the system that is to be controlled (processed, modified, e.g. moved, machined, bent, turned, heated, expanded, charged, illuminated, measured, detected, etc.).

Partially convoluted bi-(poly)-system A bi- or poly-system with a reduced number of auxiliary components.

Physical action A physical mechanism that enables performance of a specific function. For example, the function "cleaning a chemical solution from contaminants" may be based on such diverse physical actions as "moving the contaminants away from the solution," or "disintegration of the contaminants," and others.

Physical Contradiction A situation when the same component must satisfy mutually exclusive demands to its physical state, e.g. be hot and cold, electrically conductive and insulative, etc.

Poly-system A system consisting of more than two mono-systems.

Primary Function The main purpose of existence of a technological system.

Psychological inertia Predilection toward conventional ways to analyze and solve problems.

Resources Substances, fields, and other attributes of a technological system (e.g. time of functioning, occupied space, etc.) as well as of its environment and of an overall system that can be utilized to improve the system.

S-Curve Evolution of technological systems can be illustrated by an s-shaped curve reflecting changes of the system's main performance characteristics (or its benefit-to-cost ratio, Degree of Ideality) with time since its inception.

Separation Principles The approaches to resolving PCs; Separation of opposite properties in space, separation of opposite properties in time, separation of opposite properties between a system and its components.

Substance In the SFA, an element of a sufield, a technological system of any degree of complexity participating in performance of a function (physical action).

Substance-Field (Sufield) Analysis A branch of TRIZ studying transformation and evolution of sufield structures.

Sufield A model of technological system consisting of substances and fields.

Super-effect An unexpected benefit of invention.

Supersystem (Higher-level system) System that includes the system under consideration as a sub-system.

System Conflict (Engineering Contradiction, Technical Contradiction) An interaction between system's parts when the useful function/action simultaneously causes a harmful effect, introduction (intensification) of the useful function/action, or elimination or reduction of the harmful function/action causes deterioration or unacceptable complication of one of the parts or of the entire system.

Tool A component having direct physical interaction with an object (i.e. controlling the object).

TRIZ Technology Forecasting (Directed Evolution, Guided Technology Evolution) A systematic TRIZ approach to conceptual development of next-generation products and processes.

Typical System Conflicts (Typical Engineering Contradictions, Typical Technical Contradictions) Despite immense diversity of technological systems, there is a finite number of typical System Conflicts such as productivity vs. accuracy, reliability vs. complexity, shape vs. speed, etc. These System Conflicts can usually be resolved by application of the Inventive Principles.

Useful function/action A function/action that contributes to the performance of the Primary Function.

6.A TRIZ Contradiction Table

Improving Feature \ Worsening Feature		1 Weight of mobile object	2 Weight of stationary object	3 Length of mobile object	4 Length of stationary object	5 Area of mobile object	6 Area of stationary object	7 Volume of mobile object	8 Volume of stationary object	9 Rate of change, speed	10 Force exerted by object	11 Stress, pressure exerted upon object	12 Shape of object	13 Stability of object's composition	
1	Weight of mobile object			15, 8, 29,34		29,17 38,34		29,2, 40,28		2,8, 15,38	8,10, 18,37	10,36 37,40	10,14 35,40	1,35, 19,39	1
2	Weight of stationary object				10,1, 29,35		35,30 13,2		5,35, 14,2		8,10, 19,35	13,29 10,18	13,10 29,14	26,39 1,40	2
3	Length of mobile object	8, 15, 29,34				15,17 4		7,17, 4,35		13,4, 8	17,10 4	1,8, 35	1,8, 10,29	1,8, 15,34	3
4	Length of stationary object		35,28 40,29				17,7, 10,40		35,8, 2,14		28,10	1,14, 35	13,14 15,7	39,37 35	4
5	Area of mobile object	2,17, 29,4		14,15 18,4				7,14, 17,4		29,30 4,34	19,30 35,2	10,15 36,28	5,34, 29,4	11,2, 13,39	5
6	Area of stationary object		30,2, 14,18		26,7, 9,39						1,18, 35,36	10,15 36,37		2,38	6
7	Volume of mobile object	2,26, 29,40		1,7, 4,35		1,7, 4,17				29,4, 38,34	15,35 36,37	6,35, 36,37	1,15, 29,4	28,10 1,39	7
8	Volume of stationary object		35,10 19,14	19,14	35,8, 2,14						2,18, 37	24,35	7,2, 35	34,28 35,40	8
9	Rate of change, speed	2,28, 13,38		13,14 8		29,30 34		7,29, 34			13,28 15,19	6,18, 38,40	35,15 18,34	28,33 1,18	9
10	Force exerted by object	8,1, 37,18	18,13 1,28	17,19 9,36	28,10	19,10 15	1,18, 36,37	15,9, 12,37	2,36, 18,37	13,28 15,12		18,21 11	10,35 40,34	35,10 21	10
11	Stress, pressure exerted upon object	10,36 37,40	13,29 10,18	35,10 36	35,1, 14,16	10,15 36,28	10,15 36,37	6,35, 10	35,24	6,35, 36	36,35 21		35,4, 15,10	35,33 2,40	11
12	Shape of object	8,10, 29,40	15,10 26,3	29,34 5,4	13,14 10,7	5,34, 4,10		14,4, 15,22	7,2, 35	35,15 34,18	35,10 37,40	34,15 10,14		33,1, 18,4	12
13	Stability of object's composition	21,35 2,39	26,39 1,40	13,15 1,28	37	2,11, 13	39	28,10 19,39	34,28 35,40	33,15 28,18	10,35 21,16	2,35, 40	22,1, 18,4		13
14	Strength of object	1,8, 40,15	40,26 27,1	1,15, 8,35	15,14 28,26	3,34, 40,29	9,40, 28	10,15 14,7	9,14, 17,15	8,13, 26,14	10,18 3,14	10,3, 18,40	10,30 35,40	13,17 35	14
15	Durability of mobile object	19,5, 34,31		2,19, 9		3,17, 19		10,2, 19,30		3,35, 5	19,2, 16	19,3, 27	14,26 28,25	13,3, 35	15
16	Durability of stationary object		6,27, 19,16		1,40, 35				35,34 38					39,3, 35,23	16
17	Temperature of object	36,22 6,38	22,35 32	15,19 9	15,19 9	3,35, 39,18	35,38	34,39 40,18	35,6, 4	2,28, 36,30	35,10 3,21	35,39 19,2	14,22 19,32	1,35, 32	17
18	Illumination of object	19,1, 32	2,35, 32	19,32 16		19,32 26		2,13, 10		10,13 19	26,19 6		32,30	32,3, 27	18
19	Energy consumption by mobile object	12,18 28,31		12,28		15,19 25		35,13 18		8,15, 35	16,26 21,2	23,14 25	12,2, 29	19,13 17,24	19
20	Energy consumption by stationary object		19,9, 6,27								36,37			27,4, 29,18	20
		1	2	3	4	5	6	7	8	9	10	11	12	13	CT-1

Improving Feature \ Worsening Feature		14	15	16	17	18	19	20	21	22	23	24	25	26	
		Strength of object	Durability of mobile object	Durability of stationary object	Temperature of object	Illumination of object	Energy consumption by mobile object	Energy consumption by stationary object	Power supplied or consumed by object	Energy loss by object	Substance loss by object	Information loss	Time loss	Quantity of matter	
	Weight of mobile object	28,27 18,40	5,34, 31,35		6,29, 4,38	19,1, 32	35,12 34,31		12,36 18,31	6,2, 34,19	5,35, 3,31	10,24 35	10,35 20,28	3,26, 18,31	1
2	Weight of stationary object	28,2, 10,27		2,27, 19,6	28,19 32,22	19,32 35		18,19 28,1	15,19 18,22	18,19 28,15	5,8, 13,30	10,15 35	10,20 35,26	19,6, 18,26	2
3	Length of mobile object	8,35, 29,34	19		10,15 19	32	8,35, 24		1,35	7,2,3 5,39	4,29, 23,10	1,24	15,2, 29	29,35	3
4	Length of stationary object	15,14 28,26		1,40, 35	3,35, 38,18	3,25			12,8	6,28	10,28 24,35	24,26	30,29 14		4
5	Area of mobile object	3,15, 40,14	6, 3		2,15, 16	15,32 19,13	19,32		19,10 32,18	15,17 30,26	10,35 2,39	30,26	26,4	29,30 6,13	5
6	Area of stationary object	40		2,10, 19,30	35,39 38				17,32	17,7, 30	10,14 18,39	30,16	10,35 4,18	2,18, 40,4	6
7	Volume of mobile object	9,14, 15,7	6,35, 4		34,39 10,18	2,13, 10	35		35,6, 13,18	7,15, 13,16	36,39 34,10	2,22	2,6, 34,10	29,30 7	7
8	Volume of stationary object	9,14, 17,15		35,34 38	35,6, 4				30,6		10,39 35,34		35,16 32,18	35,3	8
9	Rate of change, speed	8,3, 26,14	3,19, 35,5		28,30 36,2	10,13 19	8,15, 35,38		19,35 38,2	14,20 19,35	10,13 28,38	13,26		10,19 29,38	9
10	Force exerted by object	35,10 14,27	19,2		35,10 21		19,17 10	1,16, 36,37	19,35 18,37	14,15	8,35, 40,5		10,37 36	14,29 18,36	10
11	Stress, pressure exerted upon object	9,18, 3,40	19,3, 27		35,39 19,2		14,24 10,37		10,53 14	2,36, 25	10,36 3,37		37,36 4	10,14 36	11
12	Shape of object	30,14 10,40	14,26 9,25		22,14 19,32	13,15 32	2,6, 34,14		4,6,2	14	35,29 3,5		14,10 34,17	36,22	12
13	Stability of object's composition	17,9, 15	13,27 10,35	39,3, 35,23	35,1, 32	32,3, 27,15	13,19	27,4, 29,18	32,35 27,31	14,2, 39,6	2,14, 30,40		35,27	15,32 35	13
14	Strength of object		27,3, 26		30,10 40	35,19	19,35 10	35	10,26 35,28	35	35,28 31,40		29,3, 28,10	29,10 27	14
15	Durability of mobile object	27,3, 10			19,35 39	2,19, 4,35	28,6, 35,18		19,10 35,38		28,27 3,18	10	20,10 28,18	3,35, 10,40	15
16	Durability of stationary object				19,18 36,40				16		27,16 18,38	10	28,20 10,16	3,35, 31	16
17	Temperature of object	10,30 22,40	19,13 39	19,18 36,40		32,30 21,16	19,15 3,17		2,14, 17,25	21,17 35,38	21,36 29,31		35,28 21,18	3,17, 30,39	17
18	Illumination of object	35,19	2,19, 6		32,35 19		32,1, 19	32,35 1,15	32	13,16 1,6	13,1	1,6	19,1, 26,17	1,19	18
19	Energy consumption by mobile object	5,19, 9,35	28,35 6,18		19,24 3,14	2,15, 19			6,19, 37,18	12,22 15,24	35,24 18,5		35,38 19,18	34,23 16,18	19
20	Energy consumption by stationary object	35				19,2, 35,32					28,27 18,31			3,35, 31	20
		14	15	16	17	18	19	20	21	22	23	24	25	26	CT-2

Worsening Feature / Improving Feature		27 Reliability of object	28 Accuracy of measurement	29 Precision of production	30 Harmful influence of object's environment	31 Harmful effects caused by object	32 Ease of production	33 Convenience of use	34 Ease of repair and maintenance	35 Adaptability, versatility of object	36 Complexity of object	37 Difficulties in measuring, inspection,	38 Level of automation	39 Production rate	
1	Weight of mobile object	3,11, 1,27	28,27 35,26	28,35 26,18	22,21 18,27	22,35 31,39	27,28 1,36	35,3, 2,24	2,27, 28,11	29,5, 15,8	26,30 36,34	28,29 26,32	26,35 18,19	35,3, 24,37	1
2	Weight of stationary object	10,28 8,3	18,26 28	10,1, 35,17	2,19, 22,37	35,22 1,39	28,1, 9	6,13, 1,32	2,27, 28,11	19,15 29	1,10, 26,39	25,28 17,15	2.26, 35	1,28, 15,35	2
3	Length of mobile object	10,14 29,40	28,32 4	10,28 29,37	1,15, 17,24	17,15	1,29, 17	15,29 35,4	1,28, 10	14,15 1,16	1,19, 26,24	35,1, 26,24	17,24 26,16	14,4, 28,29	3
4	Length of stationary object	15,29 28	32,28 3	2,32, 10	1,18		15,17 27	2,25	3	1,35	1,26	26		30,14 7,26	4
5	Area of mobile object	29,9	26,28 32,3	2,32	22,33 28,1	17,2, 18,39	13,1, 26,24	15,17 13,16	15,13 10,1	15,30	14,1, 13	2,36, 26,18	14,30 28,23	10,26 34,2	5
6	Area of stationary object	32,35 40,4	26,28 32,3	2,29, 18,36	27,2, 39,35	22,1, 40	40,16	16,4	16	15,16	1,18, 36	2,35, 30,18	23	10,15 17,7	6
7	Volume of mobile object	14,1, 40,11	25,26 28	25,28 2,16	22,21 27,35	17,2, 40,1	29,1, 40	15,13 30,12	10	15,29	26,1	29,26 4	35,34 16,24	10,6, 2,34	7
8	Volume of stationary object	2,35, 16		35,10 25	34,39 19,27	30,18 35,4	35		1		1,31	2,17, 26		35,37 10,2	8
9	Rate of change, speed	11,35 27,28	28,32 1,24	10,28 32,25	1,28, 35,23	2,24, 35,21	35,13 8,1	32,28 13,12	34,2, 28,27	15,10 26	10,28 4,34	3,34, 27,16	10,18		9
10	Force exerted by object	3,35, 13,21	35,10 23,24	28,29 37,36	1,35, 40,18	13,3, 36,24	15,37 18,1	1,28, 3,25	15,1, 11	15,17 18,20	26,35 10,18	36,37 10,19	2,35	3,28, 35,37	10
11	Stress, pressure exerted upon object	10,13 19,35	6,28, 25	3,35	22,2, 37	2,33, 27,18	1,35, 16	11	2	35	19,1, 35	2,36, 37	35,24	10,14 35,37	11
12	Shape of object	10,40 16	28,32 1	32,30 40	22,1, 2,35	35,1	1,32, 17,28	32,15 26	2,13, 1	1,15, 29	16,29 1,28	15,13 39	15,1, 32	17,26 34,10	12
13	Stability of object's composition		13	18	35,24 30,18	35,40 27,39	35,19	32,35 30	2,35, 10,16	35,30 34,2	2,35, 22,26	35,22 39,23	1,8, 35	23,35 40,3	13
14	Strength of object	11,3	3,27, 16	3,27	18,35 37,1	15,35 22,2	11,3, 10,32	32,40 28,2	27,11 3	15,3, 32	2,13, 28	27,3, 15,40	15	29,35 10,14	14
15	Durability of mobile object	11,2, 13	3	3,27, 16,40	22,15 33,28	21,39 16,22	27,1, 4	12,27	29,10 27	1,35, 13	10,4, 29,15	19,29 39,35	6,10	35,17 14,19	15
16	Durability of stationary object	34,27 6,40	10,26 24		17,1, 40,33	22	35,10	1	1	2		25,34 6,35	1	20,10 16,38	16
17	Temperature of object	19,35 3,10	32,19 24	24	22,33 35,2	22,35 2,24	26,27	26,27	4,10, 16	2,18, 27	2,17, 16	3,27, 35,31	26,2, 19,16	15,28 35	17
18	Illumination of object		11,15 32	3,32	15,19	35,19 32,39	19,35 28,26	28,26 19	15,17 13,16	15,10 1,19	6,32, 13	32,15	2,26, 10	2,25, 16	18
19	Energy consumption by mobile object	19,21 11,27	3,1, 32		1,35, 6,27	2,35, 6	28,26 30	19,35	1,15, 1728	15,17 13,16	2,29, 27,28	35,38	32,2	12,28 35	19
20	Energy consumption by stationary object	10,36 23			10,2, 22,37	19,22 18	1,4					19,35 16,25		1,6	20
		27	28	29	30	31	32	33	34	35	36	37	38	39	CT-3

Worsening Feature / Improving Feature		1 Weight of mobile object	2 Weight of stationary object	3 Length of mobile object	4 Length of stationary object	5 Area of mobile object	6 Area of stationary object	7 Volume of mobile object	8 Volume of stationary object	9 Rate of change, speed	10 Force exerted by object	11 Stress, pressure exerted upon object	12 Shape of object	13 Stability of object's composition	
21	Pow er supplied or consumed by object	8,36, 38,31	19,26 17,27	1,10, 35,37		19,38	17,32 13,38	35,6, 38	30,6, 25	15,35 2	26,2, 36,35	22,10 35	29,14 2,40	35,32 15,31	21
22	Energy loss by object	15,6, 19,28	19,6, 18,9	7,2, 6,13	6,38, 7	15,26 17,30	17,7, 30,18	7,18, 23	7	16,35 38	36,38			14,2, 39,6	22
23	Substance loss by object	35,6, 23,40	35,6, 22,32	14,29 10,39	10,28 24	35,2, 10,31	10,18 39,31	1,20, 30,36	3,39, 18,31	10,13 28,38	14,15 18,40	3,36, 37,10	29,35 3,5	2,14, 30,40	23
24	Information loss	10,24 35	10,35 5	1,26	26	30,26	30,16		22,2	26,32					24
25	Time loss	10,20 37,35	10,20 26,5	15,2, 29	30,24 14,5	26,4, 5,16	10,35 17,4	2,5, 34,10	35,16 32,18		10,37 36,5	37,36 4	4,10, 34,17	35,3, 22,5	25
26	Quantity of matter	35,6, 18,31	27,26 18,35	29,14 35,18		15,14 29	2,18, 40,4	15,20 29		35,29 34,28	35,14 3	10,36 14,3	35,14	15,2, 17,40	26
27	Reliability of object	3,8, 10,40	3,10, 8,28	15,9, 14,4	15,29 28,11	17,10 14,16	32,35 40,4	3,10, 14,24	2,35, 24	21,35 11,28	8,28, 10,3	10,24 35,19	35,1, 16,11		27
28	Accuracy of measurement	32,35 26,28	28,35 25,26	28,26 5,16	32,28 3,16	26,28 32,3	26,28 32,3	32,13 6		28,13 32,24	32,2	6,28, 32	6,28, 32	32,35 13	28
29	Precision of production	28,32 13,18	28,35 27,9	10,28 29,37	2,32, 10	28,33 29,32	2,29, 18,36	32,28 2	25,10 35	10,28 32	28,19 34,36	3,35	32,30 40	30,18	29
30	Harmful influence of object's environment	22,21 27,39	2,22, 13,24	17,1, 39,4	1,18	22,1, 33,28	27,2, 39,35	22,23 37,35	34,39 19,27	21,22 35,28	13,35 39,18	22,2, 37	22,1, 3,35	35,24 30,18	30
31	Harmful effects caused by object	19,22 15,39	35,22 1,39	17,15 16,22		17,2, 18,39	22,1, 40	17,2, 40	30,18 35,4	35,28 3,23	35,28 1,40	2,33, 27,18	35,1	35,40 27,39	31
32	Ease of production	28,29 15,16	1,27, 36,13	1,29, 13,17	15,17 27	13,1, 26,12	16,40	13,29 1,40	35	35,13 8,1	35,12	35,19 1,37	1,28, 13,27	11,13 1	32
33	Convenience of use	25,2, 13,15	6,13, 1,25	1,17, 13,12		1,17, 13,16	18,16 15,39	1,16, 35,15	4,18, 39,31	18,13 34	28,13 35	2,32, 12	15,34 29,28	32,35 30	33
34	Ease of repair and maintenance	2,27, 35,11	2,27, 35,11	1,28, 10,25	3,18, 31	15,13 32	16,25	25,2, 35,11	1	34,9	1,11, 10	13	1,13, 2,4	2,35	34
35	Adaptability, versatility of object	1,6, 15,8	19,15 29,16	35,1, 29,2	1,35, 16	35,30 29,7	15,16	15,35 29		35,10 14	15,17 20	35,16	15,37 1,8	35,30 14	35
36	Complexity of object	26,30 34,36	2,26, 35,39	1,19, 26,24	26	14,1, 13,16	6,36	34,26 6	1,16	34,10 28	26,16	19,1, 35	29,13 28,15	2,22, 17,19	36
37	Difficulties in measuring, inspection,	27,26 28,13	6,13, 28,1	16,17 26,24	26	2,13, 18,17	2,39, 30,16	29,1, 4,16	2,18, 26,31	3,4, 16,35	36,28 40,19	35,36 37,32	27,13 1,39	11,22 39,30	37
38	Level of automation	28,26 18,35	28,26 35,10	14,13 17,28	23	17,14 13		35,13 16		28,10	2,35	13,35	15,32 1,13	18,1	38
39	Production rate	35,26 24,37	28,27 15,3	18,4, 28,38	30,7, 14,26	10,26 34,31	10,35 17,7	2,6, 34,10	35,37 10,2		28,15 10,36	10,37 14	14,10 34,40	35,3, 22,39	39
		1	2	3	4	5	6	7	8	9	10	11	12	13	CT-4

Improving Feature \ Worsening Feature		14 Strength of object	15 Durability of mobile object	16 Durability of stationary object	17 Temperature of object	18 Illumination of object	19 Energy consumption by mobile object	20 Energy consumption by stationary object	21 Power supplied or consumed by object	22 Energy loss by object	23 Substance loss by object	24 Information loss	25 Time loss	26 Quantity of matter	
21	Pow er supplied or consumed by object	26,10 28	19,35 10,38	16	2,14, 17,25	16,6, 19	16,6, 19,37			10,35 38	28,27 18,38	10,19	35,20 10,6	4,34, 19	21
22	Energy loss by object	26			19,38 7	1,13, 32,15			3,38		35,27 2,37	19,10	10,18 32,7	7,18, 25	22
23	Substance loss by object	35,28 31,40	28,27 3,18	27,16 18,38	21,36 39,31	1,6, 13	35,18 24,5	28,27 12,31	28,27 18,38	35,27 2,31			15,18 35,10	6,3, 10,24	23
24	Information loss		10	10		19			10,19	19,10			24,26 28,32	24,28 35	24
25	Time loss	29,3, 28,28	20,10 28,18	28,20 10,16	35,29 21,18	1,19, 26,17	35,38 19,18	1	35,20 10,6	10,5, 18,32	35,18 10,39	24,26 28,32		35,38 18,16	25
26	Quantity of matter	14,35 34,10	3,35, 10,40	3,35, 31	3,17, 39		34,29 16,18	3,35, 31	35	7,18, 25	6,3, 10,24	24,28 35	35,38 18,16		26
27	Reliability of object	11,28	2,35, 3,25	34,27 6,40	3,35, 10	11,32 13	21,11 27,19	36,23	21,11 26,31	10,11 35	10,35 29,39	10,28	10,30 4	21,28 40,3	27
28	Accuracy of measurement	28,6, 32	28,6, 32	10,26 24	6,19, 28,24	6,1, 32	3,6, 32		3,6, 32	26,32 27	10,16 31,28		24,34 28,32	2,6, 32	28
29	Precision of production	3,27	3,27, 40		19,26	3,32	32,2		32,2	13,32 2	35,31 10,24		32,26 28,18	32,30	29
30	Harmful influence of object's environment	18,35 37,1	22,15 33,28	17,1, 40,33	22,33 35,2	1,19, 32,13	1,24, 6,27	10,2, 22,37	19,22 31,2	21,22 35,2	33,22 19,40	22,10 2	35,18 34	35,33 29,31	30
31	Harmful effects caused by object	15,35 22,2	15,22 33,31	21,39 16,22	22,35 2,24	19,24 39,32	2.35, 6	19,22 18	2,35, 18	21,35 2,22	10,1, 34	10,21 29	1,22	3,24, 39,1	31
32	Ease of production	1,3, 10,32	27,1, 4	35,16	27,26 18	28,24 27,1	28,26 27,1	1,4	27,1, 12,24	19,35	15,34 33	32,24 18,16	35,28 34,4	35,23 1,24	32
33	Convenience of use	32,40 3,28	29,3, 8,25	1,16, 25	26,27 13	13,17 1,24	1,13, 24		35,34 2,10	2,19, 13	28,32 2,24	4,10, 27,22	4,28, 10,34	12,35	33
34	Ease of repair and maintenance	11,1, 2,9	11,29 28,27	1	4,10	15,1, 13	15,1, 28,16		15,10 32,2	15,1, 32,19	2,35, 34,27		32,1, 10,25	2,28, 10,25	34
35	Adaptability, versatility of object	35,3, 32,6	13,1, 35	2,16	27,2, 3,35	6,22, 26,1	19,35 29,13		19,1, 29	18,15 1	15,10 2,13		35,28	3,35, 15	35
36	Complexity of object	2,13, 28	10,4, 28,15		2,17, 13	24,17 13	27,2, 29,28		20,19 30,34	10,35 13,2	35,10 28,29		6,29	13,3, 27,10	36
37	Difficulties in measuring, inspection,	27,3, 15,28	19,29 39,25	25,34 6,35	3,27, 35,16	2,24, 26	35,38	19,35 16	19,1, 16,10	35,3, 15,19	1,18, 10,24	35,33 27,22	18,28 32,9	3,27, 29,18	37
38	Level of automation	25,13	6,9		26,2, 19	8,32, 19	2,32, 13		28,2, 27	23,28	35,10 18,5	35,33	24,28 35,30	35,13	38
39	Production rate	29,28 10,18	35,10 2,18	20,10 16,38	35,21 28,10	26,17 19,1	35,10 38,19	1	35,20 10	28,10 29,35	28,10 35,23	13,15 23		35,38	39
		14	15	16	17	18	19	20	21	22	23	24	25	26	CT-5

Improving Feature \ Worsening Feature		27	28	29	30	31	32	33	34	35	36	37	38	39	
		Reliability of object	Accuracy of measurement	Precision of production	Harmful influence of object's environment	Harmful effects caused by object	Ease of production	Convenience of use	Ease of repair and maintenance	Adaptability, versatility of object	Complexity of object	Difficulties in measuring, inspection,	Level of automation	Production rate	
21	Pow er supplied or consumed by object	19,24 26,31	32,15 2	32,2	19,22 31,2	2,35, 18	26,10 34	26,35 10	35,2, 10,34	19,17 34	20,19 30,34	19,35 16	28,2, 17	28,35 34	21
22	Energy loss by object	11,10 35	32		21,22 35,2	21,35 2,22		35,32 1	2,19		7,23	35,3, 15,23	2	28,10 29,35	22
23	Substance loss by object	10,29 39,35	16,34 31,28	35,10 24,31	33,22 30,40	10,1, 34,29	15,34 33	32,28 2,24	2,35, 34,27	15,10 2	35,10 28,24	35,18 10,13	35,10 18	28,35 10,23	23
24	Information loss	10,28 23			22,10 1	10,21 22	32	27,22				35,33	35	13,23 15	24
25	Time loss	10,30 4	24,34 28,32	24,26 28,18	35,18 34	35,22 18,39	35,28 34,4	4,28, 10,34	32,1, 10	35,28	6,29	18,28 32,10	24,28 35,30		25
26	Quantity of matter	18,3, 28,40	3,2, 28	33,30	35,33 29,31	3,35, 40,39	29,1, 35,27	35,29 25,10	2,32, 10,25	15,3, 29	3,13, 27,10	3,27, 29,18	8,35	13,29 3,27	26
27	Reliability of object		32,3, 11,23	11,32 1	27,35 2,40	35,2, 40,26		27,17 40	1,11	13,35 8,24	13,35 1	27,40 28	11,13 27	1,35, 29,38	27
28	Accuracy of measurement	5,11, 1,23			28,24 22,26	3,33, 39,10	6,35, 25,18	1,13, 17,34	1,32, 13,11	13,35 2	27,35 10,34	26,24 32,28	28,2, 10,34	10,34 28,32	28
29	Precision of production	11,32 1			26,28 10,36	4,17, 34,26		1,32, 35,23	25,10		26,2, 18		26,28 18,23	10,18 32,39	29
30	Harmful influence of object's environment	27,24 2,40	28,33 23,26	26,28 10,18			24,35 2	2,25, 28,39	35,10 2	35,11 22,31	22,19 29,40	22,19 29,40	33,3, 34	22,35 13,24	30
31	Harmful effects caused by object	24,2, 40,39	3,33, 26	4,17, 34,26							19,1, 31	2,21, 27,1	2	22,35 18,39	31
32	Ease of production		1,35, 12,18		24,2			2,5, 13,16	35,1, 11,9	2,13, 15	27,26 1	6,28, 11,1	8,28, 1	35,1, 10,28	32
33	Convenience of use	17,27 8,40	25,13 2,34	1,32, 35,23	2,25, 28,39		2,5, 12		12,26 1,32	15,34 1,16	32,26 12,17		1,34. 12,3	15,1, 28	33
34	Ease of repair and maintenance	11,10 1,16	10,2, 13	25,10	35,10 2,16		1,35, 11,10	1,12, 26,15		7,1, 4,16	35,1, 13,11		34,35 7,13	1,32, 10	34
35	Adaptability, versatility of object	35,13 8,24	35,5, 1,10		35,11 32,31		1,13, 31	15,34 1,16	1,16, 7,4		15,19 37,28	1	27,34 35	35,28 6,37	35
36	Complexity of an object	13,35 1	2,26, 10,34	26,24 32	22,19 29,40	19,1	27,26 1,13	27,9, 26,24	1,13	29,15 28,37		15,10 37,28	15,1, 24	12,17 28	36
37	Difficulties in measuring, inspection,	27,40 28,8	26,24 32,28		22,19 29,28	2,21	5,28, 11,29	2,5	12,26	1,15	15,10 37,28		34,21	35,18	37
38	Level of automation	11,27 32	28,26 10,34	28,26 18,23	2,33	2	1,26, 13	1,12, 34,3	1,35, 13	27,4, 1,35	15,24 10	34,27 25		5,12, 35,26	38
39	Production rate	1,35, 10,38	1,10, 34,28	18,10 32,1	22,35 13,24	35,22 18,39	35,28 2,24	1,28, 7,19	1,32, 10,25	1,35, 28,37	12,17 28,24	35,18 27,2	5,12, 35,26		39
		27	28	29	30	31	32	33	34	35	36	37	38	39	CT-6

References

1 Hua, Z., Yang, J., Coulibaly, S. et al. (2006). Integration TRIZ with problem-solving tools: A literature review from 1995 to 2006. *International Journal of Business Innovation and Research* 1 (1–2): 111–128.

2 Rao Vallabhaneni, S. (2019). *Wiley CIA Exam Review 2020 Focus Notes, Part 3: Business Knowledge for Internal Auditing Front Cover.* Wiley.

3 The Altshuller Institute for TRIZ Studies, https://www.aitriz.org/triz.

4 Barry, K., Domb, E., and Slocum, M.S. *TRIZ – What is TRIZ.* http://Triz-journal.com. Real Innovation Network. Archived from the original on 26 September 2010. Retrieved 2 October 2010.

5 What is TRIZ?, Oxford Creativity, www.triz.co.uk/what-is-triz.

6 What is TRIZ and How can it be used in Problem Solving or Brainstorming?, InnovationManagement.se.

7 «TFLIZ» LTD, Mashinostroiteley Str.1, 40020, Sumy, Ukraine, http://triz-ltd.com/en.

8 Orloff, M.A. (2016). *ABC–TRIZ: Introduction to Creative Design Thinking with Modern TRIZ Modeling.* Spinger.

9 Rantanen, K., Conley, D.W., and Domb, E.R. (2018). *Simplified TRIZ, New Problem Solving Applications for Technical and Business Professionals*, 3e. Productivity Press.

10 Department of Chemical Engineering, University of Michigan, Ann Arbor, Using the TRIZ Crossover QMS Android Application to Improve Quality , 25/1/2013. http://umich.edu/~scps/html/07chap/html/powerpoint pics triz/Chapter%207%20 TRIZ.pdf.

11 Altshuller, G. (1999). *Innovation Algorithm: TRIZ, Systematic Innovation and Technical Creativity*, vol. 122. Worcester, MA: Technical Innovation Center, Inc.

12 Altshuller, G. (2005). *40 Principles Extended Edition: TRIZ Keys to Technical Innovation.* Worcester, MA: Technical Innovation Center Inc.

13 Orloff, M.A. (2006). *Inventive Thinking through TRIZ: A Practical Guide*, 2e, 351. Heidelberg, Germany: Springer.

14 Czinki, A. and Hentschel, C. (2016). *Solving Complex Problems and TRIZ. In TRIZ Future 2015 (TFC 2015)*, 27–32. Berlin, Germany: Elsevier.

15 Silverstein, D., DeCarlo, N., and Slocum, M. (2007). *Insourcing Innovation: How to Achieve Competitive Excellence Using TRIZ*, 4. Auerbach Publications.

16 Gadd, K. (2011). *TRIZ for Engineers: Enabling Inventive Problem Solving.* Wiley ISBN: 0470741880.

17 Yang, K. and El-Haik, B. (2003). *Design for Six Sigma: A Roadmap for Product Development*, 624. NY: McGraw-Hill.

18 What is TRIZ? *The TRIZ Journal.* http://www.TRIZ-journal.com.

19 Ekmekci, I. (2015). Methodology and an application example for product development. *Social and Behavioral Science* 95: 2689–2698. https://www.sciencedirect.com/science/article/pii/S1877042815039609.

20 Nakagawa, T. (2001). Essence of TRIZ in 50 words, Osaka Gakuin University. *The TRIZ Journal* https://triz-journal.com/essence-triz-50-words.
21 Altshuller, G. (1999). *The Innovation Algorithm*, 2e. Worcester, MA: Moscow Worker Publishing House, 1973 (in Russian). Translated into English by Lev Shulyak: Technical Innovation Center.
22 Nakagawa, T. TRIZ: Theory of Inventive Problem Solving – Understanding and Introducing It, Bulletin of Cultural and Natural Sciences in Osaka Gakuin University, No. 37 September 1998.
23 El-Haik, B., Roy, D.M., and Hart, M.A. (2006). *Service Design for Six Sigma: A Roadmap for Excellence*. American Society Quality.
24 Mishra, U. *The father of TRIZ – As we know him; A short biography of Genrich Altshuller Collection*. Bangalore, www.Trizsite.com.
25 Zhuravleva, V. (2005). *Ballad of the Stars: Stories of Science Fiction, Ultra Imagination, and TRIZ*. Technical Innovation Center, Inc.
26 Panthalookaran, V. *The TRIZ of it All*, July 02, 2018. http://people.rajagiritech.ac.in/varghese/publications/triz-it-all.
27 Smith, S.L.R. *TRIZ and Why it is Important for Six Sigma and Design for Six*. https://fdocuments.in/document/triz-and-why-it-is-important-for-six-sigma-and-design-for-and-why-it-is-important.html.
28 Jongen, W.M.F. and Muhlenberg, M.T.G. (2005). *Innovation in Agri-Food Systems*. Wageningen Academic Publishers.
29 Oxford Creativity; www.triz.co.uk/learning-centre-innovation-materials.
30 Akay, D., Demiray, A., and Kurt, M. (2008). Collaborative tool for solving human factors problems in the manufacturing environment: the theory of inventive problem solving technique (TRIZ) method. *International Journal of Production Research* 46: 2913–2925.
31 Yan, X.-T., Jiang, C.-Y., and Juster, N.P. (2004). *Perspectives from Europe and Asia on Engineering Design and Manufacture*. Springer https://www.springer.com/gp/book/9781402022111.
32 Creating Minds; http://creatingminds.org/tools/triz/triz_contradictions.htm.
33 Tan, R. (2005). *A Macro Process Model for Product Innovation Using TRIZ*.
34 Basirzadeh, M. (2019). *TRIZ 40 Inventive Principles for Innovation*, Loh-e-Mahfooz.
35 Juneja, P. *What is TRIZ Matrix and How to Use it?* Management Study Guide Content Team, https://www.managementstudyguide.com/triz-matrix.htm.
36 Altschuller, G., (1984). *Erfinden – Wege zur Lösung technischer Probleme*. Berlin, VEB Verlag Technik.
37 Dave, H. P., TRIZ: 40 principles and their ranking by contradiction matrix, 2017, 2nd International Conference for Convergence in Technology (I2CT).
38 Sitta, E. (2019). *TRIZ methodology to overtake contradictions*, 19/06/2019.
39 Williams, T. and Domb, E. (1998). *Reversibility of the 40 Principles of Problem Solving*, 02 May 1998.

40 Nakagawa, T. (2007). A new paradigm for problem solving. *The TRIZ Journal* https://triz-journal.com/a-new-paradigm-for-creative-problem-solving-six-box-scheme-in-usit.

41 Lippert, K. and Cloutier, R., TRIZ for digital systems engineering: New characteristics and principles redefined, Systems, Volume 7, Issue 3, 2019. https://www.mdpi.com/2079-8954/7/3/39.

42 Technical and physical contradictions. *The TRIZ Journal*, December 2006. http://www.triz-journal.com.

43 Duepen, E. (2004). *Ideality and the Artist*. University of Colorado. https://triz-journal.com/ideality-artist/.

44 Engel, A. (2018). *Practical Creativity and Innovation in Systems Engineering*. Wiley Series in Systems Engineering and Management.

45 Burkhardt, V. *Interview with Jack Hipple; author of The Ideal Result: What It Is and How to Achieve It*. https://www.ideaconnection.com/interviews/00358-Inventive-Principles-Illustrated-Part-5.html.

46 Mishra, U. *Finding and Solving Contradictions of Contradictions of Contradictions of Contradictions of False Positives in Virus Scanning*, Bangalore, India, https://arxiv.org/abs/1306.4652.

47 Haines-Gadd, L. *Defining Ideality as a TRIZ Concept*, https://www.dummies.com/business/management/triz-for-dummies-cheat-sheet.

48 Mishra, U. *The Concept of Resources in TRIZ*, Bangalore, India, https://arxiv.org/abs/1306.4652.

49 Souchkov, V. *Breakthrough Thinking with TRIZ for Business and Management: An Overview, ICG Training & Consulting*. https://www.researchgate.net/publication/228377110_Breakthrough_Thinking_with_TRIZ_for_Business_and_Management_An_Overview.

50 Seredinski,A. (2005). *Ideality and progress*. Proceedings of the international conference MATRIZ Fest – 2005. Saint Petersburg, Russia, July, 3th–4th 2005, pp. 228–230.

51 Petrov, V. and Seredinski, A. (2006). *Progress and Ideality*, 27 Feb, TRIZ Association of Israel, (ed. Avraam Seredinski), Head of TRIZ chair, ESICS, France, https://triz-journal.com/progress-ideality.

52 Trebar, A. Using the TRIZ Crossover QMS Android Application to Improve Quality, http://www.creativ-plus.si/triz.

53 Mann, D. (2004). Fan Technology: Evolutionary potential and evolutionary limits. *The TRIZ Journal* https://triz-journal.com/fan-technology-evolutionary-potential-evolutionary-limits.

54 LEAN PROBLEM SOLVING, www.ideas4.biz.

55 Cudney, E.A. and Agustiady, T.K. (2016). *Design for Six SIGMA: A Practical Approach through Innovation*. Boca Raton, FL: CRC Press https://publicism.info/business/sigma/12.html.

56 Hoon, D. (2015). *Trends of Engineering System Evolution*, May 5.

57 Mann, D. (2000). *Application of Triz Tools in a Non-Technical Problem*. http://systematic-innovation.com.

58 Mann, D. and Bradshaw, C. (2005). *Design for Wow 2 – Music*, http://triz-live.kobus.eu/design-wow-2-music/.

59 Domb, E. and Slocum, M. (2003). *Hands-on Systematic Innovation*,. https://triz-journal.com/hands-systematic-innovation/.

60 Tan, J. (2019). Advances in Mechanical Design, Proceedings of the 2019 International Conference on Mechanical Design (2019 ICMD).

61 Frenklach, G. (1998). *Input–Output–Trimming operator (I–O–T). The TRIZ Journal* https://triz-journal.com/input-output-trimming-operator-o-t/.

62 Haines-Gadd, L. (2016). *TRIZ for Dummies*. Wiley.

63 Souchkov, V. (2006). *Annotated List of Key TRIZ Components and Techniques, ICG Training & Consulting*; www.xtriz.com, Nov 2006, updated Nov 2011.

64 Terninko, Z., Zusman, A., and Zlotin, B. (1998). *Systematic Innovation: An Introduction to TRIZ*. Boca Raton, FL: St. Lucie Press.

65 Pannenbaecker, T. (2001). *Methodisches Erfinden in Unternehmen: Bedarf, Konzept, Perspektiven für TRIZ-basierte Erfolge*. Wiesbaden, Germany: Gabler.

66 Souchkov, V., Hoeboer, R., and van Zutphen, M. (2007). *RCA+ to solve business problems*. https://triz-journal.com/application-of-rca-to-solve-business-problems.

67 Royzen, Z. (1999). *The TRIZ Journal*. https://triz-journal.com/tool-object-product-top-function-analysis.

68 Kowalick, J. (1998). Psychological inertia. *The TRIZ Journal* https://triz-journal.com/psychological-inertia.

69 Cameron, G. (2010). *TRIZICS: Teach yourself TRIZ, How to Invent, Innovate and Solve "Impossible" Technical Problems Systematically*, vol. 86. CreateSpace.

70 Kowalick, J. (1998). Psychological inertia. *TRIZ Journal* https://triz-journal.com/psychological-inertia `(accessed 22 April 2002).

71 Savransicy, S.D. (2000). *Engineering of Creativity. Introduction to TRIZ Methodology of Inventive Problem Solving*. Boca Raton, FL: CRC Press.

72 Belski I. and Skiadopolous, A. (2017). *Theory of Inventive Problem Solving: Size Time Cost Operator*. Retrieved from www.edisons21.com.

73 Gadd, K. (2011). *TRIZ for Engineers: Enabling Inventive Problem Solving*, John Wiley & Sons.

74 Tue, N. K. (2015). *Advanced Multi-Screen Scheme of Powerful Thinking*, https://otsm-triz.org/en/content/n-khomenko-advanced-multi-screen-scheme-powerful-thinking.

75 Fey, V. (Mar 2001). *Glossary of TRIZ*. *The TRIZ Journal* https://triz-journal.com/glossary-triz.

7

Environmental Impacts of Corrosion and Assessment Strategies

Reza Javaherdashti

Eninco Engineering B.V., Hengelo, The Netherlands

7.1 Introduction

Corrosion is a multidimensional matter; on one hand it is based on electrochemistry and materials science, on the other hand it is an umbrella term for electrical engineering, economy, management, and environment. In other words, corrosion is not the simple "rusting" that one may think as being an ordinary matter that can be easily rectified and cured by a team of corrosion scientists and corrosion engineers. Corrosion has undeniable impacts on the national economy and environment; in fact, the annual global cost of corrosion – despite all the intrinsic ambiguity and uncertainties that such models inherit from the assumptions made and the mathematics involved – is much higher than all the costs imposed on humanity by natural disasters from earthquakes to tsunamis and volcanic eruptions. This and similar data have been repeatedly mentioned by corrosion specialists and scientists all around the world in all of their corrosion gatherings and conferences. Despite all this, and with a few exceptions, it is really a far-fetched idea to have expected that many educated professionals and academics take corrosion as an issue that is multifaceted. They argue that if a conference is about stainless steels, why should we have a section in corrosion? They just do not understand that one of the main reasons for having a conference about stainless steels is that these ferrous materials are one of the options for the treatment of corrosion (material selection).

One of the aspects to which the least attention has been paid is creating a systematic approach toward assessment of corrosion within an environmental

Corrosion Policy Decision Making: Science, Engineering, Management, and Economy, First Edition.
Edited by Reza Javaherdashti.
© 2022 John Wiley & Sons, Inc. Published 2022 by John Wiley & Sons, Inc.

context. In fact, while we are now living in an era where multidisciplinary sciences are quite acceptable and are even encouraged, an interdisciplinary area that encompasses corrosion–environment interactions is still a farfetched dream.[1] In his several works that expand to at least a decade, this author has always focused on the environmental aspects of corrosion [1–5]. One of the models we proposed to assess is an accident that has been caused by corrosion and has changed one or all components of the surrounding environment[2] drastically.

7.1.1 Characterization of the Disaster

In this chapter we will explain this model with new terminologies and in more details, but we need to know the definitions [1] we use; corrosion effect, corrosion impact, boundary, and dimension.

- **Corrosion Effect:** those natural causes of corrosion that can be expressed by corrosion-related terminology and within a corrosion-related context.
- **Corrosion Impact:** those natural results of corrosion that can be expressed in a quantitative way, but without using technical language related to corrosion literature (including items related to economical costs and measurable ecological/community risks). In this context, corrosion impacts can be regarded as unwanted, quantifiable aspects of those damages caused by corrosion and imposed on the components of environment.
- **Boundary:** defines the domain of specialty and skilled work.[3]
- **Dimension:** the level of concern and knowledge within the body of an industrial operator regarding corrosion management. An industrial operator is a

1 One term that is highly likely to contribute to confusion is the so-called "environmentally assisted cracking" or "environmentally-assisted corrosion." Contrary to how it may sound to environment experts, neither of the two have anything to do with environmental assessment and the related terminology. These terms just serve to highlight the role of electrolytes in accelerating corrosion of materials and, based on the "environment" (that is, the electrolyte) and the material exposed to it, can be the subject of much on-going research.

2 Our understanding is that the components of environment are water, soil, and air. Anything that changes the natural chemical properties of each of these components in a way that can be assessed as being detrimental to the overall health of the living (from micro- to macro-organisms including animals and humans) is referred to as a pollutant.

3 An oil refinery can be differentiated and distinguished from a mining plant or, within an organization, the Human Resources Department can be distinguished from the Finance Department by their boundaries.

company/plant that runs certain industrial activities that involve corrosion and thus requires corrosion control.[4]

When a disaster occurs, the first question that needs to be answered is about the boundary and dimensions involved; we have to be sure about the domain and the boundary of the operating company to decide the disaster's legal obligation to be put on the client, on the contractor, or the operators. An example in this context can be very useful; let's look at the 2013 Qingdao oil pipeline explosion that happened in Qingdao, China, where due to extensive corrosion of an oil pipeline passing under a street of this tourist-destination city, the pipeline ruptured and oil leaked for at least 15 minutes and was followed by fire and explosion. This disaster lead to the death of at least 62 people. Readers are strongly advised to see graphical documents of the blast [6].

The parties involved in this disaster apparently were[5] (i) the city's municipality as the client, (ii) the operator (Sinopec–China Petroleum & Chemical Corporation), and (iii) the contractor(s) who did the actual job. From this perspective, the boundaries of each party involved can be easily determined; client, operator, and contractor. Based on that, one can also find out about the dimensions of each of the parties. From a corrosion management point of view that insists on dealing with the risk of corrosion, it seems quite unlikely that the last two parties, the client and the contractor(s), had a certain knowledge about corrosion, its monitoring as well as treatment strategies. It can then follow that while boundaries are clear and appreciable, the dimensions are not.

What is the role of dimensions? As per our definition, dimension is the level of knowledge and skill in dealing with corrosion, in terms of both corrosion prevention and corrosion control, as well as its inspection and monitoring technologies. The lower the dimension, the more responsible is the party's contribution to the disaster. The global understanding about contractor(s) is that they basically do the physical part of the job. They must be in a robust technical information exchange and guidance by their clients so that the client must monitor what the contractor is doing.

Another example is what recently (31 December 2020) happened in South Iran, when there was oil leakage from a heavily rusted (corroded) underground out-of-service pipeline, Figure 7.1.

---

4 A copper refinery and an offshore oil rig are both operators as both are running certain industrial activities such as copper refining and oil production, and both of these operations do involve corrosion and do require corrosion control. Their activities – directly or indirectly – do require corrosion management.

5 What will be discussed here is not and should not be considered to be carrying any legal responsibility. It is just a viewpoint that has been developed to shine more light on the subject.

Figure 7.1 A leaking underground oil pipeline [7].

Reportedly, the oil that has been sitting idle within the pipe had leaked, and it had found its way through the soil to a nearby river 500 m below the site. This leak had been controlled immediately as reported by the local authorities. Based on our approach, we can identify three parties involved: the client, who is the National Oil Company; the contractor, who has installed the pipe; and the operator, who is the local technical authority. Based on the domain and dimension analysis, responsibility mainly rests on the operator. Examples could be numerous.[6] Therefore, the first step to be taken in characterizing the environmental impacts of corrosion is to realize the corrosion effect and corrosion impact, as well as the domain and dimensions of the involved parties.

The procedure mentioned above can be schematically shown in Figure 7.2. The rule of thumb in this context is that the higher the dimension in the party whose domain is the most related, the higher is its responsibility in dealing with being careful in issues that have the likelihood to become a disaster with corrosion impacts.

7.1.2 Why Environment?

Perhaps as a legacy from Marxism, some professionals still believe that economics is the main factor in explaining events. However, it may appear that as the time goes by, in the confrontation between economics and ecology, it is the ecology that will have the upper hand. It seems that we are now living in a paradigm shift era where in fact just the reverse has happened as documented in Figure 7.3.

6 A recent sewage spill (June 2020) in Florida, USA has been attributed to corrosion (https://www.materialsperformance.com/news/2021/01/researchers-say-corrosion-contributed-to-recent-florida-sewage-spill, last visited 1 February 2021).

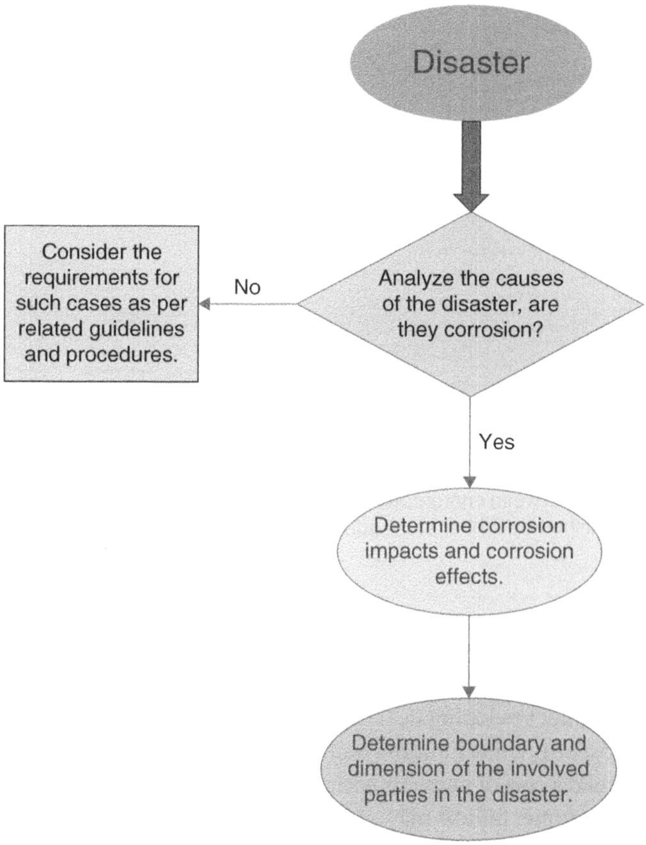

Figure 7.2 A flow chart to determine the responsibility domain of parties involved in an environmental disaster that proves to be related to corrosion.

Figure 7.3 clearly shows since almost a decade ago that the trend for renewables has started to show gradual improvement over conventional, fossil energies. It is important to bear in mind that one of the most serious critics against renewables has always been that in comparison with fossil fuels, the former has not been economically a good choice. Environmental considerations and field data, though, show just the opposite.

Indirect costs related to corrosion can be listed verbatim as below [9]:

1) Loss of a product from a container that has corroded.
2) Damage caused by leakage of a material such as oil or water from a corroded vessel.

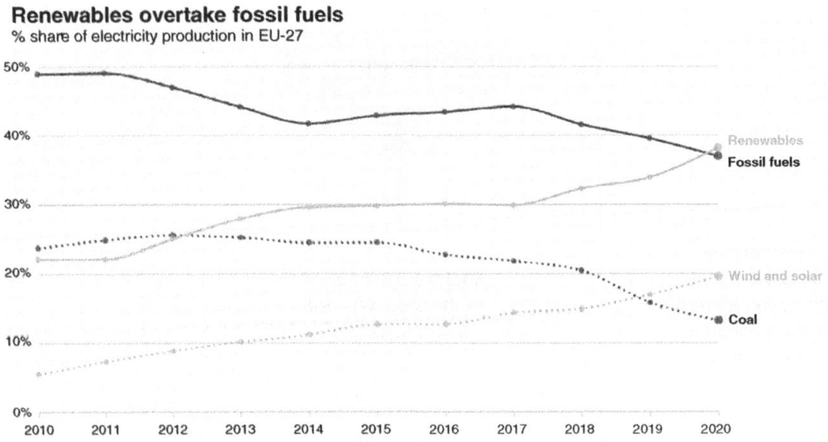

Figure 7.3 Competing trends between energies from fossil fuel sources in comparison with renewables [8] (*Source:* Published by Ember and Agora Energiewende, 25 January 2021).

3) Contamination of a product; for example traces of metal may alter the color of dyes, or food may be spoiled in corroded metal containers.
4) Loss of production because of a corrosion failure or loss of efficiency, for example, due to corrosion product reducing heat transfer. Replacement of a corroded tube in a plant may cost only a few hundred dollars, but shutdown while repairs are underway may mean a cost of many thousands of dollars in lost production.
5) Effect on safety; for example, failure may cause fire, explosion, or release of a dangerous product.
6) Appearance; a corroded material is unpleasant to the eye, so commercial establishments may use expensive trim for the sake of appearances.

Of the six items above, items 2 and 5 can be related directly to what we mean as corrosion impact in the context of this chapter. The other four items can be referred to as proprietary damages where the product itself or the factory in which production takes place are affected.

In other words, the items 2 and 5 signify the areas in which the impacts of corrosion go far from locality of a given production plant; these two items of indirect corrosion costs can affect one or all environmental components, and due to this very nature of them, their careful studies need to be prioritized in relation to the other four items listed.

It may seem more logical, then, to categorize costs of corrosion into direct and indirect, and classify them as "proprietary" and "universal" depending on the damage range they create. By this approach, the damage impact created by the corrosion impacts on universal costs, one of which being the environment will make sense.

For a balance between economy-based and ecology-based proprietary costs, our personal/professional choice is to concentrate on the ecology-based proprietary costs. Our reason is simple; you can always make money and recover from financial loss, but you can never remake the environment again and recover it.

7.1.3 Corrosion Impact and Corrosion Effect

As previously mentioned, corrosion effect differs from corrosion impact mainly in that the former is a natural interaction that occurs between the asset and the surrounding environment due to the thermodynamic nature of corrosion, whereas the latter is the way corrosion changes the environment in an unwanted way. One way to look at both proprietary and universal costs of corrosion is by studying corrosion impacts should the damage of interest be related to corrosion.

If the proprietary damage is related to corrosion in any way, it follows that one of the main mechanisms that has to be considered to answer why the damage happened is to include corrosion science and engineering. However, it is also possible that the cause of the disaster has nothing to do with corrosion; for example, the disaster could be an earthquake or the like, what are collectively referred to as natural disasters. In this category, any natural disaster that occurs must be considered within its own technological and professional nomenclature and context. However, it may also be possible that even natural disasters that seemingly show no relevance to corrosion, can in one way or another be indeed interrelated. For example, corrosion can be an instrumental tool in weakening the bond between the concrete and the rebar in reinforced concrete (RC) structures; typically for every percentage increase in corrosion level, there can exist a corresponding 16% reduction in the bond strength [10]. Therefore, even in seemingly irrelevant issues, if corrosion can be traced back to be one of the contributing factors to the damage, it needs to be determined how the damage (of proprietary or universal nature) must be studied.

Figure 7.4 shows a serious flaw in the concrete wall of a cooling tower that is near the sea, and most probably caused by the breeze containing chloride ions from seawater toward the concrete structure. We must state again that any factor in inducing proprietary costs must be considered carefully, first as per the flow diagram shown in Figure 7.5 to be relevant to corrosion to make it eligible for further corrosion-related studies. In a team composed of corrosion and environment specialists, the corrosion expert must do a root-cause analysis to first clear the air of any conclusion related to the involvement of corrosion in any damage or disasters that has occurred and been categorized as having an "environmental impact."

Figure 7.4 Damaged concrete, probably the corrosive impact of chlorides coming from the sea on a cooling tower structure built very near to the water (*Source:* Dr. R. Javaherdashti's collection).

It is possible to employ Pourbaix diagrams related to highlighting the detrimental role of chlorides on concrete structures (Figure 7.5).

As seen, existence of some corrosive ions such as chlorides, can somehow change the domains. As schematically shown in Figure 7.5, with and without chloride, the domains' shapes change dramatically, as do the values of potential and pH. Therefore, in the case above, a damage that can be categorized as proprietary damage (to a large extent item 5 and to a lesser extent item 6) will be analyzed.

7.1.4 Modeling Environmental Impacts

Modeling environmental impacts is not a new issue and there are numerous examples for it [12, 13]. To the best knowledge of this author, there is no specific model developed for studying corrosion impacts on the environment. In one of our previous works [14], we have extensively considered an example of the way corrosion

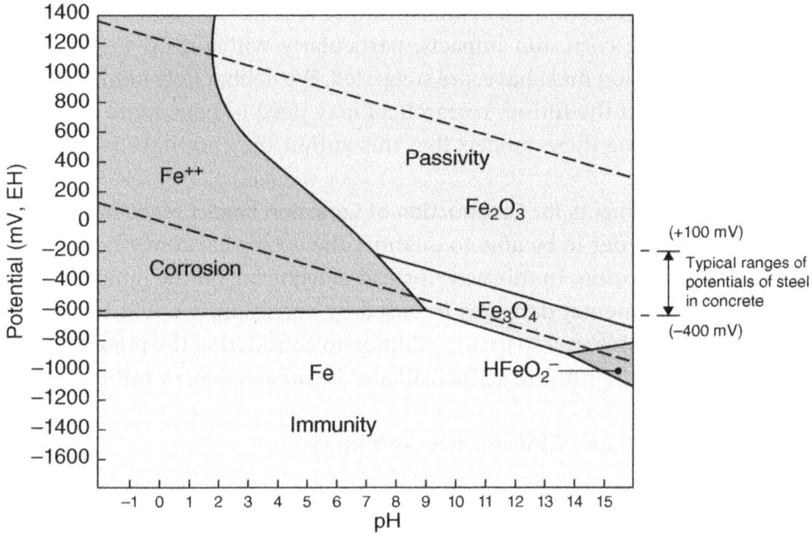

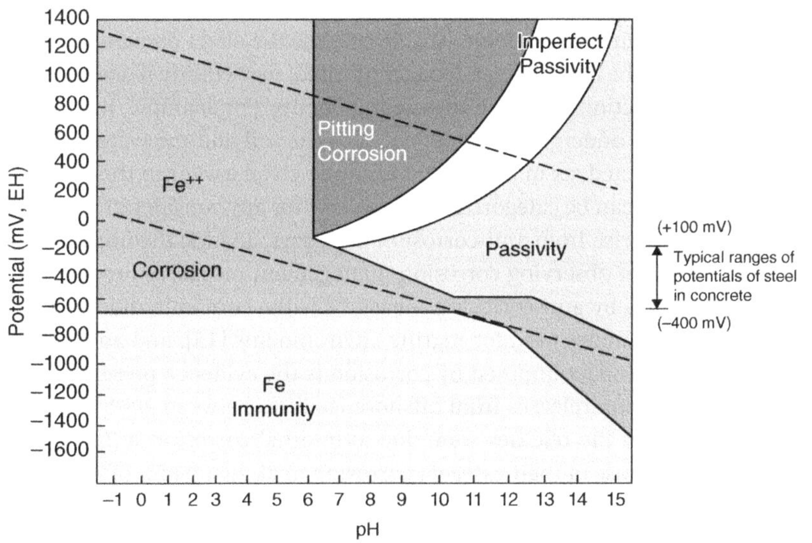

Figure 7.5 A Pourbaix diagram for iron–water system with the presence of chloride, at 25°C (top), the same diagram at the same temperature without the presence of chloride [11] (bottom).

impacts can be assessed. However, any mathematical modeling of ecological aspects of universal impacts of corrosion, to this date, has not been formulated.

While no formulated model for corrosion impacts modeling (CIM) has yet been proposed, this must not prevent us from trying to envisage some features that such models need to have. In the section below, some of the features that we believe any model for considering corrosion impacts, particularly with regard to involving management of corrosion must have, are suggested. We deliberately used the word "suggested" because in the future, researchers may need to have more elements involved in constructing these models that this author may not have included.

7.1.4.1 Necessary Elements for Construction of Corrosion Impact Modeling

CIM is necessary in order to be able to quantify the ecological component of the universal costs of corrosion. In this way, instead of lecturing on the importance of the so-called environmental damages imposed by corrosion, we will be able to express it in a measurable way so that in addition to considering the present status of corrosion impacts, their future status will also be predictable, or rather, largely predictable.

The main features of any CIM must be as follows:

1) A CIM should consider at least one component of environment (water, soil, and air) to deal with. In other words, it is better for any CIM to be environment–component specific so that the corrosion impact model in water will be differentiable from that in the soil or air. This is because the effect domain of corrosion impacts could be too wide and too complicated, particularly if one wants to consider the interactions between these components. For example, in the case of leakage from an underground pipeline, both the soil and the water components could be affected, as in the second example given earlier in this chapter.
2) Corrosion impacts can be categorized in two ways for any consideration; (i) the impacts that may arise from anti-corrosion measures, and (ii) the impacts that may result from not observing corrosion management measures properly. An example of impacts by anti-corrosion measures is the contamination induced by zinc sacrificial anodes used for marine environments [15], and an example of environmental impact imposed by corrosion is the evidence presented by a study on hydrocarbon releases from offshore facilities between 2001 and 2007 that showed 50% of the releases were due to internal corrosion in the equipment and 16% because of their external corrosion, making up 66% of hydrocarbon released from these facilities due to corrosion.
3) There is a significant difference between corrosion as a phenomena and corrosion impacts; corrosion takes place on an asset/material via series and/or parallel corrosion reactions(s), but corrosion impacts can be extended effects on one or all three components of the environment. It is this feature of corrosion impacts that makes it a very complex matter to consider and model. Pipeline leakages are good examples of the complexity involved in environmental impacts.

No official reports could be found to clearly show that corrosion was indeed the main cause of the leakage, however there are reports about the role of corrosion in disasters such as what occurred in the Niger Delta Area of Nigeria between 1999 and 2010 [16]. Reported by the authors of reference [15], from Chris Ikporukpo in 1998, the amount of corrosion in pipeline leaks is over 13%.

4) Any CIM must have a mathematical backbone to let it justify the name of a quantitative model. In other words, a mathematical framework for a CIM is needed to enable it to predict the extent of the damage of which corrosion-induced environmental impacts are capable. Many options may be available, from using Fick's laws of diffusion to applying static or dynamic sensitivity analysis techniques [17].

Another instrument that can be useful in quantifying CIM is the concept we developed in one of our previous publications [13], that is environmental risk threshold (ERT). Where $ERT = CR_{Polluted} - CR_{Unpolluted}$.

In this formula, CR stands for corrosion rate, $CR_{Polluted}$ is the average corrosion rate of the material of interest in the given environmental component when it is polluted, and $CR_{unpolluted}$ is the average corrosion rate of the material of interest in the given environmental component where it is polluted and where it is not polluted.

Based on the ERT we have developed, three possibilities can be imagined:

Possibility I: $ERT = 0$;
Possibility II: $ERT > 0$; and
Possibility III: $ERT < 0$.

The first and third possibilities (ERT is zero or it has a negative value) in our interpretation, means that pollution has no measurable effect on the severity of corrosion. From a corrosion management point of view, a positive ERT is indicative of how the pollution can act as a corrosion-enhancing factor. Examples of detrimental effects of pollution on corrosion has been the topic of many research activities [18–20]. There are some websites with free access that lay claim to software for quantifying environmental factors.[7]

5) Corrosion of the environment or environment of corrosion?

---

7 Some examples are summarized by Dr. Sanjeev Saraf for using in safety reliability projects: AIChE Chemical reactivity worksheet: https://www.aiche.org/ccps/resources/chemical-reactivity-worksheet; EPA Consequence modeling software (empirical models), ALOHA: https://www.epa.gov/cameo/aloha-software; NIST CFD modeling for fire and smoke, Fire dynamic simulator: https://pages.nist.gov/fds-smv; Kenexis Open PHA, HAZOP, LOPA: https://www.kenexis.com/software/openpha; ALD Fault Tree Analysis: https://www.fault-tree-analysis-software.com; ALD Failure rate and MTBF estimator: https://aldservice.com/Reliability-Software/free-mtbf-calculator.html; ThinkReliability Root cause analysis excel template: https://offers.thinkreliability.com/root-cause-analysis-excel-template; TNO books for consequence modeling and risk analysis: https://www.gexcon.com/products-services/coloured-books/ProSIS; SIL verification: https://www.prosisfse.com/sil-verification.

We have briefly touched on the second part of the above title, but because it could be a significant source of confusion, we think it is worth of getting back to it. As we previously mentioned, one can safely assume that what we call as environment seems to be composed of three components; air, water, and soil. If a metallic piece is placed in any of these components, due to the thermodynamics involved in corrosion as well as the electrochemistry associated with corrosion processes, the metal starts to lose the electrons that have been imposed on it during extractive metallurgy processes. However, what really affects the environmental component is not the electrons thus liberated but the ions thus formed in the components, as the component will act like an electrolyte. We will get back to this later.

The formed and then added ions are capable of changing some important properties and features of the environmental component drastically. These features could range from change of pH (acidity) of the component to adding some harmful compounds that are formed as the result of combination of the ions formed and transferred through the component-then-electrolyte.

It is in this context that corrosion of environment can make sense; it is the chemical–biochemical unbalance that is created via the progress of corrosion in an environmental component, whether air, water, or soil.

The examples of how corrosion can affect environmental components are numerous; from a leaking underground pipeline to enhancing microbial corrosion (MIC) due to altering atmospheric conditions, so-called MIC-induced atmospheric corrosion [14]. It is in this respect that environmental impact of corrosion will make sense.

A very well-known terminology in the literature of corrosion is environmentally-induced corrosion or environmental cracking. In this context, though, the environment will have another meaning; it is the electrolyte that we proposed to select to examine its effect on the acceleration of corrosion of a given material and as demonstrated by specific morphologies.

Corrosion of environment, in the sense we defined above, can have a wide and very complex domain of interactions with unwanted effects on environment. Figure 7.6 is a simple example of corrosion impacts and how they can be entangled with various components of the environment.

The complex interactions simplified and shown in Figure 7.6 can serve to summarize various interactions that a seemingly "simple" crack in the underground pipe will cause in environmental components. This figure also shows the domains that corrosion experts and environmental experts separate from each other but are still related; the crack developed could well be caused by the effect of corrosion (either internally or externally) and should be dealt with by a corrosion expert. When the crack developed and the leak occurred, it was the starting point of an environmental effect as imposed by corrosion. The contamination thus created will affect all the environmental components, particularly soil and water.

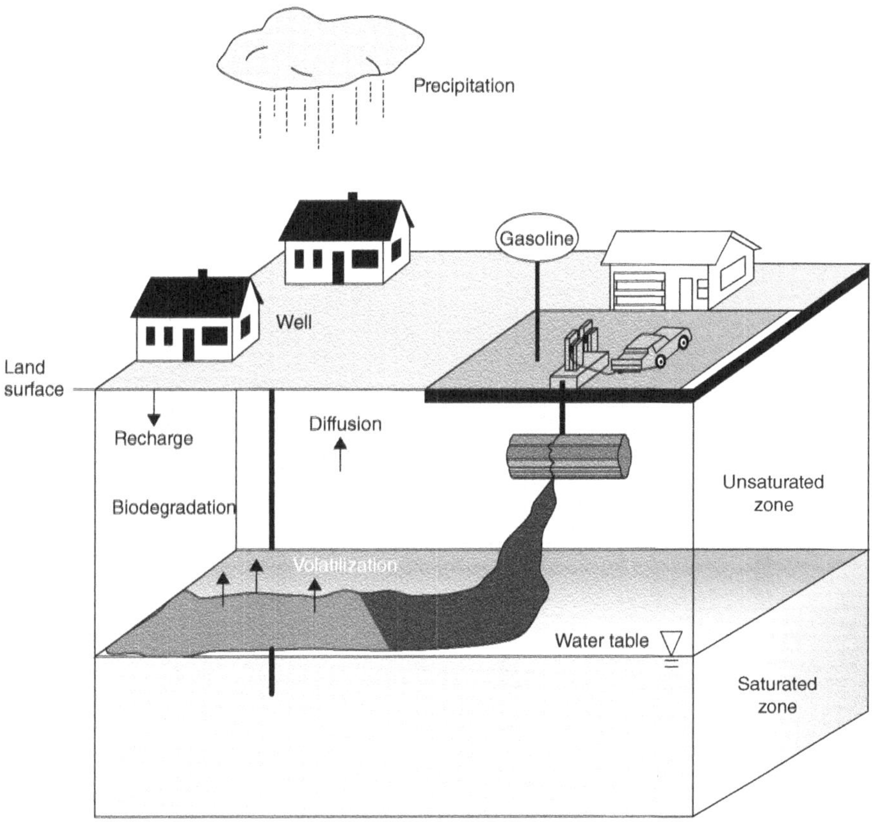

Figure 7.6 Conceptualization of the fate of petroleum hydrocarbons in a groundwater system (*Source:* Fact Sheet FS-019-98 [21]).

Obviously, as the environment does not consist of sequestered units and it is a unified totality, anything happening with one component is highly likely to affect the other.

Figure 7.7 shows toxicity anatomy diagram as caused by corrosion.

As it appears from Figure 7.7, the ions produced via corrosion can adversely affect the environment components; in water, they are mixed with water as dissolved compounds and can shift its chemistry toward (possibly) ERT positive values.

6) A suggestion: Rule 365

We will base our suggestion on MIC and it's applicable to corrosion in general. Minimum direct loss imposed by MIC is about 20% of all the damage corrosion

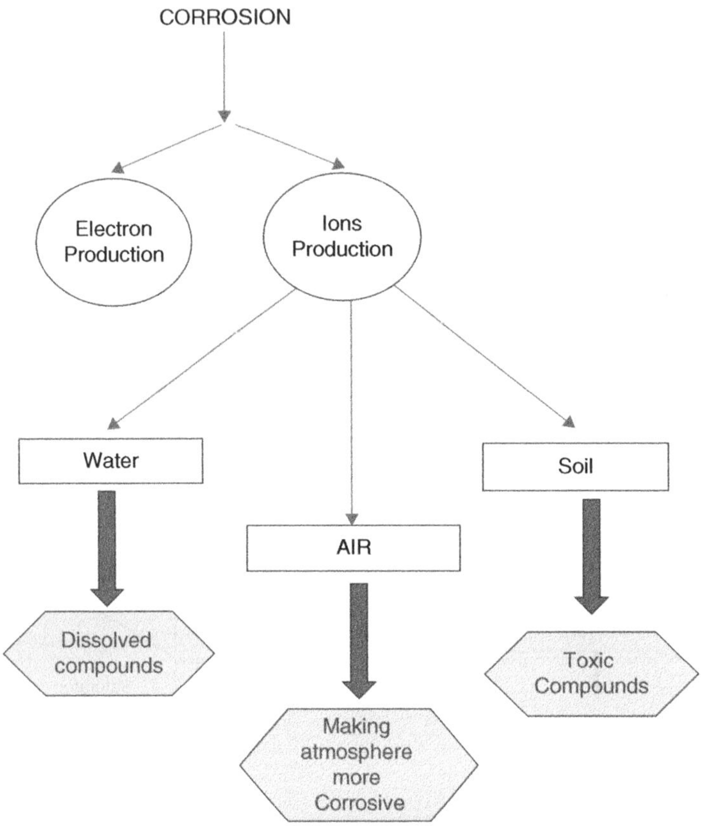

Figure 7.7 Toxicity anatomy of corrosion effects.

causes [22, 23]. We need a convention that will let us quantify necessary resources to mitigate MIC. To be on the safe side and to consider the worst possibilities, let us assume that cost of MIC is at least 10% of all the damage caused by corrosion; now we will explain a "convention" that may be called "the least required action" or Rule 365.[8] This rule states that:

- The least required action is 10%;
- In a unit of action, the least amount of action is one.

8 "Abdjad system" is a gematria (numerical values of letters). According to the abdjad system, which attributes a numerical value to each letter and vice versa, the sentence "the least required action" has an equivalent numerical value of 365. The abdjad system has been preferred over using initials due to advantages such as being easy to remember and that the number reminds us of the number of days in a year, implying the rule must be applied every day.

Action in this context means "any kintyped of activity and/or modification that is applied to the resources in order to lower losses due to MIC." It simply states, for example, of 10 corrosion engineers to be employed, at least one person (i.e. 10% of them) must have hands-on experience in MIC. Similarly, if the industrial information unit (technical library) is ordering, say, 100 books about corrosion, at least 10% of them (that is to say 10 books), must be on the topic of MIC. Equivalently, out of 10 seminars on corrosion, at least one must be completely related to MIC. The least amount of action being one means that, in case the number of engineers to be employed or books to buy or seminars to be given is less than 10, (whose 10% is one), then the least amount of action is applied so that at least one engineer will be employed or one book on MIC will be bought and so on.

7.2 Some Uses of Rule 365

- Although it may seem that rule 365 is too simple, in fact when it is applied to decision-making procedures, it will prove to be a practical, easy-to-calculate way for evaluating the existing condition(s) of resources.
- Resource evaluation by using rule 365 not only tells about present "in-hand" conditions of resources and their probable limitations and dead capacities (in terms of necessities for fighting against MIC), but also it allows us to design and shed light on future needs.
- Resource allocation in terms of time and money will become much easier so that unnecessary expenses for maintenance/operation programs will be eliminated.
- Applying rule 365 will create a massive drag force for doing more serious work in MIC.
- By applying rule 365 to managerial considerations about resources, the manager can decide about job policies and whatever is related to it with more confidence.
- Rule 365 cannot only be applicable to MIC related problems, because it also has the potential to be applied to other types of corrosion. This is true because the main feature of this convention is that it focuses on the importance of resources in corrosion mitigation problems where inadequate personnel, scarcity of finance, and similar problems may affect quality of corrosion mitigation programs quite adversely. Therefore, by making application of rule 365 a regular pattern in industrial applications, the industry will discover how much it can save in terms of timed corrosion mitigation and prevention.

7.2.1 Application of Rule 365 to Assess Corrosion Effects

In this chapter we have tried to find ways to analyze and perhaps predict the extent of corrosion effects, which are an important part of universal costs particularly related to the environment.

It must be noted that corrosion effects have a nature that seems to be gradual and will not show their damaging impact in a short time. Due to the continuous nature of corrosion processes, their effects also have the same nature. Rule 365 can be interpreted, for example, as looking at 100 possible corrosion effects, selecting 10 of them, and only concentrating on one. This could highly reduce the intrinsic complications involved in the network of corrosion effects on environmental components and continue this 24/7.

Based on the procedure proposed in Figure 7.8, it is first necessary to have an idea about the size of the corrosion effect; to do that, we need to have a model. In previous sections some of the features that such a model must have were

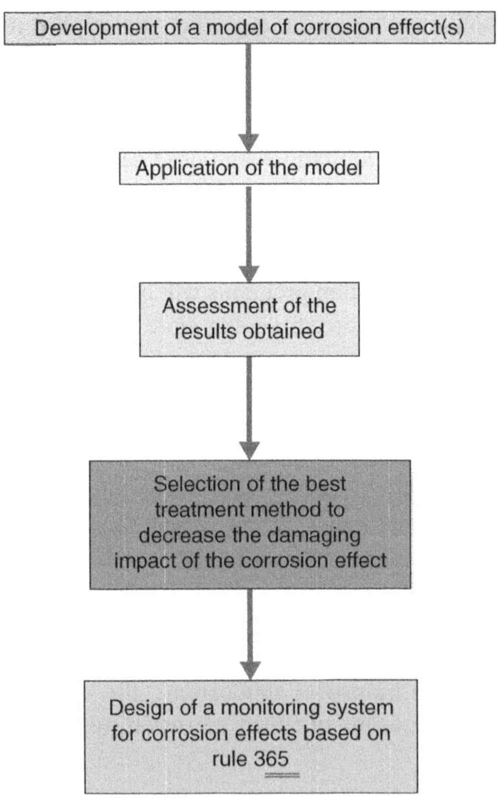

Figure 7.8 A suggested model to apply Rule 365 to deal with corrosion effects.

indicated. As soon as the model is prepared, we need to apply it; in other words, we need to find out what corrosion effects are important for us to look at. We need to remember that corrosion effects are capable of affecting all three components of the environment, and therefore it is necessary to specify what environmental component(s) are the most appropriate background for the model. It is here that an environment–corrosion interaction monitoring system can be designed that needs to be in operation 24/7 and just focused on the target(s) set by the model; we also need a certain percentage (our suggestion is 10%) of the total budget provided for corrosion management in the given plant.

Undoubtedly, as the issue of studying corrosion effects is rather a new approach toward the way corrosion can lead to environmental disasters, it is also a serious requirement to search for various ways by which new concepts such as Rule 365 can be launched and effectively applied.

7.3 Conclusions

The extent of environmental effects caused by corrosion is a discipline that to the best of our knowledge has not yet been studied in depth. In this chapter, we developed necessary basic concepts to create a dialogue between corrosion and environmental studies that are seemingly two distant disciplines from each other. In this effort, we defined the difference between corrosion impact and corrosion effect in addition to boundary and domain, to enable us to define an environmental disaster within a new context. In addition, we mentioned some of the main features of a model that attempts to explain corrosion effects within a corrosion–engineering framework.

We also introduced a new concept that we call Rule 365 to give a straightforward approach toward quantifying work and budgets required to define practical aspects of corrosion effects models.

References

1 Javaherdashti, R. and Nikraz, H. (2010). *A Global Warning on Corrosions and Environment: A New Look at Existing Technical and Managerial Strategies and Tactics*. VDM Germany.

2 Javaherdashti, R. (2000). How corrosion affects industry and life. *Anti-Corrosion Methods and Materials* 47 (1): 30–34.

3 Javaherdashti, R. (2002). How to manage corrosion control without a corrosion background. *Materials Performance* 41 (3): 30–32.

4 Javaherdashti, R. (2006). Using corrosion management to protect the environment. *Materials Performance* 45 (6): 52–54.

5 Javaherdashti, R. (2019). *A quick review of the role that microbial corrosion can play in the engineering and management of fire.* In Andreas Gergely (invited editor), Handbook tentatively entitled as "Corrosion Protection and Prevention: Methods, Applications and Technology" on behalf of the Nova Science Publishers, Inc. https://novapublishers.com/shop/phenomena-and-theories-in-corrosion-science-methods-of-prevention

6 https://qz.com/150669/china-just-changed-the-way-it-responds-to-deadly-industrial-accidents, last visited 9 January 2021.

7 https://www.shana.ir/news/311789/%D9%86%D8%B4%D8%AA-%D9%86%D9%81%D8%AA-%D8%AF%D8%B1-%DA%AF%D9%86%D8%A7%D9%88%D9%87-%D9%85%D9%87%D8%A7%D8%B1-%D9%88-%D9%BE%D8%A7%DA%A9%D8%B3%D8%A7%D8%B2%DB%8C-%D8%B4%D8%AF, last visited 11 January 2021.

8 Graichen, P. and Jones, D. (2021). *EU Power Sector in 2020.* Agora Energiewende.

9 Australasian Corrosion Association. https://membership.corrosion.com.au/training/topic/costs-and-importance-of-corrosion, Last visited 11 January 2021.

10 Shah, S., Gaudel, S., Pandit, R. et al. *Corrosion of Rebar and Earthquake Vulnerability.* 2nd International Conference on Earthquake Engineering and Post Disaster Reconstruction Planning, 25–27 April, 2019, Bhaktapur, Nepal.

11 Ribeiro, D.V. and Cunha, M.P.T. (2014). *Deterioração das estruturas de concreto armado.* In: *Corrosão em estruturas de concreto armado: Teoria, Controle e Métodos de Análise,* 1e (ed. D.V. Ribeiro), 87–118. Rio de Janeiro: Elsevier.

12 Tukker, A., Huppes, G., Guinée, J. et al. (2006). Environmental Impact of Products (EIPRO): Analysis of the life cycle environmental impacts related to the final consumption of the EU-25 (Main report), European Communities.

13 Zhao, K., Ng, T.S., Kua, H.W. et al. (2017). Modeling environmental impacts and risk under data uncertainties. *Operations Engineering and Analytics, IISE Transactions* 49 (12): 1150–1159. https://doi.org/10.1080/24725854.2017.1342054.

14 Javaherdashti, R. and Akvan, F. (2020). *Failure Modes, Effects and Causes of Microbiologically Influenced Corrosion: Advanced Perspectives and Analysis.* Elsevier.

15 Rousseau, C., Baraud, F., Leleyter, L. et al. (2009). Cathodic protection by zinc sacrificial anodes: Impact on marine sediment metallic contamination. *Journal of Hazardous Materials* 167 (1–3): 953–958. https://doi.org/10.1016/j.jhazmat.2009.01.083. Epub 2009 Jan 30. PMID: 19250740.

16 Achebe, C.H., Nneke, U.C., and Anisiji, O.E. Analysis of Oil Pipeline Failures in the Oil and Gas Industries in the Niger Delta Area of Nigeria. Proceedings of the International Multiconference of Engineers and Computer sciences, Vol. II, IMECS, March 14–16, 2012, Hong Kong.

17 Pianosi, F., Beven, K., Freer, J. et al. (2016). Sensitivity analysis of environmental models: a systematic review with practical workflow. *Environmental Modelling and Software* 79: 214–232.

18 Melchers, R.E. (2007). The effects of water pollution on the immersion corrosion of mild and low alloy steels. *Corrosion Science* 49 (8): 3149–3167.

19 Ukpaka, C.P., Wami, E.N., and Amadi, S.A. (2015). Effect of pollution on metal corrosion: a case study of carbon steel metal in acidic media. *Current Science Perspectives* 1 (4): 107–111.

20 Melchers, R.E. (2015). Effect of water nutrient pollution on long-term corrosion of 90:10 copper nickel alloy. *Materials* 8: 8047–8058. https://doi.org/10.3390/ma8125443.

21 Simulating transport of volatile organic compounds in the unsaturated zone using the computer model R-UNSAT. U.S. Department of the Interior, U.S. Geological Survey, Fact Sheet FS-019-98, January 1998.

22 Flemming, H.-C. (1995). *Microbially Influenced Corrosion of Materials*. Berlin, Heidelberg: Springer-Verlag.

23 Videla, H.A. (1986). Proc. Argentine-USA workshop on Biodeteriration 7 (CONICET-NFS), Aquatec Quimica, Sao Paulo, Brazil.

Index

Corrosion Policy Decision Making: Science, Engineering, Management, and Economy, First Edition.
Edited by Reza Javaherdashti.
© 2022 John Wiley & Sons, Inc. Published 2022 by John Wiley & Sons, Inc.

Printed and bound by CPI Group (UK) Ltd, Croydon, CR0 4YY

17/09/2026

14964321-0003